Epigenetics and Human Health

Volume 18

This books series concerns the quickly expanding field of epigenetics. It evokes novel research questions and how these are linked to human health. Each of its volumes covers a particular topic that is important to clinical practice and research, and to researchers in epigenetics who study the relevant medical issues of today. The individual books of the series complement each other and underline the importance of epigenetics in health-related disciplines such as toxicology, reproduction, psychiatry, and nutritional and environmental medicine.

Epigenetics and Human Health was founded in 2011. Prof. Dr. Robert Feil, PD Dr. Mario Noyer-Weidner and Prof. Dr. Jörn Walter were the founding editors and were responsible as series editors for volumes 1-10 until 2016. In 2023, the series was relaunched together with Prof. Rasime Kalkan and Dr. Luis M. Vaschetto, who are the current series editors.

Epigenetics and Human Health is indexed in SCOPUS, EMBASE, Reaxys, EMCare, EMBiology and SCImago.

Sara C. Zapico
Editor

Forensic Science and Epigenetics

How Epigenetics Could Help Solve Crimes

Springer

Editor
Sara C. Zapico
Department of Chemistry and Environmental Sciences
New Jersey Institute of Technology
Newark, NJ, USA

ISSN 2191-2262 ISSN 2191-2270 (electronic)
Epigenetics and Human Health
ISBN 978-3-032-23084-3 ISBN 978-3-032-23085-0 (eBook)
https://doi.org/10.1007/978-3-032-23085-0

This Springer imprint is published by the registered company Springer Nature Switzerland AG
The registered company address is: Gewerbestrasse 11, 6330 Cham, Switzerland

Preface

The field of forensic epigenetics has advanced rapidly over the past decade, transforming what was once a speculative idea into a growing body of robust scientific applications. As I have watched these developments unfold, it became increasingly clear that the discipline needed a comprehensive, structured resource—one that gathers its foundational principles, emerging methods, and diverse applications in a single volume. This book arose from that need and from my own commitment to supporting the continued growth of this promising area of forensic science.

The chapters in this volume reflect both the maturity and the complexity of the field. Chapter 1 provides an overview of the conceptual and historical foundations of epigenetics in forensic contexts, while Chapter 2 outlines the primary laboratory and analytical techniques that underpin current research. The following chapters highlight specialized applications: the discrimination of monozygotic twins (Chapter 3), the identification of body fluids using methylation patterns (Chapter 4), and blood-based age estimation models (Chapter 5). Chapters 6 and 7 extend age-prediction approaches to Y-chromosome methylation and to semen and saliva. Chapter 8 expands these methodologies to skeletal tissues, offering a bridge between molecular biology and forensic anthropology. Chapter 9 examines emerging strategies for postmortem interval estimation. Finally, Chapter 10 provides a forward-looking perspective on the technological, ethical, and legal landscapes that will shape the next generation of forensic epigenetics.

What stands out most to me as editor is the collaborative spirit behind these contributions. The authors—distinguished researchers from institutions across the United States, Europe, Asia, and Latin America—bring complementary expertise in molecular biology, bioinformatics, forensic science, and anthropology. Their collective work reflects the interdisciplinary nature of epigenetics and the shared commitment to methodological rigor and scientific innovation that drives this field forward.

It is my hope that this book serves not only as a comprehensive resource for researchers, practitioners, and students but also as an invitation to further innovation in a field poised to transform forensic science. The challenges ahead are substantial, but so too are the opportunities. As epigenetics continues to reveal its forensic potential, the insights gathered in these pages will help guide the discipline toward greater accuracy, reliability, and societal impact.

Newark, NJ, USA Sara C. Zapico

Acknowledgments

I would like to express my sincere gratitude to the readers for including this book among the resources in their libraries. I hope it meets their expectations and serves as a valuable reference for their work and research. This volume would not have been possible without the dedication of the contributing authors. I thank them wholeheartedly for their willingness to prepare review chapters in their respective areas of expertise and for sharing their knowledge with the forensic and scientific communities.

I am deeply appreciative of the Editorial Team at Springer Nature, particularly Bibhuti Bhusan Sharma, for their guidance and support throughout the development of this book. Their expertise, patience, and assistance were instrumental in bringing this project to fruition.

I would like to thank all those who contributed to the refinement and revision of the chapters, helping ensure clarity and cohesion across the volume. I am especially grateful to Dr. Joe Adserias Garriga for encouraging me to undertake this project; her support and enthusiasm have been invaluable.

Finally, I dedicate this book to my Spanish and American families, whose unwavering support and encouragement sustain me in every academic and professional endeavor. Thank you very much for everything.

Contents

Editor and Contributors

About the Editor

Sara C. Zapico is an Assistant Professor in the Department of Chemistry and Environmental Sciences at New Jersey Institute of Technology (NJIT), in Newark, New Jersey. She is also a Research Associate at Smithsonian Institution, in Washington, DC. Before NJIT, she was the graduate program director of the Professional Science Master's in Forensic Science at Florida International University in Miami, Florida. She also served as an associate at the forensic unit from the International Committee of the Red Cross in Geneva, Switzerland. Her research interests focus on the application of biochemical techniques to forensic science issues, like age-at-death estimation and the determination of post-mortem interval, with implications in aging and biomedical research.

Contributors

José Alonso Aguilar-Velázquez Laboratorio de Ciencias Morfológico Forenses y Medicina Molecular, Departamento de Morfología, Centro Universitario de Ciencias de la Salud, Universidad de Guadalajara, Guadalajara, Jalisco, México

Denisse Stephania Becerra-Loaiza Laboratorio de Ciencias Morfológico Forenses y Medicina Molecular, Departamento de Morfología, Centro Universitario de Ciencias de la Salud, Universidad de Guadalajara, Guadalajara, Jalisco, México

Departamento de Aparatos y Sistemas II, Universidad Autónoma de Guadalajara, Zapopan, Jalisco, México

Bram Bekaert Forensic Biomedical Sciences, Department of Imaging & Pathology, KU Leuven, Leuven, Belgium

Department of forensic Medicine, Laboratory of Forensic Genetics, UZ Leuven, Leuven, Belgium

María Josefina Castagnola Department of Chemistry and Environmental Sciences, New Jersey Institute of Technology, Newark, NJ, USA

Mariano Guardado-Estrada Laboratorio de Genética, Escuela Nacional de Ciencias Forenses, Universidad Nacional Autónoma de México, Ciudad de México, México

Mirna Ghemrawi Department of Chemistry and Biochemistry, Florida International University, Miami, FL, USA

Global Forensic and Justice Center, Largo, FL, USA

Georgiana Gibson-Daw Department of Physical and Biological Sciences, Western New England University, Springfield, MA, USA

Sae Rom Hong Biometrology Group, Korea Research Institute of Standards and Science, Daejeon, South Korea

Manfred Kayser Department of Pathology and Clinical Bioinformatics, Erasmus MC University Medical Center Rotterdam, Rotterdam, The Netherlands

Hwan Young Lee Department of Forensic Medicine, Seoul National University College of Medicine, Seoul, South Korea

Institute of Forensic and Anthropological Science, Seoul National University College of Medicine, Seoul, South Korea

Ji Eun Lee Department of Forensic Medicine, Seoul National University College of Medicine, Seoul, South Korea

Bruce R. McCord Department of Chemistry and Biochemistry, Florida International University, Miami, FL, USA

Mariza Panoutsopoulou Department of Clinical Genetics, Maastricht University Medical Center, Maastricht, The Netherlands

Deborah S. B. S. Silva Chemistry Department, Hofstra University, Hempstead, NY, USA

Cordulle Vanthillo Forensic Biomedical Sciences, Department of Imaging & Pathology, KU Leuven, Leuven, Belgium

Athina Vidaki Department of Genetics and Cell Biology, GROW and CARIM Institutes, Maastricht University, Maastricht, The Netherlands

Sara C. Zapico Department of Chemistry and Environmental Sciences, New Jersey Institute of Technology, Newark, NJ, USA

Anthropology Department and Laboratories of Analytical Biology, National Museum of Natural History, Smithsonian Institution, Washington, DC, USA

Chapter 1
Applications of Epigenetics to Forensic Sciences. Introduction

Sara C. Zapico

Abstract Epigenetics is the study of processes that produce heritable phenotypes that do not strictly depend on the DNA sequence. Epigenetic modifications have been studied in various research areas to understand their involvement in processes such as aging and disease, among others, and to explore the potential for reversing their effects, thereby developing strategies for epigenetic reprogramming. Based on these studies, epigenetics has emerged as a significant field in forensic science, aiming to address issues encountered when analyzing biological evidence, including body fluid identification, age estimation (for both criminalistics purposes and identification of human remains), discrimination of monozygotic twins, determination of the postmortem interval, and lifestyle predictions. This chapter is an introduction to the concept of epigenetics, epigenetic modifications, influential factors, and a summary of its current applications in forensic science.

Keywords Epigenetics · DNA methylation · Chromatin remodeling and histone modifications · Non-coding RNAs · Epigenetic clocks · Forensic science

Abbreviations

CHD	Chromodomain-helicase DNA-binding protein
circRNAs	Circular RNAs
CRCs	Chromatin Remodeling Complexes
DNMT	DNA Methyltransferases
DunedinPACE	Dunedin for Pace of Aging Calculated from Epigenome

S. C. Zapico (✉)
Department of Chemistry and Environmental Sciences, New Jersey Institute of Technology, Newark, NJ, USA

Anthropology Department and Laboratories of Analytical Biology, National Museum of Natural History, Smithsonian Institution, Washington, DC, USA
e-mail: sc338@njit.edu; casado-zapicos@si.edu

S. C. Zapico (ed.), *Forensic Science and Epigenetics*, Epigenetics and Human Health 18, https://doi.org/10.1007/978-3-032-23085-0_1

DunedinPoAm	Dunedin Pace of Aging methylation
eRNAs	Enhancer RNAs
HATs	Histone Acetyltransferases
HDAC	Histone Deacetylases
HDM	Histone Demethylases
HMT	Histone Methyltransferases
INO80	Inositol-requiring 80
ISWI	Imitation Switch
lncRNAs	Long non-coding RNAs
miRNAs	MicroRNAs
MPS	Massive Parallel Sequencing
MZ	Monozygotic
NATs	Natural Antisense Transcripts
M6A	N-6-methyladenosine
ncRNA	Non-coding RNA
piRNAs	PIWI-interacting RNAs
PMI	Postmortem Interval
RT-QPCR	Reverse Transcription Quantitative Polymerase Chain Reaction
RISC	RNA-induced Silencing Complexes
RNAi	RNA interference
rRNA	Ribosomal RNA
SAM	S-adenosylmethionine
siRNAs	Small interfering RNA
snRNA	Small nuclear RNA
sncRNAs	Small non-coding RNAs
snoRNA	Small nucleolar RNA
SUMO	Small ubiquitin-related modifier
SWI/SNF	Switch/sucrose nonfermentable
tDMRs	Tissue-specific Differentially Methylated Regions
TET	Ten-eleven Translocation
TSD	Time-since-death
TR	Telomerase RNA
tRNA	Transfer RNA
VISAGE	Visible Attributes through Genomics
WBC	White Blood Cell count

1.1 Introduction

The term "epigenetics" was first introduced by Conrad Waddington (1905–1975) in the early 1940s, as a combination of the words "epigenesis," the formation of a complex organism from a single unit, and "genetics." It was defined as "the branch of biology which studies the causal interactions between genes and their products which bring the phenotype into being" (Waddington 2012). He was one of the first researchers to understand the importance of genes and environment in embryology, proposing the famous "epigenetic landscape," in which a ball, symbolizing a cell, could follow different paths due to the roughness of the surface (intra- and extracellular environmental influences) (Goldberg et al. 2007).

Waddington's hypothesis did not contain any information about where exactly a gene was located in the cell until the works of Avery, Chase, and McCloud, who discovered that DNA was the hereditary material (Griffith 1928; Meza-Menchaca et al. 2024). This culminated with the modeling of the DNA structure performed by Watson and Crick (1953).

Several researchers continued Waddington's work, particularly David Ledbetter Naney's emphasized that epigenetic mechanisms shaped by environmental influences regulated gene expression and were maintained over cell divisions, defined as "Control Systems" (Nanney 1958). However, Nanney also addressed the heritability of epigenetics indicated that "the lack of universal agreement on this view suggests that epigenetic systems vary in stability" (Nanney 1958). In fact, it was demonstrated that some epigenetic marks are very stable, while others are not. Therefore, an early definition of epigenetics included heritability, "the study of mitotically and/or meiotically heritable changes in gene function that do not entail a change in DNA sequence" (Ospelt 2022). A widely acknowledged definition is the "study of processes that produce heritable phenotypes that do not strictly depend on the DNA sequence" (Lieb et al. 2006).

Epigenetic modifications can be reversible and encompass various processes, including DNA methylation, chromatin remodeling and histone modifications, and non-coding RNA regulation, being involved in X chromosome inactivation and genomic imprinting (Fig. 1.1).

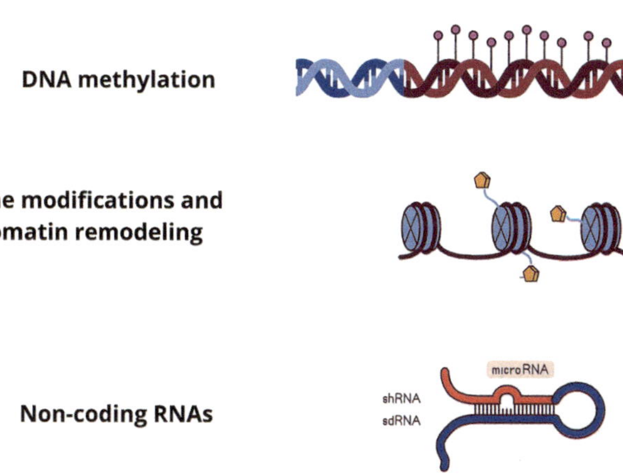

Fig. 1.1 Epigenetic modifications described in this chapter

1.2 Epigenetic Modifications

1.2.1 DNA Methylation

DNA methylation was the earliest epigenetic modification to be identified (Craddock and Magee 1965; Hotchkiss 1948; Scarano et al. 1965). This process entails the covalent attachment of a methyl group (–CH_3) to the fifth carbon of the cytosine ring, producing 5-methylcytosine (5mC). In mammalian somatic tissues, methylation occurs predominantly at 5′CpG3′ dinucleotides, affecting approximately 98% of CpG sites, whereas in embryonic stem cells roughly 75% of CpGs are methylated (Kim and Kim 2021). Although CpG sites constitute the primary context for methylation, non-CpG methylation has also been detected in human genomes (Lister et al. 2009; Yan et al. 2011). Overall, an estimated 60–90% of CpG sites in the human genome are methylated. Regions enriched in unmethylated CpG sites cluster into "CpG islands," which typically span 300–3000 bp and have a GC content exceeding 55% (Kim and Costello 2017). Approximately 70% of protein-coding gene promoters are associated with these CpG islands, and most promoter-associated islands remain largely unmethylated. In contrast, CpG islands located within intragenic or intergenic regions may be either methylated or unmethylated, contributing to substantial epigenetic heterogeneity across the genome (Cain et al. 2022; Ioshikhes and Zhang 2000; Kim and Costello 2017).

DNA methylation plays a central role in regulating gene expression. Dynamic cycles of methylation and demethylation can modulate transcriptional activity, and a well-established inverse relationship exists between promoter CpG-island methylation and gene expression, whereby increased methylation is typically associated with transcriptional silencing (Kader and Ghai 2015; Meza-Menchaca et al. 2024). However, this paradigm is not universal. In several contexts, DNA methylation shows little or no correlation with transcriptional repression and may even coincide with gene activation, particularly at non-CpG-island regions (Kader and Ghai 2015). These observations underscore the complexity of DNA methylation as a regulatory mechanism and highlight its context-dependent effects on gene activity.

Beyond its well-recognized function in transcriptional regulation, DNA methylation contributes to multiple additional cellular processes, including transcriptional elongation, maintenance of cellular homeostasis, and preservation of genomic integrity. These effects are mediated in part through methylation-dependent regulation of alternative splicing and the expression of various non-coding RNAs. DNA methylation is also essential for X-chromosome inactivation, as most CpG islands on the inactive X chromosome acquire methylation marks. Similarly, genomic imprinting relies on allele-specific DNA methylation patterns that establish monoallelic gene expression. Moreover, DNA methylation supports chromosomal stability and suppresses the activity of repetitive elements, thereby preventing genomic instability (Deniz et al. 2019; Feng and Riddle 2020; Kim and Kim 2021; Meng et al. 2015).

DNA methylation is orchestrated by a family of DNA methyltransferases (DNMTs), which catalyze the transfer of a methyl group from the donor molecule S-adenosylmethionine (SAM) to cytosine residues. DNMT1, the first enzyme of this family to be identified, functions primarily as a maintenance methyltransferase. During DNA replication, DNMT1 recognizes hemimethylated CpG sites and restores symmetrical methylation, thereby ensuring the faithful propagation of epigenetic marks to daughter cells during mitosis (Pradhan et al. 1999). Approximately a decade later, DNMT3A and DNMT3B were discovered and characterized as de novo methyltransferases. Unlike DNMT1, these enzymes introduce methylation at previously unmethylated CpG sites on both DNA strands, independently of DNA replication. Their activity is critical for the establishment of methylation patterns during embryogenesis and for the regulation of tissue-specific gene expression. Aberrant activity of DNMT3A and DNMT3B is also implicated in pathological gene silencing in various disorders, including cancer (Okano et al. 1999). Subsequent work identified DNMT3L, a regulatory cofactor that enhances the catalytic activity of de novo DNMTs and stabilizes DNMT-containing protein complexes. Although structurally related to other DNMTs, DNMT3L lacks intrinsic methyltransferase activity and functions instead as an essential mediator of proper methylation establishment, particularly during germ cell development (Aapola et al. 2000; Deplus et al. 2002; Moore et al. 2013).

DNA methylation is a reversible epigenetic modification, and two major mechanisms are responsible for its removal. Active DNA demethylation is catalyzed by the ten-eleven translocation (TET) family of dioxygenases, which sequentially oxidize 5-methylcytosine (5mC) to 5-hydroxymethylcytosine and further derivatives, ultimately enabling replacement with unmethylated cytosine (Melamed et al. 2018). In contrast, passive DNA demethylation occurs during DNA replication when maintenance methylation is impaired or absent, resulting in the progressive dilution of methylation marks across successive cell divisions (Ito et al. 2010; Kohli and Zhang 2013; Ramchandani et al. 1999).

In addition, growing evidence highlights extensive cooperation between DNA methylation and histone modifications in the regulation of gene repression. Characterization of multiple chromatin-modifying complexes has revealed a tightly coordinated crosstalk between these epigenetic systems, underscoring the integrated nature of epigenomic regulation (Peixoto et al. 2020).

1.2.2 Chromatin Remodeling and Histone Modifications

In eukaryotic cells, DNA is packaged within the nucleus by associating with proteins—primarily histones—to form chromatin. This organization exists in two major structural states. Heterochromatin represents a densely compacted form of chromatin that facilitates efficient genomic packaging but restricts access of transcriptional machinery and other regulatory proteins to promoter regions. In contrast, euchromatin denotes a more relaxed and accessible configuration that supports active transcription (Kornberg 1974). Chromatin remodeling refers to the dynamic transition between these condensed and open states, enabling regulatory protein complexes to access genomic DNA and modulate gene expression. This process is driven by a variety of chromatin-remodeling mechanisms and protein complexes that reposition, restructure, or modify nucleosomes, thereby orchestrating transcriptional regulation in response to cellular and environmental cues (Clapier and Cairns 2009; Henikoff and Smith 2015; Tyagi et al. 2016).

The fundamental unit of chromatin organization is the nucleosome, which consists of DNA wrapped around an octamer of histone proteins—the core structural components of chromatin. This octamer contains two copies, each of histones H2A, H2B, H3, and H4, forming a stable protein core around which approximately 147 base pairs of DNA are wound. An additional histone, known as the linker histone H1, binds to the nucleosome exterior and plays a critical role in higher-order chromatin compaction (Kim and Kim 2021).

Chromatin remodeling is governed by two major mechanisms. The first involves covalent modification of chromatin components, primarily through histone post-translational modifications and DNA chemical modifications such as methylation. The second comprises non-covalent ATP-dependent remodeling carried out by chromatin-remodeling complexes (CRCs), which use the energy from ATP hydrolysis to alter nucleosome positioning, structure, or composition (Gerra et al. 2024).

1.2.2.1 Histone Modifications

The concept that histones undergo post-translational modifications (PTMs) with functional consequences was first reported in 1964, revealing that these chemical modifications influence chromatin structure and gene activity (Allfrey et al. 1964). Among the numerous PTMs identified to date, histone acetylation and methylation remain the most extensively studied.

Histone acetylation occurs on lysine residues located in the N-terminal tails of histones and is catalyzed by histone acetyltransferases (HATs). The addition of an acetyl group neutralizes the positive charge of lysine, weakening the electrostatic attraction between the negatively charged DNA backbone and the histone core. This charge neutralization results in a more relaxed chromatin state, facilitating greater DNA accessibility to transcription factors. In addition to its structural effects, histone acetylation generates specific recognition motifs that are bound by diverse effector proteins, enabling the recruitment of chromatin-modifying complexes that either promote or suppress transcriptional activity (Chen et al. 2017; Gavazzo et al. 1997; Melamed et al. 2018). The removal of acetyl groups is mediated by histone deacetylases (HDACs), which restore the positive charge of lysine residues and promote chromatin compaction, generally leading to transcriptional repression (Peixoto et al. 2020; Yun et al. 2011).

Histone methylation occurs on lysine and arginine residues within histone tails and is catalyzed by histone methyltransferases (HMTs), which transfer methyl groups from the donor molecule S-adenosylmethionine (SAM). Removal of these marks is mediated by histone demethylases (HDMs). Depending on the residue and degree of methylation, this modification can promote either transcriptional activation or repression by altering chromatin accessibility and the binding of transcriptional regulators (Cao and Zhang 2004; Melamed et al. 2018; Yun et al. 2011).

In addition to acetylation and methylation, several other post-translational histone modifications contribute to chromatin regulation. Histone ubiquitination participates in transcriptional control and plays a central role in DNA damage responses. Sumoylation—the conjugation of small ubiquitin-related modifier (SUMO) proteins to histones—has been linked to modulation of gene expression and chromatin architecture. Histones may also undergo ADP-ribosylation at aspartate or glutamate residues, a modification required for efficient DNA repair processes (Peixoto et al. 2020).

Collectively, these post-translational modifications constitute the foundation of the "histone code," a conceptual framework proposing that specific combinations of histone marks encode regulatory information that influences chromatin structure, including the balance between euchromatin and heterochromatin. This epigenetic code plays essential roles in cell differentiation, mitotic progression, DNA repair, and X-chromosome inactivation, and governs transcriptional activation or repression depending on the context (Henikoff 2005; Kondo 2009; Strahl and Allis 2000).

Histone diversity is further expanded by the presence of histone variants. For instance, the H2A family includes variants such as H2AX and H2AZ, which contribute to specialized functions including DNA damage signaling and repair (Redon et al. 2002; Sevilla and Binda 2014; West and Bonner 1980). Beyond naturally occurring variants, numerous histone mutations—collectively termed "oncohistones"—have been identified in human cancers. These mutant histones alter chromatin dynamics and transcriptional programs, thereby contributing to oncogenic transformation and tumor progression (Justin et al. 2016).

1.2.2.2 Chromatin Remodeling Complexes

ATP-dependent chromatin-remodeling complexes (CRCs) are evolutionarily conserved, multisubunit assemblies that utilize ATP hydrolysis to reorganize nucleosome positioning and thereby regulate chromatin accessibility. Based on the sequence and structural features of their ATPase subunits, CRCs are classified into four major families: switch/sucrose nonfermentable (SWI/SNF), inositol-requiring 80 (INO80), imitation switch (ISWI), and chromodomain-helicase DNA-binding (CHD) complexes. Each family encompasses distinct subunits and remodeling activities that define specialized regulatory pathways. All CRCs contain a conserved ATPase core that converts the energy from ATP hydrolysis into mechanical force, enabling sliding, eviction, assembly, or disassembly of nucleosomes, as well as the exchange of specific histone variants. Through these versatile remodeling activities, CRCs dynamically modulate chromatin structure to influence transcription, replication, DNA repair, and genome integrity. Despite substantial progress in structural and biochemical studies, the detailed molecular mechanisms governing CRC subunit organization, assembly pathways, and specific modes of nucleosome recognition remain incompletely understood (Bartholomew 2014; Clapier and Cairns 2009; Jiang et al. 2023a).

1.2.3 Non-coding RNAs

Non-coding RNAs (ncRNAs) comprise a broad class of RNA molecules that are not translated into proteins. They are generally categorized into two major groups. The first includes "housekeeping ncRNAs," which support essential cellular processes and encompass ribosomal RNAs (rRNAs), transfer RNAs (tRNAs), small nuclear RNAs (snRNAs), small nucleolar RNAs (snoRNAs), and telomerase RNAs (TR). The second category consists of regulatory ncRNAs, which modulate gene expression through interactions with DNA, RNA, or chromatin. These interactions can alter transcriptional activity and chromatin architecture without modifying the underlying DNA sequence, placing regulatory ncRNAs firmly within the realm of epigenetic control (Hombach and Kretz 2016). Regulatory ncRNAs include both long non-coding RNAs (lncRNAs) and a diverse group of small non-coding RNAs (sncRNAs). The latter encompasses microRNAs (miRNAs), small interfering RNAs (siRNAs), PIWI-interacting RNAs (piRNAs), circular RNAs (circRNAs), antisense RNAs, and enhancer RNAs (eRNAs), each fulfilling specialized roles in post-transcriptional regulation, chromatin remodeling, or genome defense (Song et al. 2024).

Long non-coding RNAs (lncRNAs) are defined as transcripts exceeding 200 base pairs in length and represent an exceptionally heterogeneous class of molecules, both in sequence and in origin. Some lncRNAs serve as precursors for small interfering RNAs (siRNAs), while others exert diverse regulatory functions. These include modulating transcription, directing chromatin remodeling, and acting as molecular scaffolds that recruit multiple proteins required for the formation of ribonucleoprotein complexes. Through these interactions, lncRNAs can influence histone-tail modifications and orchestrate a variety of other epigenetic processes (Statello et al. 2021; Winkle et al. 2021).

MicroRNAs (miRNAs) are small RNA molecules approximately 20 nucleotides long that function as key components of the RNA interference (RNAi) pathway. After processing, miRNAs are incorporated into the RNA-induced silencing complex (RISC), where they guide the complex to target mRNAs through sequence complementarity. Depending on the degree of complementarity between the miRNA and its target, RISC induces either mRNA degradation or translational repression. It is estimated that miRNAs regulate roughly 30% of human protein-coding genes, highlighting their broad impact on gene expression networks (Hutvagner and Simard 2008; Jinek and Doudna 2009; Winkle et al. 2021).

Small interfering RNAs (siRNAs), typically fewer than 50 nucleotides in length, also participate in RNA-mediated silencing pathways. siRNAs act in both the cytoplasm and nucleus, where they can trigger stable and potentially heritable epigenetic silencing by directing chromatin modifications or transcriptional repression at target loci (Verdel et al. 2009).

Circular RNAs (circRNAs) are covalently closed, exon-derived non-coding RNAs formed through back-splicing that links the 5′ and 3′ ends to create a continuous loop. They are widely expressed in mammalian cells and are increasingly recognized as important endogenous regulatory molecules. In addition to functioning within the cell of origin, circRNAs can act in trans through their incorporation into extracellular vesicles. These molecules participate in the epigenetic regulation of gene expression at both transcriptional and post-transcriptional levels, often through interactions with miRNAs and various RNA-binding proteins (Zeng et al. 2021; Zhou et al. 2020).

Antisense transcripts are RNA molecules transcribed from the strand opposite to canonical sense genes. These natural antisense transcripts (NATs) are broadly distributed across the genome and exhibit a wide spectrum of regulatory activities. Beyond forming complementary RNA–RNA duplexes, NATs can indirectly influence gene expression by interacting with key chromatin-modifying enzymes such as DNA methyltransferases, histone acetyltransferases, and histone deacetylases. Over time, NATs have emerged as significant epigenetic regulators, modulating gene expression through diverse transcriptional and post-transcriptional mechanisms (Su et al. 2010).

PIWI-interacting RNAs (piRNAs) are typically 21–35 nucleotides in length, making them longer than miRNAs. They constitute a highly heterogeneous class of small non-coding RNAs and are therefore difficult to categorize alongside other ncRNA families. piRNAs guide PIWI proteins to complementary RNA sequences, leading to targeted RNA degradation. In addition to post-transcriptional silencing, piRNA–PIWI complexes can direct DNA methylation, modulate histone modifications, and contribute to chromatin remodeling. Although piRNAs regulate many evolutionarily conserved genes, they also possess the capacity to adaptively target novel sequences (Carninci 2009; Zhang et al. 2011c).

Enhancer RNAs (eRNAs) are medium-length transcripts, generally shorter than 2 kb, and differ structurally and functionally from other ncRNAs such as miRNAs and circRNAs. eRNAs participate primarily in transcriptional regulation and are thought to play a key role in facilitating communication between enhancers and promoters. Although their mechanisms of action are not yet fully understood, accumulating evidence suggests that eRNAs modulate promoter activity and contribute to enhancer-driven transcriptional activation (Kim et al. 2018; Rahnamoun et al. 2020).

1.2.4 RNA Methylation

More than 160 chemical modifications have been identified on RNA molecules, with RNA methylation representing the most abundant and extensively studied category. Among these, N-6-methyladenosine (m6A) is the predominant mark, accounting for over 80% of all methylated residues in RNA. M6A functions in part as a docking platform for specific RNA-binding proteins, which, upon recognition of this modification, can recruit additional protein partners to regulate RNA processing, stability, translation, and localization. Although several other RNA methylation marks have been discovered, the precise biological roles of many of these modifications remain incompletely characterized (Bokar et al. 1997; Dominissini et al. 2012; Fu et al. 2014; Konno et al. 2019; Lin and Gregory 2014; Wang et al. 2014).

1.3 Influential Factors

1.3.1 Aging

In recent years, research across a wide range of organisms has demonstrated that aging is accompanied by characteristic epigenetic alterations. These age-related changes include shifts in chromatin organization—such as modifications in DNA methylation patterns and histone marks—as well as alterations in the expression profiles of both coding and non-coding RNAs (Berger et al. 2009; Inukai et al. 2012; Kato et al. 2011; Wood et al. 2010).

The relationship between aging and the epigenome—particularly DNA methylation—has been illustrated most clearly through studies of monozygotic twins, who share identical genetic sequences. In early life, twins exhibit remarkably similar methylation profiles; however, as they age, their epigenetic patterns progressively diverge, reflecting the cumulative influence of environmental exposures, lifestyle differences, and stochastic biological processes (Fraga et al. 2005; Vidaki et al. 2013). Numerous studies have shown that aging is accompanied by a global decline in DNA methylation across the genome, coupled with site-specific hypermethylation at particular gene promoters (Calvanese et al. 2009; Zhang et al. 2011a). These alterations disrupt normal gene expression and protein production, contributing to the emergence of aging-related phenotypes.

These early observations laid the foundation for the development of "epigenetic clocks." These models emerged from genome-wide analyses examining age-associated DNA methylation patterns at specific CpG sites, as well as their relationships with biological sex, disease status, lifestyle factors, and other physiological traits (Hannum et al. 2013; Horvath 2013; Horvath and Raj 2018; Koch and Wagner 2011). Data derived from these clocks reveal that DNA methylation undergoes pronounced shifts during early life, characterized by rapid and widespread changes in childhood followed by a stabilization phase that persists throughout adulthood (Horvath 2013; Horvath et al. 2012).

Over the years, numerous epigenetic clocks have been developed, broadly classified into chronological and biological clocks. Chronological clocks represent the first generation of these models. They identify CpG sites whose methylation levels correlate strongly with chronological age—defined as the time elapsed since birth—and are capable of predicting age-related outcomes, including all-cause mortality risk. The earliest and most influential examples include the Hannum clock, which was developed using blood-derived DNA, and the Horvath multi-tissue clock, designed to function across a wide range of human tissues (Hannum et al. 2013; Horvath 2013).

The second generation of epigenetic clocks—referred to as biological clocks—aims to estimate an individual's biological or physiological age by integrating both intrinsic biological states and environmental influences (Horvath and Raj 2018). Although these models partially overlap with earlier chronological clocks, they incorporate additional clinical biomarkers to better capture health status and disease risk. For example, the PhenoAge clock includes variables such as creatinine levels, C-reactive protein, and white blood cell counts, among others (Levine et al. 2018). By integrating these clinical indicators with DNA methylation data, biological clocks offer improved predictive power for a wide range of health outcomes, including morbidity and mortality, thereby demonstrating their value in diagnostic and prognostic applications (Levine et al. 2018).

Another widely used biological clock is GrimAge, recently updated to a second version known as GrimAge2 (Lu et al. 2019, 2022). This model incorporates DNA-methylation surrogates for various plasma proteins—such as high-sensitivity C-reactive protein and hemoglobin A1c in GrimAge2—along with chronological age, biological sex, and smoking-related markers to estimate time to death. The second version exhibits improved predictive accuracy for mortality and stronger associations with age-related conditions, including type 2 diabetes.

Longitudinal epigenetic clocks have also been developed to quantify the rate of biological aging. One of the first was the DunedinPoAm (Dunedin Pace of Aging methylation) clock, constructed from a birth-cohort study that tracked changes in 18 blood-based biomarkers over a 12-year period (Belsky et al. 2020). A subsequent model, DunedinPACE (Dunedin Pace of Aging Calculated from the Epigenome), expanded on this approach by evaluating 19 physiological indicators measured repeatedly over two decades in the same cohort. DunedinPACE has demonstrated robust associations with morbidity, disability, and mortality, among other health outcomes (Belsky et al. 2022).

Taken together, these findings highlight that aging is accompanied by extensive alterations in the epigenome—including shifts in gene expression profiles, chromatin landscapes, and genome architecture—underscoring the central role of epigenetic regulation in the aging process (Brunet and Berger 2014).

1.3.2 Biological Sex and Ancestry

As described previously, X inactivation in females is related to epigenetic modifications. Several studies assessed mostly DNA methylation differences among males and females (Shen et al. 2008). The study developed by Boks et al. (2009)

confirmed these assertions, finding a higher methylation in the X-chromosome sites in females than males. Moreover, numerous autosomal loci of CpGs were also found to be differentially methylated in females, as previously pointed out by El-Maarri et al. (2007). In this study, the variation in levels of methylation between sexes was not too high, but still considerable. Particularly, it was found in LINE repetitive elements, being higher in males. This was confirmed by Zhang et al. (2011a) and Zhu et al. (2012). In these works, males showed higher methylation than females, with the only exceptions being tDMRs (tissue-specific differentially methylated regions) at the imprinted genes, which demonstrated similar methylation levels in both sexes, though with an inclination toward a lower methylation in females.

These findings have been reflected when applying epigenetic clocks. Hannum found that men age approximately 4% faster than women (Hannum et al. 2013). These results were supported by other epigenetic clock studies, like Horvath et al. (2016) and others (Tajuddin et al. 2019).

Other studies assessed methylation differences among populations. The study of Fraser et al. (2012) pointed out differential methylation profiles of approximately 30% of CpG sites in blood between Northern European and West African ancestries. Heyn et al. (2013) evaluated these patterns among African Americans, Caucasian Americans, and Han-Chinese Americans, finding 439 CpGs differentially methylated. A more focused study carried out in New York City, assessing methylation differences among Blacks, Whites, and Hispanics (Terry et al. 2008), demonstrated lower levels of methylation in Blacks with respect to Whites and Hispanics, confirmed by Zhang later on (Zhang et al. 2011a). This ancestry differential methylation should be considered for future studies.

1.3.3 Tissue Specificity

Genome-wide association studies have demonstrated that DNA methylation patterns exhibit strong tissue specificity. Horvath's original epigenetic clock was designed as a multi-tissue model, explicitly accounting for the methylation differences that exist among various human tissues (Horvath 2013; Horvath and Raj 2018). Consistent with this, multiple chromosomal regions—referred to as tissue-specific differentially methylated regions (tDMRs)—have been identified and are known to display distinct methylation profiles depending on the tissue or cell type examined (An et al. 2013; Choi et al. 2014; Lee et al. 2012; Rakyan et al. 2008).

Comprehensive genome-wide mapping has shown that the majority of CpG islands are largely unmethylated. Nevertheless, studies indicate that approximately 4–17% of CpG sites exhibit variable methylation across different tissues and cellular contexts (Cohen et al. 2011). Two primary models have been proposed to explain these methylation differences within CpG islands. One model suggests that bound proteins—such as transcription factors or chromatin-associated proteins—restrict access to specific DNA regions, thereby limiting the ability of methyltransferases to modify certain CpG sites. The second model posits that sequence-specific DNA-binding proteins can actively recruit or guide methylation machinery to targeted genomic regions, establishing localized methylation patterns through a directed, methylation-targeting mechanism.

tDMRs are often enriched at the peripheries of CpG islands, where both CpG density and overall G/C content are lower than in the island cores. CpG islands are thought to act as developmental regulatory switches, establishing epigenetic memory through the formation of cell-type-specific hypo- and hypermethylation patterns. Consistent with this function, tDMRs have been strongly implicated in mammalian development and tissue differentiation. These regions are believed to exert their regulatory influence by either inhibiting or facilitating the binding of specific transcriptional or chromatin-associated factors in a methylation-dependent manner (Igarashi et al. 2008; Ohgane et al. 2008; Vidaki et al. 2013). Correlations between tDMR methylation status and transcriptional activity have been documented, suggesting functional involvement in gene regulation. However, the precise mechanistic role of tDMR methylation in directing transcriptional outcomes remains incompletely understood. Despite this uncertainty, substantial evidence supports a prominent relationship between tDMR hypermethylation and gene silencing (Kader and Ghai 2015).

1.3.4 Diseases

The potential involvement of epigenetic mechanisms in human disease was first proposed by Robin Holliday (Holliday and Pugh 1975). Since then, alterations in epigenetic marks—particularly changes in DNA methylation—have been documented across a wide range of chronic conditions, including neurodegenerative disorders, autoimmune rheumatic and musculoskeletal diseases, and type 1 diabetes (Ballestar and Li 2017; Berson et al. 2018; Mazzone et al. 2019). Analyses of disease-associated epimutations in these conditions have revealed that the affected genes and signaling pathways play central roles in disease development and progression, highlighting the pathogenic significance of epigenetic dysregulation (Ospelt 2022).

In addition to these findings, analyses of epigenetic age using DNA methylation–based clocks have revealed accelerated biological aging in individuals living with HIV compared with healthy controls (Gross et al. 2016; Horvath and Levine 2015). Similar associations have been reported in other infectious diseases: infections such as Helicobacter pylori and Cytomegalovirus have likewise been linked to increased epigenetic age estimates, suggesting that chronic infection and immune activation may contribute to epigenetic aging (Gao et al. 2017; Kananen et al. 2015).

Aberrant DNA methylation patterns have also been widely reported across a broad spectrum of cancers, including those of the blood, colon, stomach, kidney, lung, prostate, urothelium, and breast, among others (Ambatipudi et al. 2017; Dugue et al. 2018; Horvath 2013; Perna et al. 2016; Spolnicka et al. 2018; Zheng et al.

2016). These alterations often involve both global hypomethylation and site-specific hypermethylation, reflecting disrupted epigenetic regulation as a hallmark of tumorigenesis.

Collectively, these observations underscore the potential of epigenetic analyses to advance our understanding of disease mechanisms and to inform the development of therapeutic strategies aimed at epigenetic reprogramming.

1.3.5 Environment and Lifestyle

As epigenetics refers to the "interaction between genes and environment," many studies have been carried out assessing the effects of exogenous factors on epigenetic modifications, mostly focused on DNA methylation.

Tobacco As it has been demonstrated, smoking causes DNA damage and telomere shortening (Huang et al. 2013). Different works assessed the epigenetic effects of tobacco, finding some CpG sites associated with smoking (Breitling et al. 2011; Satta et al. 2008; Wan et al. 2012; Zhang et al. 2016). GrimAge epigenetic clock considers smoking for its prediction of chronological age (Lu et al. 2019, 2022). Thus, some studies report increased epigenetic age estimates in smokers (Fiorito et al. 2019; McCartney et al. 2018), and also demonstrated an epigenetic acceleration associated with cumulative smoking (pack-years) (Yang et al. 2019).

Alcohol Differential DNA methylation has been shown in regular alcohol consumers versus non-drinkers, finding a significant increase in global blood methylation in chronic alcoholics (Bonsch et al. 2004; Hillemacher et al. 2009; Nieratschker et al. 2014; Zhang et al. 2013). Moderate alcohol consumption (1–7 drinks/week) has been shown to decrease epigenetic age (Quach et al. 2017), though habitual alcohol consumption and alcohol dependence were associated with an increase in epigenetic age (Fiorito et al. 2019; Luo et al. 2020; Rosen et al. 2018).

Diet Various dietary bioactive food components have been observed to alter gene expression via changes in DNA methylation (Hardy and Tollefsbol 2011; Park et al. 2012). Vitamins and folates influence the activity of enzymes that participate in cellular methylation processes and very much influence the rate of disease symptoms (Jacob 1999; Jaenisch and Bird 2003). Epigallocatechin-3-gallate, which is a primary polyphenol in green tea, was found to reduce global DNA methylation in cancer cell lines via competitive inhibition of DNA methyltransferases (Fang et al. 2003; Kato et al. 2008). This points out diet as a potential regulator of DNA methylation to be considered as an adjuvant treatment in certain diseases.

Exercise Physical exercise can impact the epigenome and regulate gene expression (Ntanasis-Stathopoulos et al. 2013; Voisin et al. 2015). Also, it reduces genetic effects on individuals' body mass index (BMI) (Wang et al. 2016). Though different works in this area are not in agreement regarding the potential benefits at the epigenetic level of exercise (Luttropp et al. 2013; White et al. 2013; Zeng et al. 2012; Zhang et al. 2011b). Additional long-term studies should be carried out to explore these impacts further.

Environmental Toxin Exposure Exposure to chemical and environmental pollutants induces changes in DNA methylation, resulting in epimutation-associated phenotypes. These demonstrated to accelerate epigenetic age, particularly after exposure to fine particulate matter and nitrogen oxide (Ward-Caviness et al. 2016), black carbon or fine particulate matter (Nwanaji-Enwerem et al. 2016), and pesticides (Lind et al. 2018).

Life Experiences Altered DNA methylation patterns have also been associated with psycho-social factors such as stress and early life socio-economic status (Borghol et al. 2012; Lam et al. 2012; Terry et al. 2008). This research is still ongoing, trying to elucidate the impact of these factors on methylation and their relationship with diseases.

1.4 Forensic Applications

Most applications of epigenetics to forensic science have been based on DNA methylation. Non-coding RNAs are also emerging as potential tools to improve evidence analysis, with more extensive research focused on body fluid identification and postmortem Interval (PMI) estimation. This section will give an overview of the applications of epigenetics to forensic science, with more details in the different chapters of this book.

1.4.1 Body Fluid Identification

Identification of body fluids at crime scenes is of utmost importance to reconstruct the sequence of events and identify the potential perpetrator, and in some cases, the victim (Medina-Paz et al. 2024). There are presumptive and confirmatory techniques to detect body fluids, including immunochromatographic tests, based on antigen-antibody reaction (Zapico and Roca 2024). These tests are not exempt from limitations; for that reason, forensic scientists have been looking for other methods to clearly identify body fluids.

As described previously, each tissue/cell/fluid has tissue-specific differentially methylated regions (tDMRs) (Lee et al. 2012), which could be used to identify these body fluids. In 2006, Eckhardt et al. (2006) demonstrated that methylation profiles at tDMRs are specific and stable, thus it can be used for body fluid identification. These regions are unmethylated in particular tissue/cell/fluid, and/or showing various degrees of methylation in others (Choi et al. 2014; Sugimoto et al. 2009). Since this discovery, several attempts have been made to apply this technique to differentiate forensically relevant body fluids: blood, semen, and saliva (Forat et al. 2016; Frumkin et al. 2011; Huang et al. 2022; Konrad et al. 2023; Park et al. 2014; Tian et al. 2020), but also for other less common body fluids like sweat (Fernandez-Tejero et al. 2023), vaginal fluid, and menstrual blood (Lee et al. 2016). This led to the identification and validation of suitable tissue-specific CpG markers, including multiplex assays (Jung et al. 2016).

miRNA has been widely studied as a promising biomarker in the identification of tissues and body fluids based on its small molecular weight and rich tissue specificity (Glynn 2020). Most of the studies in non-coding RNAs are focused on miRNA (Altmeyer et al. 2024; Layne et al. 2019; Rhodes et al. 2023). CircRNAs are abundant, stable, and they are conservatively expressed throughout the life cycle, being highly tissue-specific and resistant to degradation by ribonuclease R (RNase R), thanks to their closed-loop structure. Some studies point out the potential of circRNAs for the identification of relevant body fluids (Jeck et al. 2013; Memczak et al. 2013; Zhang et al. 2018), also in combination with mRNA and lncRNA (Liu et al. 2024), demonstrating their usefulness to apply to forensic practice. Another non-coding RNA to consider is piRNA. It is short in length, and it has tissue-specific expression, which could also make it a promising epigenetic marker. Different studies were carried out to assess this, demonstrating its utility to identify body fluids (Wang et al. 2019a, b).

Chapter 4 of this book will describe in more depth different approaches to body fluid identification applying epigenetic techniques.

1.4.2 Age Estimation

Age estimation is probably the most studied application of epigenetics in forensic science. First, it is necessary to distinguish between age estimation from the point of view of criminalistics versus forensic anthropology.

In criminalistics, age estimation of a biological sample left at the crime scene could help with the identification of the perpetrator if there is no match in forensic databases. In forensic anthropology, age estimation is part of the biological profile for the identification of human remains (determination of biological sex, population affinity, height, and age). In this scenario, age constitutes the most challenging parameter to estimate by anthropological means in adult individuals (Zapico and Ubelaker 2013). Moreover, age estimation can be applied in civil cases to determine if someone is actually a minor or an adult (Williams 2018).

There have been more epigenetic studies assessing the possibility of estimating the age for criminalistics purposes based on body fluids left at the crime scene. The main reason is the development of epigenetic clocks, particularly the chronological clocks (Hannum et al. 2013; Horvath 2013), which included these body fluids in their analysis, providing a starting point for differentially methylated regions based on age. As a result, CpGs were selected depending on the body fluid, applying different combinations and statistical analysis to reach variable accuracies, between 2.5 and 8 years with respect to chronological age, with better accuracies obtained with blood samples (Bekaert et al. 2015; Correia Dias et al. 2020a, c; Dias et al. 2020; Hamano et al. 2017; Heidegger et al. 2020; Lee et al. 2018; Weidner et al. 2014; Zbiec-Piekarska et al. 2015a, b).

In recent years, the study of DNA methylation of the Y chromosome has been growing, particularly in terms of age estimation. Identifying the Y chromosome in a profile will indicate a male perpetrator, performing Y-STRs. Then, if age is estimated, it would be possible to distinguish between males from the same lineage (Castagnola et al. 2024). There are a few studies assessing DNA methylation of Y chromosome for forensic purposes (Jiang et al. 2023b; Vidaki et al. 2021), but the results are encouraging. Further research and validation are needed for the potential application of this technique to forensic casework.

Age estimation based on DNA methylation for forensic anthropology has been less developed than criminalistics research, as the main tissues remaining are bones and teeth, which were not considered in the epigenetic clocks. Current research in this field applies previously identified CpG sites in blood, semen, and saliva to these samples, with moderate accuracy depending on the tissue of election, CpG sites selected, statistical analysis, and methodological approach. The majority of the studies were carried out in tooth samples, using the whole tooth (Marquez-Ruiz et al. 2020), or separating the layers (Bekaert et al. 2015; Zapico et al. 2021; Dominguez and Zapico 2025; Giuliani et al. 2016), with a few in bone samples (Correia Dias et al. 2020a, b, c; Lee et al. 2020).

Furthermore, some groups have tried to create universal tools for age estimation in both scenarios, criminalistics and forensic anthropology, like the VISAGE (Visible Attributes through Genomics) group, for blood, buccal cells, and bones (Wozniak et al. 2021), and Correia-Dias et al. with a blood-bone-tooth model prediction tool (Correia Dias et al. 2021), both models with moderate accuracy for age prediction.

A more in-depth description of the current DNA methylation approaches for age estimation will be depicted in Chap. 5 (blood), Chap. 6 (Y-chromosome), Chap. 7 (semen and saliva), and Chap. 8 (forensic anthropology).

There have been a few studies assessing the possibility of estimating age based on non-coding RNAs. These studies were mostly focused on miRNAs (Huan et al. 2018; Noren Hooten et al. 2010), circRNAs (Wang et al. 2022a), and their combination (Wang et al. 2022b). The findings are promising, but further research needs to be carried out to determine the usefulness of non-coding RNAs for age estimation.

1.4.3 Epigenetic Discrimination of Monozygotic Twins

Monozygotic (MZ) twins possess the same DNA sequence, making it a challenge to discriminate them in forensic cases from biological evidence. In this regard, DNA methylation has emerged as a potential solution to this problem (Planterose Jimenez et al. 2021). Different studies have been carried out trying to identify specific CpG markers being able to distinguish MZ twins with varying degrees of success, and mostly in blood samples (Li et al. 2013; Planterose Jimenez et al. 2021; Vidaki et al. 2017). Another line of research has focused on the use of non-coding RNAs to discriminate between MZ, mostly applying miRNA (Fang et al. 2019; Wang et al. 2025; Xiao et al. 2019) and circRNA (Wang et al. 2024). Further studies are needed to identify potential DNA methylation and/or miRNA/circRNA markers to clearly distinguish between MZ twins.

Chapter 3 of this book will give an overview of the current research on this topic.

1.4.4 Postmortem Interval Estimation

The determination of postmortem interval (PMI) or time-since-death (TSD) is one of the most challenging parameters to estimate in forensic science, particularly during longer PMIs (Martinez et al. 2019; Zapico et al. 2014). There have been different approaches to improve this estimate based on molecular biology and microbial techniques (Zapico and Adserias-Garriga 2022). In terms of epigenetics, different studies point out the potential usefulness of non-coding RNAs toward this purpose, particularly miRNAs (Guardado-Estrada et al. 2023; Lee et al. 2024; Na 2020; Pasaribu et al. 2023), circRNAs (Tu et al. 2019), and/or a combination of non-coding RNAs (Lv et al. 2014, 2017; Tu et al. 2018).

Chapter 9 of this book will describe current research in this topic.

1.4.5 Lifestyle

Lifestyle, like consumption of alcohol and tobacco, has an impact on our epigenetic signature, particularly DNA methylation. When facing biological evidence at the crime scene, if the DNA profile does not match forensic databases, it is imperative to get as much information as possible from the evidence, such as the age of the perpetrator or whether the perpetrator is a smoker or an alcohol consumer. There have been some works trying to identify specific CpG markers to predict the tobacco/alcohol status of the sample (Alghanim et al. 2018; Liu et al. 2018; Vidaki et al. 2023). Although the results are promising, further research needs to be developed to undoubtedly confirm the utility of these markers toward this goal, as well as to identify other lifestyle markers like diet, exercise, and socio-economic status, among others.

1.4.6 Epigenetic Techniques

DNA methylation is the most widely studied epigenetic modification for forensic purposes. The gold standard technique to assess DNA methylation is pyrosequencing (Ghemrawi et al. 2023). However, many researchers applied SNaPshot or Sanger sequencing (Correia Dias et al. 2020a, b, c, 2021) because the equipment is already available for forensic laboratories. Current trends are guided toward the use of massive parallel sequencing (MPS) (Wozniak et al. 2021), based on its advantages regarding sample preparation and analysis.

With respect to the analysis of non-coding RNAs, the majority of the studies used reverse transcription quantitative PCR (RT-QPCR) (Layne et al. 2019; Rhodes et al. 2023; Wang et al. 2019b, 2024), or applied microarrays (Guardado-Estrada et al. 2023; Wang et al. 2022b) and MPS (Liu et al. 2024).

An overview of different epigenetic methods applied to forensic science will be described in Chap. 2.

1.5 Conclusion

Epigenetics is an emerging field in forensic science aiming to improve the analysis of biological evidence. Among epigenetic modifications, the most widely studied in forensic science is DNA methylation, with most of the research focused on age estimation based on the development of epigenetic clocks. Furthermore, tDMRs also open the door to improving body fluid identification at crime scenes. Other areas of interest are the discrimination of MZ twins as well as determining lifestyle habits from the evidence, though further studies are required to evaluate the utility of DNA methylation toward this goal. Non-coding RNAs are also raised as potential tools to address some of the issues in forensic sciences, like body fluid identification, discrimination of MZ twins, postmortem interval determination, and, to a lesser extent, age estimation. Figure 1.2 summarizes forensic applications of epigenetics. Although this chapter also describes chromatin remodeling and histone modifications, there have not yet been applications to forensic science, probably due to the technical

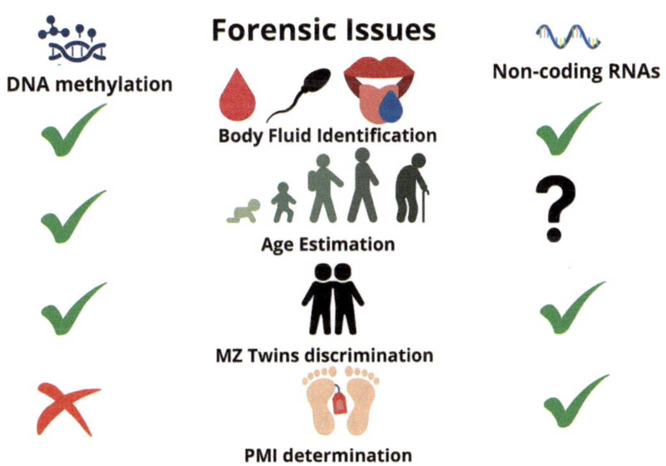

Fig. 1.2 Summary of epigenetic forensic applications. A green check mark indicates the potential usefulness of DNA methylation and non-coding RNAs in overcoming forensic science issues

difficulties in their analysis, and the fact that it will require the isolation of proteins from the biological evidence. Current research is aimed at improving and expanding DNA methylation and non-coding RNAs applications to forensic science, considering the different issues encountered during the analysis of biological evidence.

Compliance with Ethical Standards Sara C. Zapico declares that she has no conflict of interest. This chapter is a review of previously published accounts, as such, no animal or human studies were performed.

References

Aapola U, Kawasaki K, Scott HS, Ollila J, Vihinen M, Heino M, Shintani A, Kawasaki K, Minoshima S, Krohn K, Antonarakis SE, Shimizu N, Kudoh J, Peterson P (2000) Isolation and initial characterization of a novel zinc finger gene, DNMT3L, on 21q22.3, related to the cytosine-5-methyltransferase 3 gene family. Genomics 65(3):293–298. https://doi.org/10.1006/geno.2000.6168

Alghanim H, Wu W, McCord B (2018) DNA methylation assay based on pyrosequencing for determination of smoking status. Electrophoresis 39(21):2806–2814. https://doi.org/10.1002/elps.201800098

Allfrey VG, Faulkner R, Mirsky AE (1964) Acetylation and methylation of histones and their possible role in the regulation of RNA synthesis. Proc Natl Acad Sci USA 51(5):786–794. https://doi.org/10.1073/pnas.51.5.786

Altmeyer L, Baumer K, Hall D (2024) Differentiation of five forensically relevant body fluids using a small set of microRNA markers. Electrophoresis 45(19–20):1785–1795. https://doi.org/10.1002/elps.202400089

Ambatipudi S, Horvath S, Perrier F, Cuenin C, Hernandez-Vargas H, Le Calvez-Kelm F, Durand G, Byrnes G, Ferrari P, Bouaoun L, Sklias A, Chajes V, Overvad K, Severi G, Baglietto L, Clavel-Chapelon F, Kaaks R, Barrdahl M, Boeing H et al (2017) DNA methylome analysis identifies accelerated epigenetic ageing associated with postmenopausal breast cancer susceptibility. Eur J Cancer 75:299–307. https://doi.org/10.1016/j.ejca.2017.01.014

An JH, Choi A, Shin KJ, Yang WI, Lee HY (2013) DNA methylation-specific multiplex assays for body fluid identification. Int J Legal Med 127(1):35–43. https://doi.org/10.1007/s00414-012-0719-1

Ballestar E, Li T (2017) New insights into the epigenetics of inflammatory rheumatic diseases. Nat Rev Rheumatol 13(10):593–605. https://doi.org/10.1038/nrrheum.2017.147

Bartholomew B (2014) Regulating the chromatin landscape: structural and mechanistic perspectives. Annu Rev Biochem 83:671–696. https://doi.org/10.1146/annurev-biochem-051810-093157

Bekaert B, Kamalandua A, Zapico SC, Van de Voorde W, Decorte R (2015) Improved age determination of blood and teeth samples using a selected set of DNA methylation markers. Epigenetics 10(10):922–930. https://doi.org/10.1080/15592294.2015.1080413

Belsky DW, Caspi A, Arseneault L, Baccarelli A, Corcoran DL, Gao X, Hannon E, Harrington HL, Rasmussen LJ, Houts R, Huffman K, Kraus WE, Kwon D, Mill J, Pieper CF, Prinz JA, Poulton R, Schwartz J, Sugden K et al (2020) Quantification of the pace of biological aging in humans through a blood test, the DunedinPoAm DNA methylation algorithm. elife 9:e54870. https://doi.org/10.7554/eLife.54870

Belsky DW, Caspi A, Corcoran DL, Sugden K, Poulton R, Arseneault L, Baccarelli A, Chamarti K, Gao X, Hannon E, Harrington HL, Houts R, Kothari M, Kwon D, Mill J, Schwartz J, Vokonas P, Wang C, Williams BS, Moffitt TE (2022) DunedinPACE, a DNA methylation biomarker of the pace of aging. elife 11:e73420. https://doi.org/10.7554/eLife.73420

Berger SL, Kouzarides T, Shiekhattar R, Shilatifard A (2009) An operational definition of epigenetics. Genes Dev 23(7):781–783. https://doi.org/10.1101/gad.1787609

Berson A, Nativio R, Berger SL, Bonini NM (2018) Epigenetic regulation in neurodegenerative diseases. Trends Neurosci 41(9):587–598. https://doi.org/10.1016/j.tins.2018.05.005

Bokar JA, Shambaugh ME, Polayes D, Matera AG, Rottman FM (1997) Purification and cDNA cloning of the AdoMet-binding subunit of the human mRNA (N6-adenosine)-methyltransferase. RNA 3(11):1233–1247. https://www.ncbi.nlm.nih.gov/pubmed/9409616

Boks MP, Derks EM, Weisenberger DJ, Strengman E, Janson E, Sommer IE, Kahn RS, Ophoff RA (2009) The relationship of DNA methylation with age, gender and genotype in twins and healthy controls. PLoS One 4(8):e6767. https://doi.org/10.1371/journal.pone.0006767

Bonsch D, Lenz B, Reulbach U, Kornhuber J, Bleich S (2004) Homocysteine associated genomic DNA hypermethylation in patients with chronic alcoholism. J Neural Transm (Vienna) 111(12):1611–1616. https://doi.org/10.1007/s00702-004-0232-x

Borghol N, Suderman M, McArdle W, Racine A, Hallett M, Pembrey M, Hertzman C, Power C, Szyf M (2012) Associations with early-life socio-economic position in adult DNA methylation. Int J Epidemiol 41(1):62–74. https://doi.org/10.1093/ije/dyr147

Breitling LP, Yang R, Korn B, Burwinkel B, Brenner H (2011) Tobacco-smoking-related differential DNA methylation: 27K discovery and replication. Am J Hum Genet 88(4):450–457. https://doi.org/10.1016/j.ajhg.2011.03.003

Brunet A, Berger SL (2014) Epigenetics of aging and aging-related disease. J Gerontol A Biol Sci Med Sci 69(Suppl 1):S17–S20. https://doi.org/10.1093/gerona/glu042

Cain JA, Montibus B, Oakey RJ (2022) Intragenic CpG islands and their impact on gene regulation. Front Cell Dev Biol 10:832348. https://doi.org/10.3389/fcell.2022.832348

Calvanese V, Lara E, Kahn A, Fraga MF (2009) The role of epigenetics in aging and age-related diseases [Research Support, N.I.H., Extramural Research Support, Non-U.S. Gov't Review]. Ageing Res Rev 8(4):268–276. https://doi.org/10.1016/j.arr.2009.03.004

Cao R, Zhang Y (2004) SUZ12 is required for both the histone methyltransferase activity and the silencing function of the EED-EZH2 complex. Mol Cell 15(1):57–67. https://doi.org/10.1016/j.molcel.2004.06.020

Carninci P (2009) Molecular biology: the long and short of RNAs. Nature 457(7232):974–975. https://doi.org/10.1038/457974b

Castagnola MJ, Medina-Paz F, Zapico SC (2024) Uncovering forensic evidence: a path to age estimation through DNA methylation. Int J Mol Sci 25(9):4917. https://doi.org/10.3390/ijms25094917

Chen Z, Li S, Subramaniam S, Shyy JY, Chien S (2017) Epigenetic regulation: a new frontier for biomedical engineers. Annu Rev Biomed Eng 19:195–219. https://doi.org/10.1146/annurev-bioeng-071516-044720

Choi A, Shin KJ, Yang WI, Lee HY (2014) Body fluid identification by integrated analysis of DNA methylation and body fluid-specific microbial DNA. Int J Legal Med 128(1):33–41. https://doi.org/10.1007/s00414-013-0918-4

Clapier CR, Cairns BR (2009) The biology of chromatin remodeling complexes. Annu Rev Biochem 78:273–304. https://doi.org/10.1146/annurev.biochem.77.062706.153223

Cohen NM, Kenigsberg E, Tanay A (2011) Primate CpG islands are maintained by heterogeneous evolutionary regimes involving minimal selection. Cell 145(5):773–786. https://doi.org/10.1016/j.cell.2011.04.024

Correia Dias H, Cordeiro C, Corte Real F, Cunha E, Manco L (2020a) Age estimation based on DNA methylation using blood samples from deceased individuals. J Forensic Sci 65(2):465–470. https://doi.org/10.1111/1556-4029.14185

Correia Dias H, Corte-Real F, Cunha E, Manco L (2020b) DNA methylation age estimation from human bone and teeth. Aust J Forensic Sci 54(2):163–176. https://doi.org/10.1080/00450618.2020.180511

Correia Dias H, Cunha E, Corte Real F, Manco L (2020c) Age prediction in living: forensic epigenetic age estimation based on blood samples. Leg Med (Tokyo) 47:101763. https://doi.org/10.1016/j.legalmed.2020.101763

Correia Dias H, Manco L, Corte Real F, Cunha E (2021) A blood-bone-tooth model for age prediction in forensic contexts. Biology (Basel) 10(12):1312. https://doi.org/10.3390/biology10121312

Craddock VM, Magee PN (1965) Methylation of liver DNA in the intact animal by the carcinogen dimethylnitrosamine during carcinogenesis. Biochim Biophys Acta 95:677–678. https://doi.org/10.1016/0005-2787(65)90526-5

Deniz O, Frost JM, Branco MR (2019) Regulation of transposable elements by DNA modifications. Nat Rev Genet 20(7):417–431. https://doi.org/10.1038/s41576-019-0106-6

Deplus R, Brenner C, Burgers WA, Putmans P, Kouzarides T, de Launoit Y, Fuks F (2002) Dnmt3L is a transcriptional repressor that recruits histone deacetylase. Nucleic Acids Res 30(17):3831–3838. https://doi.org/10.1093/nar/gkf509

Dias HC, Cordeiro C, Pereira J, Pinto C, Real FC, Cunha E, Manco L (2020) DNA methylation age estimation in blood samples of living and deceased individuals using a multiplex SNaPshot assay. Forensic Sci Int 311:110267. https://doi.org/10.1016/j.forsciint.2020.110267

Dominguez VM, Zapico SC (2025) Epi and histo: partners in improving age-at-death estimation. American Academy of Forensic Science, Baltimore

Dominissini D, Moshitch-Moshkovitz S, Schwartz S, Salmon-Divon M, Ungar L, Osenberg S, Cesarkas K, Jacob-Hirsch J, Amariglio N, Kupiec M, Sorek R, Rechavi G (2012) Topology of the human and mouse m6A RNA methylomes revealed by m6A-seq. Nature 485(7397):201–206. https://doi.org/10.1038/nature11112

Dugue PA, Bassett JK, Joo JE, Jung CH, Ming Wong E, Moreno-Betancur M, Schmidt D, Makalic E, Li S, Severi G, Hodge AM, Buchanan DD, English DR, Hopper JL, Southey MC, Giles GG, Milne RL (2018) DNA methylation-based biological aging and cancer risk and survival: pooled analysis of seven prospective studies. Int J Cancer 142(8):1611–1619. https://doi.org/10.1002/ijc.31189

Eckhardt F, Lewin J, Cortese R, Rakyan VK, Attwood J, Burger M, Burton J, Cox TV, Davies R, Down TA, Haefliger C, Horton R, Howe K, Jackson DK, Kunde J, Koenig C, Liddle J, Niblett D, Otto T et al (2006) DNA methylation profiling of human chromosomes 6, 20 and 22. Nat Genet 38(12):1378–1385. https://doi.org/10.1038/ng1909

El-Maarri O, Becker T, Junen J, Manzoor SS, Diaz-Lacava A, Schwaab R, Wienker T, Oldenburg J (2007) Gender specific differences in levels of DNA methylation at selected loci from human total blood: a tendency toward higher methylation levels in males. Hum Genet 122(5):505–514. https://doi.org/10.1007/s00439-007-0430-3

Fang MZ, Wang Y, Ai N, Hou Z, Sun Y, Lu H, Welsh W, Yang CS (2003) Tea polyphenol (-)-epigallocatechin-3-gallate inhibits DNA methyltransferase and reactivates methylation-silenced genes in cancer cell lines. Cancer Res 63(22):7563–7570. https://www.ncbi.nlm.nih.gov/pubmed/14633667

Fang C, Zhao J, Liu X, Zhang J, Cao Y, Yang Y, Yu C, Zhang X, Qian J, Liu W, Wu H, Yan J (2019) MicroRNA profile analysis for discrimination of monozygotic twins using massively parallel sequencing and real-time PCR. Forensic Sci Int Genet 38:23–31. https://doi.org/10.1016/j.fsigen.2018.09.011

Feng JX, Riddle NC (2020) Epigenetics and genome stability. Mamm Genome 31(5–6):181–195. https://doi.org/10.1007/s00335-020-09836-2

Fernandez-Tejero N, Gauthier Q, Cho S, McCord BR (2023) High-resolution melt analysis for the detection of skin/sweat via DNA methylation. Electrophoresis 44(3–4):371–377. https://doi.org/10.1002/elps.202200120

Fiorito G, McCrory C, Robinson O, Carmeli C, Ochoa-Rosales C, Zhang Y, Colicino E, Dugue PA, Artaud F, McKay GJ, Jeong A, Mishra PP, Nost TH, Krogh V, Panico S, Sacerdote C, Tumino R, Palli D, Matullo G et al (2019) Socioeconomic position, lifestyle habits and biomarkers of epigenetic aging: a multi-cohort analysis. Aging (Albany NY) 11(7):2045–2070. https://doi.org/10.18632/aging.101900

Forat S, Huettel B, Reinhardt R, Fimmers R, Haidl G, Denschlag D, Olek K (2016) Methylation markers for the identification of body fluids and tissues from forensic trace evidence. PLoS One 11(2):e0147973. https://doi.org/10.1371/journal.pone.0147973

Fraga MF, Ballestar E, Paz MF, Ropero S, Setien F, Ballestar ML, Heine-Suner D, Cigudosa JC, Urioste M, Benitez J, Boix-Chornet M, Sanchez-Aguilera A, Ling C, Carlsson E, Poulsen P, Vaag A, Stephan Z, Spector TD, Wu YZ et al (2005) Epigenetic differences arise during the lifetime of monozygotic twins. Proc Natl Acad Sci USA 102(30):10604–10609. https://doi.org/10.1073/pnas.0500398102

Fraser HB, Lam LL, Neumann SM, Kobor MS (2012) Population-specificity of human DNA methylation. Genome Biol 13(2):R8. https://doi.org/10.1186/gb-2012-13-2-r8

Frumkin D, Wasserstrom A, Budowle B, Davidson A (2011) DNA methylation-based forensic tissue identification. Forensic Sci Int Genet 5(5):517–524. https://doi.org/10.1016/j.fsigen.2010.12.001

Fu Y, Dominissini D, Rechavi G, He C (2014) Gene expression regulation mediated through reversible m(6)A RNA methylation. Nat Rev Genet 15(5):293–306. https://doi.org/10.1038/nrg3724

Gao X, Zhang Y, Brenner H (2017) Associations of *Helicobacter pylori* infection and chronic atrophic gastritis with accelerated epigenetic ageing in older adults. Br J Cancer 117(8):1211–1214. https://doi.org/10.1038/bjc.2017.314

Gavazzo P, Vergani L, Mascetti GC, Nicolini C (1997) Effects of histone acetylation on chromatin structure. J Cell Biochem 64(3):466–475. https://doi.org/10.1002/(sici)1097-4644(19970301)64:3<466::aid-jcb13>3.0.co;2-e

Gerra MC, Dallabona C, Cecchi R (2024) Epigenetic analyses in forensic medicine: future and challenges. Int J Legal Med 138(3):701–719. https://doi.org/10.1007/s00414-024-03165-8

Ghemrawi M, Tejero NF, Duncan G, McCord B (2023) Pyrosequencing: current forensic methodology and future applications—a review. Electrophoresis 44(1–2):298–312. https://doi.org/10.1002/elps.202200177

Giuliani C, Cilli E, Bacalini MG, Pirazzini C, Sazzini M, Gruppioni G, Franceschi C, Garagnani P, Luiselli D (2016) Inferring chronological age from DNA methylation patterns of human teeth. Am J Phys Anthropol 159(4):585–595. https://doi.org/10.1002/ajpa.22921

Glynn CL (2020) Potential applications of microRNA profiling to forensic investigations. RNA 26(1):1–9. https://doi.org/10.1261/rna.072173.119

Goldberg AD, Allis CD, Bernstein E (2007) Epigenetics: a landscape takes shape. Cell 128(4):635–638. https://doi.org/10.1016/j.cell.2007.02.006

Griffith F (1928) The significance of pneumococcal types. J Hyg (Lond) 27(2):113–159. https://doi.org/10.1017/s0022172400031879

Gross AM, Jaeger PA, Kreisberg JF, Licon K, Jepsen KL, Khosroheidari M, Morsey BM, Swindells S, Shen H, Ng CT, Flagg K, Chen D, Zhang K, Fox HS, Ideker T (2016) Methylome-wide analysis of chronic HIV infection reveals five-year increase in biological age and epigenetic targeting of HLA. Mol Cell 62(2):157–168. https://doi.org/10.1016/j.molcel.2016.03.019

Guardado-Estrada M, Cardenas-Monroy CA, Martinez-Rivera V, Cortez F, Pedraza-Lara C, Millan-Catalan O, Perez-Plasencia C (2023) A miRNome analysis at the early postmortem interval. PeerJ 11:e15409. https://doi.org/10.7717/peerj.15409

Hamano Y, Manabe S, Morimoto C, Fujimoto S, Tamaki K (2017) Forensic age prediction for saliva samples using methylation-sensitive high resolution melting: exploratory application for cigarette butts. Sci Rep 7(1):10444. https://doi.org/10.1038/s41598-017-10752-w

Hannum G, Guinney J, Zhao L, Zhang L, Hughes G, Sadda S, Klotzle B, Bibikova M, Fan JB, Gao Y, Deconde R, Chen M, Rajapakse I, Friend S, Ideker T, Zhang K (2013) Genome-wide methylation profiles reveal quantitative views of human aging rates. Mol Cell 49(2):359–367. https://doi.org/10.1016/j.molcel.2012.10.016

Hardy TM, Tollefsbol TO (2011) Epigenetic diet: impact on the epigenome and cancer. Epigenomics 3(4):503–518. https://doi.org/10.2217/epi.11.71

Heidegger A, Xavier C, Niederstatter H, de la Puente M, Pospiech E, Pisarek A, Kayser M, Branicki W, Parson W, Consortium V (2020) Development and optimization of the VISAGE basic prototype tool for forensic age estimation. Forensic Sci Int Genet 48:102322. https://doi.org/10.1016/j.fsigen.2020.102322

Henikoff S (2005) Histone modifications: combinatorial complexity or cumulative simplicity? Proc Natl Acad Sci USA 102(15):5308–5309. https://doi.org/10.1073/pnas.0501853102

Henikoff S, Smith MM (2015) Histone variants and epigenetics. Cold Spring Harb Perspect Biol 7(1):a019364. https://doi.org/10.1101/cshperspect.a019364

Heyn H, Moran S, Hernando-Herraez I, Sayols S, Gomez A, Sandoval J, Monk D, Hata K, Marques-Bonet T, Wang L, Esteller M (2013) DNA methylation contributes to natural human variation. Genome Res 23(9):1363–1372. https://doi.org/10.1101/gr.154187.112

Hillemacher T, Frieling H, Hartl T, Wilhelm J, Kornhuber J, Bleich S (2009) Promoter specific methylation of the dopamine transporter gene is altered in alcohol dependence and associated with craving. J Psychiatr Res 43(4):388–392. https://doi.org/10.1016/j.jpsychires.2008.04.006

Holliday R, Pugh JE (1975) DNA modification mechanisms and gene activity during development. Science 187(4173):226–232. https://www.ncbi.nlm.nih.gov/pubmed/1111098

Hombach S, Kretz M (2016) Non-coding RNAs: classification, biology and functioning. Adv Exp Med Biol 937:3–17. https://doi.org/10.1007/978-3-319-42059-2_1

Horvath S (2013) DNA methylation age of human tissues and cell types. Genome Biol 14(10):R115. https://doi.org/10.1186/gb-2013-14-10-r115

Horvath S, Levine AJ (2015) HIV-1 infection accelerates age according to the epigenetic clock. J Infect Dis 212(10):1563–1573. https://doi.org/10.1093/infdis/jiv277

Horvath S, Raj K (2018) DNA methylation-based biomarkers and the epigenetic clock theory of ageing. Nat Rev Genet 19(6):371–384. https://doi.org/10.1038/s41576-018-0004-3

Horvath S, Zhang Y, Langfelder P, Kahn RS, Boks MP, van Eijk K, van den Berg LH, Ophoff RA (2012) Aging effects on DNA methylation modules in human brain and blood tissue. Genome Biol 13(10):R97. https://doi.org/10.1186/gb-2012-13-10-r97

Horvath S, Gurven M, Levine ME, Trumble BC, Kaplan H, Allayee H, Ritz BR, Chen B, Lu AT, Rickabaugh TM, Jamieson BD, Sun D, Li S, Chen W, Quintana-Murci L, Fagny M, Kobor MS, Tsao PS, Reiner AP et al (2016) An epigenetic clock analysis of race/ethnicity, sex, and coronary heart disease. Genome Biol 17(1):171. https://doi.org/10.1186/s13059-016-1030-0

Hotchkiss RD (1948) The quantitative separation of purines, pyrimidines, and nucleosides by paper chromatography. J Biol Chem 175(1):315–332. https://www.ncbi.nlm.nih.gov/pubmed/18873306

Huan T, Chen G, Liu C, Bhattacharya A, Rong J, Chen BH, Seshadri S, Tanriverdi K, Freedman JE, Larson MG, Murabito JM, Levy D (2018) Age-associated microRNA expression in human peripheral blood is associated with all-cause mortality and age-related traits. Aging Cell 17(1):e12687. https://doi.org/10.1111/acel.12687

Huang J, Okuka M, Lu W, Tsibris JC, McLean MP, Keefe DL, Liu L (2013) Telomere shortening and DNA damage of embryonic stem cells induced by cigarette smoke. Reprod Toxicol 35:89–95. https://doi.org/10.1016/j.reprotox.2012.07.003

Huang H, Liu X, Cheng J, Xu L, He X, Xiao C, Huang D, Yi S (2022) A novel multiplex assay system based on 10 methylation markers for forensic identification of body fluids. J Forensic Sci 67(1):136–148. https://doi.org/10.1111/1556-4029.14872

Hutvagner G, Simard MJ (2008) Argonaute proteins: key players in RNA silencing. Nat Rev Mol Cell Biol 9(1):22–32. https://doi.org/10.1038/nrm2321

Igarashi J, Muroi S, Kawashima H, Wang X, Shinojima Y, Kitamura E, Oinuma T, Nemoto N, Song F, Ghosh S, Held WA, Nagase H (2008) Quantitative analysis of human tissue-specific differences in methylation. Biochem Biophys Res Commun 376(4):658–664. https://doi.org/10.1016/j.bbrc.2008.09.044

Inukai S, de Lencastre A, Turner M, Slack F (2012) Novel microRNAs differentially expressed during aging in the mouse brain. PLoS One 7(7):e40028. https://doi.org/10.1371/journal.pone.0040028

Ioshikhes IP, Zhang MQ (2000) Large-scale human promoter mapping using CpG islands. Nat Genet 26(1):61–63. https://doi.org/10.1038/79189

Ito S, D'Alessio AC, Taranova OV, Hong K, Sowers LC, Zhang Y (2010) Role of Tet proteins in 5mC to 5hmC conversion, ES-cell self-renewal and inner cell mass specification. Nature 466(7310):1129–1133. https://doi.org/10.1038/nature09303

Jacob RA (1999) The role of micronutrients in DNA synthesis and maintenance. Adv Exp Med Biol 472:101–113. https://doi.org/10.1007/978-1-4757-3230-6_10
Jaenisch R, Bird A (2003) Epigenetic regulation of gene expression: how the genome integrates intrinsic and environmental signals. Nat Genet 33(Suppl):245–254. https://doi.org/10.1038/ng1089
Jeck WR, Sorrentino JA, Wang K, Slevin MK, Burd CE, Liu J, Marzluff WF, Sharpless NE (2013) Circular RNAs are abundant, conserved, and associated with ALU repeats. RNA 19(2):141–157. https://doi.org/10.1261/rna.035667.112
Jiang D, Li T, Guo C, Tang TS, Liu H (2023a) Small molecule modulators of chromatin remodeling: from neurodevelopment to neurodegeneration. Cell Biosci 13(1):10. https://doi.org/10.1186/s13578-023-00953-4
Jiang L, Zhang K, Wei X, Li J, Wang S, Wang Z, Zhou Y, Zha L, Luo H, Song F (2023b) Developing a male-specific age predictive model based on Y-CpGs for forensic analysis. Forensic Sci Int 343:111566. https://doi.org/10.1016/j.forsciint.2023.111566
Jinek M, Doudna JA (2009) A three-dimensional view of the molecular machinery of RNA interference. Nature 457(7228):405–412. https://doi.org/10.1038/nature07755
Jung SE, Cho S, Antunes J, Gomes I, Uchimoto ML, Oh YN, Di Giacomo L, Schneider PM, Park MS, van der Meer D, Williams G, McCord B, Ahn HJ, Choi DH, Lee YH, Lee SD, Lee HY (2016) A collaborative exercise on DNA methylation based body fluid typing. Electrophoresis 37(21):2759–2766. https://doi.org/10.1002/elps.201600256
Justin N, Zhang Y, Tarricone C, Martin SR, Chen S, Underwood E, De Marco V, Haire LF, Walker PA, Reinberg D, Wilson JR, Gamblin SJ (2016) Structural basis of oncogenic histone H3K27M inhibition of human polycomb repressive complex 2. Nat Commun 7:11316. https://doi.org/10.1038/ncomms11316
Kader F, Ghai M (2015) DNA methylation and application in forensic sciences. Forensic Sci Int 249:255–265. https://doi.org/10.1016/j.forsciint.2015.01.037
Kananen L, Nevalainen T, Jylhava J, Marttila S, Hervonen A, Jylha M, Hurme M (2015) Cytomegalovirus infection accelerates epigenetic aging. Exp Gerontol 72:227–229. https://doi.org/10.1016/j.exger.2015.10.008
Kato K, Long NK, Makita H, Toida M, Yamashita T, Hatakeyama D, Hara A, Mori H, Shibata T (2008) Effects of green tea polyphenol on methylation status of RECK gene and cancer cell invasion in oral squamous cell carcinoma cells. Br J Cancer 99(4):647–654. https://doi.org/10.1038/sj.bjc.6604521
Kato M, Chen X, Inukai S, Zhao H, Slack FJ (2011) Age-associated changes in expression of small, noncoding RNAs, including microRNAs, in *C. elegans*. RNA 17(10):1804–1820. https://doi.org/10.1261/rna.2714411
Kim M, Costello J (2017) DNA methylation: an epigenetic mark of cellular memory. Exp Mol Med 49(4):e322. https://doi.org/10.1038/emm.2017.10
Kim DY, Kim JM (2021) Multi-omics integration strategies for animal epigenetic studies—a review. Anim Biosci 34(8):1271–1282. https://doi.org/10.5713/ab.21.0042
Kim YJ, Xie P, Cao L, Zhang MQ, Kim TH (2018) Global transcriptional activity dynamics reveal functional enhancer RNAs. Genome Res 28(12):1799–1811. https://doi.org/10.1101/gr.233486.117
Koch CM, Wagner W (2011) Epigenetic-aging-signature to determine age in different tissues. Aging (Albany NY) 3(10):1018–1027. https://doi.org/10.18632/aging.100395
Kohli RM, Zhang Y (2013) TET enzymes, TDG and the dynamics of DNA demethylation. Nature 502(7472):472–479. https://doi.org/10.1038/nature12750
Kondo Y (2009) Epigenetic cross-talk between DNA methylation and histone modifications in human cancers. Yonsei Med J 50(4):455–463. https://doi.org/10.3349/ymj.2009.50.4.455
Konno M, Koseki J, Asai A, Yamagata A, Shimamura T, Motooka D, Okuzaki D, Kawamoto K, Mizushima T, Eguchi H, Takiguchi S, Satoh T, Mimori K, Ochiya T, Doki Y, Ofusa K, Mori M, Ishii H (2019) Distinct methylation levels of mature microRNAs in gastrointestinal cancers. Nat Commun 10(1):3888. https://doi.org/10.1038/s41467-019-11826-1

Konrad H, Jurgens L, Hartung B, Poetsch M (2023) More than just blood, saliva, or sperm-setup of a workflow for body fluid identification by DNA methylation analysis. Int J Legal Med 137(6):1683–1692. https://doi.org/10.1007/s00414-023-03069-z

Kornberg RD (1974) Chromatin structure: a repeating unit of histones and DNA. Science 184(4139):868–871. https://doi.org/10.1126/science.184.4139.868

Lam LL, Emberly E, Fraser HB, Neumann SM, Chen E, Miller GE, Kobor MS (2012) Factors underlying variable DNA methylation in a human community cohort. Proc Natl Acad Sci USA 109(Suppl 2):17253–17260. https://doi.org/10.1073/pnas.1121249109

Layne TR, Green RA, Lewis CA, Nogales F, Dawson Cruz TC, Zehner ZE, Seashols-Williams SJ (2019) microRNA detection in blood, urine, semen, and saliva stains after compromising treatments. J Forensic Sci 64(6):1831–1837. https://doi.org/10.1111/1556-4029.14113

Lee HY, Park MJ, Choi A, An JH, Yang WI, Shin KJ (2012) Potential forensic application of DNA methylation profiling to body fluid identification. Int J Legal Med 126(1):55–62. https://doi.org/10.1007/s00414-011-0569-2

Lee HY, Jung SE, Lee EH, Yang WI, Shin KJ (2016) DNA methylation profiling for a confirmatory test for blood, saliva, semen, vaginal fluid and menstrual blood. Forensic Sci Int Genet 24:75–82. https://doi.org/10.1016/j.fsigen.2016.06.007

Lee JW, Choung CM, Jung JY, Lee HY, Lim SK (2018) A validation study of DNA methylation-based age prediction using semen in forensic casework samples. Leg Med (Tokyo) 31:74–77. https://doi.org/10.1016/j.legalmed.2018.01.005

Lee HY, Hong SR, Lee JE, Hwang IK, Kim NY, Lee JM, Fleckhaus J, Jung SE, Lee YH (2020) Epigenetic age signatures in bones. Forensic Sci Int Genet 46:102261. https://doi.org/10.1016/j.fsigen.2020.102261

Lee EH, Jeong M, Park K, Lee DG, Lee EJ, Lee H, Kim AY, Ahn JW, Woo HJ, Kim S, Lim J, Kim J (2024) Detection of miR-133a-5p using a molecular beacon probe for investigating postmortem intervals. Noncoding RNA 10(6):58. https://doi.org/10.3390/ncrna10060058

Levine ME, Lu AT, Quach A, Chen BH, Assimes TL, Bandinelli S, Hou L, Baccarelli AA, Stewart JD, Li Y, Whitsel EA, Wilson JG, Reiner AP, Aviv A, Lohman K, Liu Y, Ferrucci L, Horvath S (2018) An epigenetic biomarker of aging for lifespan and healthspan. Aging (Albany NY) 10(4):573–591. https://doi.org/10.18632/aging.101414

Li C, Zhao S, Zhang N, Zhang S, Hou Y (2013) Differences of DNA methylation profiles between monozygotic twins' blood samples. Mol Biol Rep 40(9):5275–5280. https://doi.org/10.1007/s11033-013-2627-y

Lieb JD, Beck S, Bulyk ML, Farnham P, Hattori N, Henikoff S, Liu XS, Okumura K, Shiota K, Ushijima T, Greally JM (2006) Applying whole-genome studies of epigenetic regulation to study human disease. Cytogenet Genome Res 114(1):1–15. https://doi.org/10.1159/000091922

Lin S, Gregory RI (2014) Methyltransferases modulate RNA stability in embryonic stem cells. Nat Cell Biol 16(2):129–131. https://doi.org/10.1038/ncb2914

Lind PM, Salihovic S, Lind L (2018) High plasma organochlorine pesticide levels are related to increased biological age as calculated by DNA methylation analysis. Environ Int 113:109–113. https://doi.org/10.1016/j.envint.2018.01.019

Lister R, Pelizzola M, Dowen RH, Hawkins RD, Hon G, Tonti-Filippini J, Nery JR, Lee L, Ye Z, Ngo QM, Edsall L, Antosiewicz-Bourget J, Stewart R, Ruotti V, Millar AH, Thomson JA, Ren B, Ecker JR (2009) Human DNA methylomes at base resolution show widespread epigenomic differences. Nature 462(7271):315–322. https://doi.org/10.1038/nature08514

Liu C, Marioni RE, Hedman AK, Pfeiffer L, Tsai PC, Reynolds LM, Just AC, Duan Q, Boer CG, Tanaka T, Elks CE, Aslibekyan S, Brody JA, Kuhnel B, Herder C, Almli LM, Zhi D, Wang Y, Huan T et al (2018) A DNA methylation biomarker of alcohol consumption. Mol Psychiatry 23(2):422–433. https://doi.org/10.1038/mp.2016.192

Liu Z, Yang J, Wang N, Liu J, Geng J, Zhu J, Cong B, Sun H, Wu R (2024) Integrative lncRNA, circRNA, and mRNA analysis reveals expression profiles of six forensic body fluids/tissue. Int J Legal Med 138(3):731–742. https://doi.org/10.1007/s00414-023-03131-w

Lu AT, Quach A, Wilson JG, Reiner AP, Aviv A, Raj K, Hou L, Baccarelli AA, Li Y, Stewart JD, Whitsel EA, Assimes TL, Ferrucci L, Horvath S (2019) DNA methylation GrimAge strongly predicts lifespan and healthspan. Aging (Albany NY) 11(2):303–327. https://doi.org/10.18632/aging.101684

Lu AT, Binder AM, Zhang J, Yan Q, Reiner AP, Cox SR, Corley J, Harris SE, Kuo PL, Moore AZ, Bandinelli S, Stewart JD, Wang C, Hamlat EJ, Epel ES, Schwartz JD, Whitsel EA, Correa A, Ferrucci L et al (2022) DNA methylation GrimAge version 2. Aging (Albany NY) 14(23):9484–9549. https://doi.org/10.18632/aging.204434

Luo A, Jung J, Longley M, Rosoff DB, Charlet K, Muench C, Lee J, Hodgkinson CA, Goldman D, Horvath S, Kaminsky ZA, Lohoff FW (2020) Epigenetic aging is accelerated in alcohol use disorder and regulated by genetic variation in APOL2. Neuropsychopharmacology 45(2):327–336. https://doi.org/10.1038/s41386-019-0500-y

Luttropp K, Nordfors L, Ekstrom TJ, Lind L (2013) Physical activity is associated with decreased global DNA methylation in Swedish older individuals. Scand J Clin Lab Invest 73(2):184–185. https://doi.org/10.3109/00365513.2012.743166

Lv YH, Ma KJ, Zhang H, He M, Zhang P, Shen YW, Jiang N, Ma D, Chen L (2014) A time course study demonstrating mRNA, microRNA, 18S rRNA, and U6 snRNA changes to estimate PMI in deceased rat's spleen. J Forensic Sci 59(5):1286–1294. https://doi.org/10.1111/1556-4029.12447

Lv YH, Ma JL, Pan H, Zeng Y, Tao L, Zhang H, Li WC, Ma KJ, Chen L (2017) Estimation of the human postmortem interval using an established rat mathematical model and multi-RNA markers. Forensic Sci Med Pathol 13(1):20–27. https://doi.org/10.1007/s12024-016-9827-4

Marquez-Ruiz AB, Gonzalez-Herrera L, Luna JD, Valenzuela A (2020) DNA methylation levels and telomere length in human teeth: usefulness for age estimation. Int J Legal Med 134(2):451–459. https://doi.org/10.1007/s00414-019-02242-7

Martinez PN, Menendez ST, Villaronga MLA, Ubelaker DH, Garcia-Pedrero JM, Zapico SC (2019) The big sleep: elucidating the sequence of events in the first hours of death to determine the postmortem interval. Sci Justice 59(4):418–424. https://doi.org/10.1016/j.scijus.2019.03.001

Mazzone R, Zwergel C, Artico M, Taurone S, Ralli M, Greco A, Mai A (2019) The emerging role of epigenetics in human autoimmune disorders. Clin Epigenetics 11(1):34. https://doi.org/10.1186/s13148-019-0632-2

McCartney DL, Stevenson AJ, Walker RM, Gibson J, Morris SW, Campbell A, Murray AD, Whalley HC, Porteous DJ, McIntosh AM, Evans KL, Deary IJ, Marioni RE (2018) Investigating the relationship between DNA methylation age acceleration and risk factors for Alzheimer's disease. Alzheimers Dement (Amst) 10:429–437. https://doi.org/10.1016/j.dadm.2018.05.006

Medina-Paz F, Kuba B, Kryvorutsky E, Roca G, Zapico SC (2024) Assessment of blood and semen detection and DNA collection from swabs up to three months after deposition on five different cloth materials. Int J Mol Sci 25(6):3522. https://doi.org/10.3390/ijms25063522

Melamed P, Yosefzon Y, David C, Tsukerman A, Pnueli L (2018) Tet enzymes, variants, and differential effects on function. Front Cell Dev Biol 6:22. https://doi.org/10.3389/fcell.2018.00022

Memczak S, Jens M, Elefsinioti A, Torti F, Krueger J, Rybak A, Maier L, Mackowiak SD, Gregersen LH, Munschauer M, Loewer A, Ziebold U, Landthaler M, Kocks C, le Noble F, Rajewsky N (2013) Circular RNAs are a large class of animal RNAs with regulatory potency. Nature 495(7441):333–338. https://doi.org/10.1038/nature11928

Meng H, Cao Y, Qin J, Song X, Zhang Q, Shi Y, Cao L (2015) DNA methylation, its mediators and genome integrity. Int J Biol Sci 11(5):604–617. https://doi.org/10.7150/ijbs.11218

Meza-Menchaca T, Albores-Medina A, Heredia-Mendez AJ, Ruiz-May E, Ricano-Rodriguez J, Gallegos-Garcia V, Esquivel A, Vettoretti-Maldonado G, Campos-Parra AD (2024) Revisiting epigenetics fundamentals and its biomedical implications. Int J Mol Sci 25(14):7927. https://doi.org/10.3390/ijms25147927

Moore LD, Le T, Fan G (2013) DNA methylation and its basic function. Neuropsychopharmacology 38(1):23–38. https://doi.org/10.1038/npp.2012.112

Na JY (2020) Estimation of the post-mortem interval using microRNA in the bones. J Forensic Leg Med 75:102049. https://doi.org/10.1016/j.jflm.2020.102049
Nanney DL (1958) Epigenetic control systems. Proc Natl Acad Sci USA 44(7):712–717. https://doi.org/10.1073/pnas.44.7.712
Nieratschker V, Grosshans M, Frank J, Strohmaier J, von der Goltz C, El-Maarri O, Witt SH, Cichon S, Nothen MM, Kiefer F, Rietschel M (2014) Epigenetic alteration of the dopamine transporter gene in alcohol-dependent patients is associated with age. Addict Biol 19(2):305–311. https://doi.org/10.1111/j.1369-1600.2012.00459.x
Noren Hooten N, Abdelmohsen K, Gorospe M, Ejiogu N, Zonderman AB, Evans MK (2010) microRNA expression patterns reveal differential expression of target genes with age. PLoS One 5(5):e10724. https://doi.org/10.1371/journal.pone.0010724
Ntanasis-Stathopoulos J, Tzanninis JG, Philippou A, Koutsilieris M (2013) Epigenetic regulation on gene expression induced by physical exercise. J Musculoskelet Neuronal Interact 13(2):133–146. https://www.ncbi.nlm.nih.gov/pubmed/23728100
Nwanaji-Enwerem JC, Colicino E, Trevisi L, Kloog I, Just AC, Shen J, Brennan K, Dereix A, Hou L, Vokonas P, Schwartz J, Baccarelli AA (2016) Long-term ambient particle exposures and blood DNA methylation age: findings from the VA normative aging study. Environ Epigenet 2(2):dvw006. https://doi.org/10.1093/eep/dvw006
Ohgane J, Yagi S, Shiota K (2008) Epigenetics: the DNA methylation profile of tissue-dependent and differentially methylated regions in cells. Placenta 29 Suppl A:S29–S35. https://doi.org/10.1016/j.placenta.2007.09.011
Okano M, Bell DW, Haber DA, Li E (1999) DNA methyltransferases Dnmt3a and Dnmt3b are essential for de novo methylation and mammalian development. Cell 99(3):247–257. https://doi.org/10.1016/s0092-8674(00)81656-6
Ospelt C (2022) A brief history of epigenetics. Immunol Lett 249:1–4. https://doi.org/10.1016/j.imlet.2022.08.001
Park LK, Friso S, Choi SW (2012) Nutritional influences on epigenetics and age-related disease. Proc Nutr Soc 71(1):75–83. https://doi.org/10.1017/S0029665111003302
Park JL, Kwon OH, Kim JH, Yoo HS, Lee HC, Woo KM, Kim SY, Lee SH, Kim YS (2014) Identification of body fluid-specific DNA methylation markers for use in forensic science. Forensic Sci Int Genet 13:147–153. https://doi.org/10.1016/j.fsigen.2014.07.011
Pasaribu RS, Auerkari EI, Suhartono AW, Auerkari P (2023) A small RNA, microRNA as a potential biomolecular marker to estimate post mortem interval in forensic science: a systematic review. Int J Legal Med 137(5):1313–1325. https://doi.org/10.1007/s00414-023-03015-z
Peixoto P, Cartron PF, Serandour AA, Hervouet E (2020) From 1957 to nowadays: a brief history of epigenetics. Int J Mol Sci 21(20):7571. https://doi.org/10.3390/ijms21207571
Perna L, Zhang Y, Mons U, Holleczek B, Saum KU, Brenner H (2016) Epigenetic age acceleration predicts cancer, cardiovascular, and all-cause mortality in a German case cohort. Clin Epigenetics 8:64. https://doi.org/10.1186/s13148-016-0228-z
Planterose Jimenez B, Liu F, Caliebe A, Montiel Gonzalez D, Bell JT, Kayser M, Vidaki A (2021) Equivalent DNA methylation variation between monozygotic co-twins and unrelated individuals reveals universal epigenetic inter-individual dissimilarity. Genome Biol 22(1):18. https://doi.org/10.1186/s13059-020-02223-9
Pradhan S, Bacolla A, Wells RD, Roberts RJ (1999) Recombinant human DNA (cytosine-5) methyltransferase. I. Expression, purification, and comparison of de novo and maintenance methylation. J Biol Chem 274(46):33002–33010. https://doi.org/10.1074/jbc.274.46.33002
Quach A, Levine ME, Tanaka T, Lu AT, Chen BH, Ferrucci L, Ritz B, Bandinelli S, Neuhouser ML, Beasley JM, Snetselaar L, Wallace RB, Tsao PS, Absher D, Assimes TL, Stewart JD, Li Y, Hou L, Baccarelli AA et al (2017) Epigenetic clock analysis of diet, exercise, education, and lifestyle factors. Aging (Albany NY) 9(2):419–446. https://doi.org/10.18632/aging.101168
Rahnamoun H, Orozco P, Lauberth SM (2020) The role of enhancer RNAs in epigenetic regulation of gene expression. Transcription 11(1):19–25. https://doi.org/10.1080/21541264.2019.1698934

Rakyan VK, Down TA, Thorne NP, Flicek P, Kulesha E, Graf S, Tomazou EM, Backdahl L, Johnson N, Herberth M, Howe KL, Jackson DK, Miretti MM, Fiegler H, Marioni JC, Birney E, Hubbard TJ, Carter NP, Tavare S, Beck S (2008) An integrated resource for genome-wide identification and analysis of human tissue-specific differentially methylated regions (tDMRs). Genome Res 18(9):1518–1529. https://doi.org/10.1101/gr.077479.108

Ramchandani S, Bhattacharya SK, Cervoni N, Szyf M (1999) DNA methylation is a reversible biological signal. Proc Natl Acad Sci USA 96(11):6107–6112. https://doi.org/10.1073/pnas.96.11.6107

Redon C, Pilch D, Rogakou E, Sedelnikova O, Newrock K, Bonner W (2002) Histone H2A variants H2AX and H2AZ. Curr Opin Genet Dev 12(2):162–169. https://doi.org/10.1016/s0959-437x(02)00282-4

Rhodes C, Lewis C, Price K, Valentine A, Creighton MA, Boone E, Seashols-Williams S (2023) Evaluation and verification of a microRNA panel using quadratic discriminant analysis for the classification of human body fluids in DNA extracts. Genes (Basel) 14(5):968. https://doi.org/10.3390/genes14050968

Rosen AD, Robertson KD, Hlady RA, Muench C, Lee J, Philibert R, Horvath S, Kaminsky ZA, Lohoff FW (2018) DNA methylation age is accelerated in alcohol dependence. Transl Psychiatry 8(1):182. https://doi.org/10.1038/s41398-018-0233-4

Satta R, Maloku E, Zhubi A, Pibiri F, Hajos M, Costa E, Guidotti A (2008) Nicotine decreases DNA methyltransferase 1 expression and glutamic acid decarboxylase 67 promoter methylation in GABAergic interneurons. Proc Natl Acad Sci USA 105(42):16356–16361. https://doi.org/10.1073/pnas.0808699105

Scarano E, Iaccarino M, Grippo P, Winckelmans D (1965) On methylation of DNA during development of the sea urchin embryo. J Mol Biol 14(2):603–607. https://doi.org/10.1016/s0022-2836(65)80211-x

Sevilla A, Binda O (2014) Post-translational modifications of the histone variant H2AZ. Stem Cell Res 12(1):289–295. https://doi.org/10.1016/j.scr.2013.11.004

Shen Y, Matsuno Y, Fouse SD, Rao N, Root S, Xu R, Pellegrini M, Riggs AD, Fan G (2008) X-inactivation in female human embryonic stem cells is in a nonrandom pattern and prone to epigenetic alterations. Proc Natl Acad Sci USA 105(12):4709–4714. https://doi.org/10.1073/pnas.0712018105

Song B, Qian J, Fu J (2024) Research progress and potential application of microRNA and other non-coding RNAs in forensic medicine. Int J Legal Med 138(2):329–350. https://doi.org/10.1007/s00414-023-03091-1

Spolnicka M, Zbiec-Piekarska R, Karp M, Machnicki MM, Wlasiuk P, Makowska Z, Pieta A, Gambin T, Gasperowicz P, Branicki W, Giannopoulos K, Stoklosa T, Ploski R (2018) DNA methylation signature in blood does not predict calendar age in patients with chronic lymphocytic leukemia but may alert to the presence of disease. Forensic Sci Int Genet 34:e15–e17. https://doi.org/10.1016/j.fsigen.2018.02.004

Statello L, Guo CJ, Chen LL, Huarte M (2021) Gene regulation by long non-coding RNAs and its biological functions. Nat Rev Mol Cell Biol 22(2):96–118. https://doi.org/10.1038/s41580-020-00315-9

Strahl BD, Allis CD (2000) The language of covalent histone modifications. Nature 403(6765):41–45. https://doi.org/10.1038/47412

Su WY, Xiong H, Fang JY (2010) Natural antisense transcripts regulate gene expression in an epigenetic manner. Biochem Biophys Res Commun 396(2):177–181. https://doi.org/10.1016/j.bbrc.2010.04.147

Sugimoto K, Koh E, Sin HS, Maeda Y, Narimoto K, Izumi K, Kobori Y, Kitamura E, Nagase H, Yoshida A, Namiki M (2009) Tissue-specific differentially methylated regions of the human VASA gene are potentially associated with maturation arrest phenotype in the testis. J Hum Genet 54(8):450–456. https://doi.org/10.1038/jhg.2009.59

Tajuddin SM, Hernandez DG, Chen BH, Noren Hooten N, Mode NA, Nalls MA, Singleton AB, Ejiogu N, Chitrala KN, Zonderman AB, Evans MK (2019) Novel age-associated DNA meth-

ylation changes and epigenetic age acceleration in middle-aged African Americans and whites. Clin Epigenetics 11(1):119. https://doi.org/10.1186/s13148-019-0722-1
Terry MB, Ferris JS, Pilsner R, Flom JD, Tehranifar P, Santella RM, Gamble MV, Susser E (2008) Genomic DNA methylation among women in a multiethnic New York City birth cohort. Cancer Epidemiol Biomarkers Prev 17(9):2306–2310. https://doi.org/10.1158/1055-9965.EPI-08-0312
Tian H, Bai P, Tan Y, Li Z, Peng D, Xiao X, Zhao H, Zhou Y, Liang W, Zhang L (2020) A new method to detect methylation profiles for forensic body fluid identification combining ARMS-PCR technique and random forest model. Forensic Sci Int Genet 49:102371. https://doi.org/10.1016/j.fsigen.2020.102371
Tu C, Du T, Shao C, Liu Z, Li L, Shen Y (2018) Evaluating the potential of housekeeping genes, rRNAs, snRNAs, microRNAs and circRNAs as reference genes for the estimation of PMI. Forensic Sci Med Pathol 14(2):194–201. https://doi.org/10.1007/s12024-018-9973-y
Tu C, Du T, Ye X, Shao C, Xie J, Shen Y (2019) Using miRNAs and circRNAs to estimate PMI in advanced stage. Leg Med (Tokyo) 38:51–57. https://doi.org/10.1016/j.legalmed.2019.04.002
Tyagi M, Imam N, Verma K, Patel AK (2016) Chromatin remodelers: we are the drivers!! Nucleus 7(4):388–404. https://doi.org/10.1080/19491034.2016.1211217
Verdel A, Vavasseur A, Le Gorrec M, Touat-Todeschini L (2009) Common themes in siRNA-mediated epigenetic silencing pathways. Int J Dev Biol 53(2–3):245–257. https://doi.org/10.1387/ijdb.082691av
Vidaki A, Daniel B, Court DS (2013) Forensic DNA methylation profiling—potential opportunities and challenges. Forensic Sci Int Genet 7(5):499–507. https://doi.org/10.1016/j.fsigen.2013.05.004
Vidaki A, Diez Lopez C, Carnero-Montoro E, Ralf A, Ward K, Spector T, Bell JT, Kayser M (2017) Epigenetic discrimination of identical twins from blood under the forensic scenario. Forensic Sci Int Genet 31:67–80. https://doi.org/10.1016/j.fsigen.2017.07.014
Vidaki A, Montiel Gonzalez D, Planterose Jimenez B, Kayser M (2021) Male-specific age estimation based on Y-chromosomal DNA methylation. Aging (Albany NY) 13(5):6442–6458. https://doi.org/10.18632/aging.202775
Vidaki A, Planterose Jimenez B, Poggiali B, Kalamara V, van der Gaag KJ, Maas SCE, Consortium BIOS, Ghanbari M, Sijen T, Kayser M (2023) Targeted DNA methylation analysis and prediction of smoking habits in blood based on massively parallel sequencing. Forensic Sci Int Genet 65:102878. https://doi.org/10.1016/j.fsigen.2023.102878
Voisin S, Eynon N, Yan X, Bishop DJ (2015) Exercise training and DNA methylation in humans. Acta Physiol (Oxford) 213(1):39–59. https://doi.org/10.1111/apha.12414
Waddington CH (2012) The epigenotype. 1942. Int J Epidemiol 41(1):10–13. https://doi.org/10.1093/ije/dyr184
Wan ES, Qiu W, Baccarelli A, Carey VJ, Bacherman H, Rennard SI, Agusti A, Anderson W, Lomas DA, Demeo DL (2012) Cigarette smoking behaviors and time since quitting are associated with differential DNA methylation across the human genome. Hum Mol Genet 21(13):3073–3082. https://doi.org/10.1093/hmg/dds135
Wang Y, Li Y, Toth JI, Petroski MD, Zhang Z, Zhao JC (2014) N6-methyladenosine modification destabilizes developmental regulators in embryonic stem cells. Nat Cell Biol 16(2):191–198. https://doi.org/10.1038/ncb2902
Wang B, Gao W, Lv J, Yu C, Wang S, Pang Z, Cong L, Dong Z, Wu F, Wang H, Wu X, Jiang G, Wang X, Wang B, Cao W, Li L (2016) Physical activity attenuates genetic effects on BMI: results from a study of Chinese adult twins. Obesity (Silver Spring) 24(3):750–756. https://doi.org/10.1002/oby.21402
Wang S, Wang Z, Tao R, He G, Liu J, Li C, Hou Y (2019a) The potential use of Piwi-interacting RNA biomarkers in forensic body fluid identification: a proof-of-principle study. Forensic Sci Int Genet 39:129–135. https://doi.org/10.1016/j.fsigen.2019.01.002
Wang S, Wang Z, Tao R, Wang M, Liu J, He G, Yang Y, Xie M, Zou X, Hou Y (2019b) Expression profile analysis of piwi-interacting RNA in forensically relevant biological fluids. Forensic Sci Int Genet 42:171–180. https://doi.org/10.1016/j.fsigen.2019.07.015

Wang J, Wang C, Wei Y, Zhao Y, Wang C, Lu C, Feng J, Li S, Cong B (2022a) Circular RNA as a potential biomarker for forensic age prediction. Front Genet 13:825443. https://doi.org/10.3389/fgene.2022.825443

Wang J, Zhang H, Wang C, Fu L, Wang Q, Li S, Cong B (2022b) Forensic age estimation from human blood using age-related microRNAs and circular RNAs markers. Front Genet 13:1031806. https://doi.org/10.3389/fgene.2022.1031806

Wang J, Fu G, Wang Q, Ma G, Wang Z, Lu C, Fu L, Zhang X, Cong B, Li S (2024) Differences of circular RNA expression profiles between monozygotic twins' blood, with the forensic application in bloodstain and saliva. Forensic Sci Int Genet 69:103001. https://doi.org/10.1016/j.fsigen.2023.103001

Wang DY, Tian MH, Chen YZ, Wang SW, Xing XY, Sun ML, Liu Z, Liu Y, Wang H, Wei J, Zhong Y, Yao J (2025) Profiling microRNA expression differentiates monozygotic twins in peripherical blood by droplet digital PCR. Forensic Sci Int Genet 76:103230. https://doi.org/10.1016/j.fsigen.2025.103230

Ward-Caviness CK, Nwanaji-Enwerem JC, Wolf K, Wahl S, Colicino E, Trevisi L, Kloog I, Just AC, Vokonas P, Cyrys J, Gieger C, Schwartz J, Baccarelli AA, Schneider A, Peters A (2016) Long-term exposure to air pollution is associated with biological aging. Oncotarget 7(46):74510–74525. https://doi.org/10.18632/oncotarget.12903

Watson JD, Crick FH (1953) Molecular structure of nucleic acids; a structure for deoxyribose nucleic acid. Nature 171(4356):737–738. https://doi.org/10.1038/171737a0

Weidner CI, Lin Q, Koch CM, Eisele L, Beier F, Ziegler P, Bauerschlag DO, Jockel KH, Erbel R, Muhleisen TW, Zenke M, Brummendorf TH, Wagner W (2014) Aging of blood can be tracked by DNA methylation changes at just three CpG sites. Genome Biol 15(2):R24. https://doi.org/10.1186/gb-2014-15-2-r24

West MH, Bonner WM (1980) Histone 2A, a heteromorphous family of eight protein species. Biochemistry 19(14):3238–3245. https://doi.org/10.1021/bi00555a022

White AJ, Sandler DP, Bolick SC, Xu Z, Taylor JA, DeRoo LA (2013) Recreational and household physical activity at different time points and DNA global methylation. Eur J Cancer 49(9):2199–2206. https://doi.org/10.1016/j.ejca.2013.02.013

Williams G (2018) The emerging field of forensic epigenetics. Forensic Sci Int 290:e24–e25. https://doi.org/10.1016/j.forsciint.2018.07.019

Winkle M, El-Daly SM, Fabbri M, Calin GA (2021) Noncoding RNA therapeutics—challenges and potential solutions. Nat Rev Drug Discov 20(8):629–651. https://doi.org/10.1038/s41573-021-00219-z

Wood JG, Hillenmeyer S, Lawrence C, Chang C, Hosier S, Lightfoot W, Mukherjee E, Jiang N, Schorl C, Brodsky AS, Neretti N, Helfand SL (2010) Chromatin remodeling in the aging genome of Drosophila. Aging Cell 9(6):971–978. https://doi.org/10.1111/j.1474-9726.2010.00624.x

Wozniak A, Heidegger A, Piniewska-Rog D, Pospiech E, Xavier C, Pisarek A, Kartasinska E, Boron M, Freire-Aradas A, Wojtas M, de la Puente M, Niederstatter H, Ploski R, Spolnicka M, Kayser M, Phillips C, Parson W, Branicki W, Consortium V (2021) Development of the VISAGE enhanced tool and statistical models for epigenetic age estimation in blood, buccal cells and bones. Aging (Albany NY) 13(5):6459–6484. https://doi.org/10.18632/aging.202783

Xiao C, Pan C, Liu E, He H, Liu C, Huang Y, Yi S, Huang D (2019) Differences of microRNA expression profiles between monozygotic twins' blood samples. Forensic Sci Int Genet 41:152–158. https://doi.org/10.1016/j.fsigen.2019.05.003

Yan J, Zierath JR, Barres R (2011) Evidence for non-CpG methylation in mammals. Exp Cell Res 317(18):2555–2561. https://doi.org/10.1016/j.yexcr.2011.08.019

Yang Y, Gao X, Just AC, Colicino E, Wang C, Coull BA, Hou L, Zheng Y, Vokonas P, Schwartz J, Baccarelli AA (2019) Smoking-related DNA methylation is associated with DNA methylation phenotypic age acceleration: the veterans affairs normative aging study. Int J Environ Res Public Health 16(13):2356. https://doi.org/10.3390/ijerph16132356

Yun M, Wu J, Workman JL, Li B (2011) Readers of histone modifications. Cell Res 21(4):564–578. https://doi.org/10.1038/cr.2011.42
Zapico SC, Adserias-Garriga J (2022) Postmortem interval estimation: new approaches by the analysis of human tissues and microbial communities' changes. Forensic Sci 2:163–174
Zapico SC, Roca G (2024) Making the most of lateral flow immunochromatographic tests: an efficient protocol to recover DNA. Methods Protoc 7(1):8. https://doi.org/10.3390/mps7010008
Zapico SC, Ubelaker DH (2013) Applications of physiological bases of ageing to forensic sciences. Estimation of age-at-death [Research Support, Non-U.S. Gov't]. Ageing Res Rev 12(2):605–617. https://doi.org/10.1016/j.arr.2013.02.002
Zapico SC, Menendez ST, Nunez P (2014) Cell death proteins as markers of early postmortem interval. Cell Mol Life Sci 71(15):2957–2962. https://doi.org/10.1007/s00018-013-1531-x
Zapico SC, Gauthier Q, Antevska A, McCord BR (2021) Identifying methylation patterns in dental pulp aging: application to age-at-death estimation in forensic anthropology. Int J Mol Sci 22(7):3717. https://doi.org/10.3390/ijms22073717
Zbiec-Piekarska R, Spolnicka M, Kupiec T, Makowska Z, Spas A, Parys-Proszek A, Kucharczyk K, Ploski R, Branicki W (2015a) Examination of DNA methylation status of the ELOVL2 marker may be useful for human age prediction in forensic science. Forensic Sci Int Genet 14:161–167. https://doi.org/10.1016/j.fsigen.2014.10.002
Zbiec-Piekarska R, Spolnicka M, Kupiec T, Parys-Proszek A, Makowska Z, Paleczka A, Kucharczyk K, Ploski R, Branicki W (2015b) Development of a forensically useful age prediction method based on DNA methylation analysis. Forensic Sci Int Genet 17:173–179. https://doi.org/10.1016/j.fsigen.2015.05.001
Zeng H, Irwin ML, Lu L, Risch H, Mayne S, Mu L, Deng Q, Scarampi L, Mitidieri M, Katsaros D, Yu H (2012) Physical activity and breast cancer survival: an epigenetic link through reduced methylation of a tumor suppressor gene L3MBTL1. Breast Cancer Res Treat 133(1):127–135. https://doi.org/10.1007/s10549-011-1716-7
Zeng X, Yuan X, Cai Q, Tang C, Gao J (2021) Circular RNA as an epigenetic regulator in chronic liver diseases. Cells 10(8):1945. https://doi.org/10.3390/cells10081945
Zhang FF, Cardarelli R, Carroll J, Fulda KG, Kaur M, Gonzalez K, Vishwanatha JK, Santella RM, Morabia A (2011a) Significant differences in global genomic DNA methylation by gender and race/ethnicity in peripheral blood. Epigenetics 6(5):623–629. https://doi.org/10.4161/epi.6.5.15335
Zhang FF, Cardarelli R, Carroll J, Zhang S, Fulda KG, Gonzalez K, Vishwanatha JK, Morabia A, Santella RM (2011b) Physical activity and global genomic DNA methylation in a cancer-free population. Epigenetics 6(3):293–299. https://doi.org/10.4161/epi.6.3.14378
Zhang Y, Wang X, Kang L (2011c) A k-mer scheme to predict piRNAs and characterize locust piRNAs. Bioinformatics 27(6):771–776. https://doi.org/10.1093/bioinformatics/btr016
Zhang H, Wang F, Kranzler HR, Zhao H, Gelernter J (2013) Profiling of childhood adversity-associated DNA methylation changes in alcoholic patients and healthy controls. PLoS One 8(6):e65648. https://doi.org/10.1371/journal.pone.0065648
Zhang Y, Schottker B, Florath I, Stock C, Butterbach K, Holleczek B, Mons U, Brenner H (2016) Smoking-associated DNA methylation biomarkers and their predictive value for all-cause and cardiovascular mortality. Environ Health Perspect 124(1):67–74. https://doi.org/10.1289/ehp.1409020
Zhang Y, Liu B, Shao C, Xu H, Xue A, Zhao Z, Shen Y, Tang Q, Xie J (2018) Evaluation of the inclusion of circular RNAs in mRNA profiling in forensic body fluid identification. Int J Legal Med 132(1):43–52. https://doi.org/10.1007/s00414-017-1690-7
Zheng Y, Joyce BT, Colicino E, Liu L, Zhang W, Dai Q, Shrubsole MJ, Kibbe WA, Gao T, Zhang Z, Jafari N, Vokonas P, Schwartz J, Baccarelli AA, Hou L (2016) Blood epigenetic age may predict cancer incidence and mortality. EBioMedicine 5:68–73. https://doi.org/10.1016/j.ebiom.2016.02.008

Zhou WY, Cai ZR, Liu J, Wang DS, Ju HQ, Xu RH (2020) Circular RNA: metabolism, functions and interactions with proteins. Mol Cancer 19(1):172. https://doi.org/10.1186/s12943-020-01286-3
Zhu ZZ, Hou L, Bollati V, Tarantini L, Marinelli B, Cantone L, Yang AS, Vokonas P, Lissowska J, Fustinoni S, Pesatori AC, Bonzini M, Apostoli P, Costa G, Bertazzi PA, Chow WH, Schwartz J, Baccarelli A (2012) Predictors of global methylation levels in blood DNA of healthy subjects: a combined analysis. Int J Epidemiol 41(1):126–139. https://doi.org/10.1093/ije/dyq154

Chapter 2
Epigenetic Techniques Applied to Forensic Science

Mirna Ghemrawi and Sara C. Zapico

Abstract Epigenetic markers have emerged as valuable molecular tools in forensic science, providing information beyond traditional STR profiling. This chapter reviews the principal techniques used to analyze DNA methylation and RNA-based epigenetic signatures, emphasizing their underlying chemistries, analytical strengths, limitations, and current forensic applications. Genome-wide approaches—such as Illumina methylation microarrays, short-read sequencing, and long-read nanopore platforms—enable broad screening for age-, tissue-, and exposure-associated loci. Targeted assays, including pyrosequencing, MS-qPCR/MS-HRM, RT-qPCR for microRNAs, digital PCR, ARMS-PCR, CE-based SNaPshot, and MassARRAY EpiTyper, offer sensitive and operationally feasible alternatives for validating and implementing selected markers in casework. Across these methods, the most developed forensic applications include age estimation and body-fluid or tissue identification, with expanding work on complex trace interpretation and individualization. While challenges remain, particularly regarding input quantity, degradation, bisulfite-related DNA loss, and workflow standardization, the combined use of discovery-scale and targeted platforms continues to advance the integration of epigenetic evidence into forensic practice.

M. Ghemrawi (✉)
Department of Chemistry and Biochemistry, Florida International University, Miami, FL, USA

Global Forensic and Justice Center, Largo, FL, USA
e-mail: mghemraw@fiu.edu

S. C. Zapico
Department of Chemistry and Environmental Sciences, New Jersey Institute of Technology, Newark, NJ, USA

Anthropology Department and Laboratories of Analytical Biology, National Museum of Natural History, Smithsonian Institution, Washington, DC, USA
e-mail: sc338@njit.edu; casado-zapicos@si.edu

S. C. Zapico (ed.), *Forensic Science and Epigenetics*, Epigenetics and Human Health 18, https://doi.org/10.1007/978-3-032-23085-0_2

Keywords Forensic epigenetics · DNA methylation · MicroRNA · PCR-based assays · Digital PCR · SNaPshot · Pyrosequencing · Nanopore sequencing · Age estimation · Body-fluid identification

Abbreviations

CE	Capillary Electrophoresis
CpG	Cytosine–Phosphate–Guanine Dinucleotide
cSNP	Coding-Region Single Nucleotide Polymorphism
ddPCR	Droplet Digital PCR
DNA	Deoxyribonucleic Acid
dPCR	Digital Polymerase Chain Reaction
HRM	High-Resolution Melting
mRNA	Messenger RNA
miRNA	MicroRNA
MS-HRM	Methylation-Sensitive High-Resolution Melting
MS-qPCR	Methylation-Specific Quantitative PCR
NGS	Next-Generation Sequencing
PCR	Polymerase Chain Reaction
RT-qPCR	Reverse Transcription-Quantitative PCR
SBE	Single-Base Extension
SBS	Sequencing by Synthesis
SNaPshot	Single-Nucleotide Primer Extension Assay
STR	Short Tandem Repeat

2.1 Introduction

The study of epigenetic modifications has been carried out through different approaches, moreover, considering the type of modification and the forensic application. The most popular, thanks to its capacity to assess several CpG sites and microRNAs (miRNAs) at the same time, is the microarray (Bocklandt et al. 2011; Guardado-Estrada et al. 2023). Other next-generation sequencing (NGS) technologies, based on different chemistries, have been applied to conduct epigenetic analysis in both large and small scales (Beentjes et al. 2025; Guan et al. 2021; Hanggi et al. 2025; Zeng et al. 2025). The gold standard method for DNA methylation analysis is pyrosequencing (Alghanim et al. 2020; Ghemrawi et al. 2023), though it is not exempt from drawbacks, including the lack of multiplexing. As a result, researchers have looked for other approaches to overcome these disadvantages and increase their applicability in forensic labs, including PCR-based methodologies (Fang et al. 2023a; Wang et al. 2019a, b), and capillary-based techniques (Dias et al. 2020; Jiang et al. 2023).

This chapter will give an overview of the different techniques, their advantages and disadvantages, and their current application in forensic science.

2.2 Microarrays

Microarrays are the most popular method for genome-wide DNA methylation profiling. Epigenetic clocks were developed based on Illumina (CA, USA) methylation assays (Hannum et al. 2013; Horvath 2013; Levine et al. 2018). These platforms utilize bisulfite conversion alongside bead arrays to assess methylation at specific CpG sites, enabling multiplex genotyping (Fouse et al. 2010). Genomic DNA, after undergoing bisulfite treatment, is hybridized to arrays containing two probes tailored to each site—one targeting the methylated cytosine and the other the unmethylated site, which has been converted to uracil. A single-base extension reaction incorporates fluorescently labeled dideoxynucleoside triphosphates (ddNTPs) into the probes. The resulting fluorescence is scanned to compare signal intensities between the methylated and unmethylated probes, thereby quantifying the degree of methylation (Bibikova et al. 2011). The large-scale methylation platform from Illumina is the Infinium methylation assay. The 27K Illumina Infinium Methylation BeadChip analyzes methylation states for 27,000 CpGs, which is limited to only 1% of all CpGs. The next Chip developed, the Infinium HumanMethylation450 Bead Chip, interrogates 482,422 cytosines across the human genome, representing approximately 1.7% of all CpG sites (Stirzaker et al. 2014). However, it covers 99% of RefSeq genes, with an average of 17 CpG sites per gene region distributed across the promoter 5′UTR, first exon, gene body, and 3′UTR (Bibikova et al. 2011; Sandoval et al. 2011). It can use amounts of DNA around 500 ng, being able to perform in different types of tissues (Stirzaker et al. 2014). The new generation of Chips, the Infinium MethylationEPIC v2.0 BeadChip Kit assay, covers 99% of the human reference genome for a total of 950,000 CpG sites (Illumina 2025). It enables the use of as little as 250 ng of DNA, making it suitable for potentially degraded samples, such as those commonly encountered in forensic investigations. This chip has replaced previous versions and is currently the one used for epigenetic screening applications.

In forensic sciences, several studies have applied these Chips for previous screenings to identify methylation markers for age estimation (Lee et al. 2020; Weidner et al. 2014), even it has been applied to estimate the age of mummified human remains (Schmidt et al. 2021). Other applications have been the identification of body fluids at crime scenes (Fang et al. 2023b; Lee et al. 2015; Park et al. 2014), and discrimination of monozygotic twins (Vidaki et al. 2017b, 2018).

Despite their broad CpG coverage, Illumina methylation arrays require relatively high DNA input, which limits their direct use in forensic casework. Even the newer EPIC v2.0 array, which operates with ~250 ng of bisulfite-converted DNA, demands more material than is typically recovered from trace or degraded forensic samples. As a result, these platforms have been used mainly as screening tools to identify informative CpG markers, which are then adapted to low-input methods such as qPCR, dPCR, or targeted sequencing for practical forensic applications (Christiansen et al. 2022).

2.3 PCR-Based Techniques in Forensic Epigenetics

Polymerase chain reaction (PCR) remains a central tool in molecular biology and forensic genetics (McDonald et al. 2024), and its adaptability has enabled the interrogation of epigenetic features from limited or degraded biological traces (Hernandez et al. 2013). Although the enzymatic amplification chemistry mirrors that are used for STR profiling, epigenetic applications leverage PCR to quantify methylation states, detect allele-specific epigenetic variants, and characterize small regulatory RNAs relevant to tissue source or donor attributes (Nussbaumer et al. 2006). The analytical framework also differs: instead of generating qualitative genotypes, epigenetic PCR assays aim to measure proportions—such as the fraction of methylated templates, levels of microRNA expression, or the relative abundance of epialleles associated with aging, tissue specificity, or environmental exposure (Pajares et al. 2021). The subsections below describe how distinct PCR platforms have been adapted to resolve methylation signatures and small-RNA profiles under forensic conditions, where low quantity, degradation, and mixed biological sources are common.

2.3.1 *Quantitative PCR*

Quantitative PCR (qPCR) is one of the most accessible tools for locus-specific DNA methylation analysis and has been implemented in either methylation-specific formats (MS-qPCR) or in combination with high-resolution melt profiling (MS-HRM) (Wojdacz et al. 2008). Both approaches rely on sodium bisulfite conversion to discriminate cytosine methylation states: unmethylated cytosines are deaminated to uracil, whereas methylated cytosines remain unchanged. Subsequent amplification with either methylation-specific primers (Herman et al. 1996) or sequence-agnostic primers followed by melt curve analysis (Wojdacz and Dobrovic 2007) enables quantification of methylation levels through Ct value differentials or characteristic melting transitions.

Single-locus assays have been widely used for age estimation. The ELOVL2 locus, in particular, consistently shows one of the strongest age-associated methylation trajectories in blood and has been validated across qPCR-based platforms for forensic age inference (Ogata et al. 2022; Paparazzo et al. 2022). MS-HRM assays targeting ELOVL2 or parallel age-informative loci have demonstrated high analytical sensitivity, tolerance to moderate DNA degradation, and suitability for casework-type samples (Hamano et al. 2016).

The appeal of MS-qPCR for forensic laboratories lies in its compatibility with existing qPCR infrastructure, low per-sample cost, and rapid turnaround time. However, several methodological constraints persist. Bisulfite conversion substantially reduces DNA template integrity, incomplete conversion can introduce quantification biases, and qPCR's limited multiplexing capability restricts the number of

CpGs interrogated per reaction. Moreover, the dynamic range and precision of qPCR are lower than those achieved by digital PCR platforms or targeted sequencing (consortium 2016; Sabeeha and Hasnain 2019). For these reasons, qPCR-based methylation assays are most effective when applied to small, thoroughly validated marker panels with robust internal controls, particularly in scenarios where rapid or resource-constrained analysis is required.

2.3.2 Reverse Transcription qPCR

Reverse transcription quantitative PCR (RT-qPCR) enables the interrogation of RNA species, particularly microRNAs (miRNAs), which are short (18–24 nt), highly stable non-coding RNAs that persist even in environmentally challenged or aged forensic stains (Sauer et al. 2014). Their compact size, resistance to RNase-mediated degradation, and marked tissue-specific expression have positioned them as informative biomarkers for body fluid identification (Sauer et al. 2016). Early work by Zubakov et al. (2010) demonstrated this potential by screening miRNA expression with microarrays and subsequently validating candidate markers through RT-qPCR, identifying miR-144 and miR-451 as blood-specific and miR-135b and miR-10a as semen-associated, with detection sensitivity down to the sub-picogram level. Subsequent studies expanded these panels and confirmed their robustness in casework-like samples, including saliva-, vaginal fluid-, and menstrual blood–specific miRNAs (Courts and Madea 2011; Hanson and Ballantyne 2013).

RT-qPCR workflows typically involve total RNA extraction from the forensic matrix, poly(A) tailing or stem-loop primer reverse transcription to generate cDNA, and amplification with miRNA-specific primers (Fujimoto et al. 2019). Quantification relies on cycle threshold values normalized against endogenous small RNA controls such as RNU6B or SNORD43, though the choice of reference remains a point of methodological variability across laboratories (Kaija et al. 2020; Wei et al. 2023). The short amplicon size (<70 bp) makes RT-qPCR well-suited for degraded samples, and the method's compatibility with existing qPCR instrumentation facilitates integration into standard forensic workflows.

Despite these advantages, several limitations warrant consideration. RNA extraction from minute or environmentally exposed stains is technically challenging; inter-individual variability and physiological influences can affect miRNA expression patterns (Silva et al. 2015); and the lack of consensus on normalization strategies hinders straightforward cross-laboratory comparison (Sauer et al. 2014). Moreover, some miRNAs exhibit overlapping expression across tissues, necessitating carefully curated multiplex panels. Nonetheless, RT-qPCR-based miRNA profiling remains one of the molecular tools for body fluid identification and provides forensic analysts with an additional layer of contextual information that complements DNA profiling (Hanson et al. 2009; Soliman et al. 2018).

2.3.3 Nanoplate dPCR

Nanoplate digital PCR (dPCR) extends quantitative PCR into the digital domain by partitioning the reaction mixture into thousands of nanoliter-scale wells, enabling absolute quantification of nucleic acid molecules without the need for external calibration curves (Morley 2014). Following bisulfite conversion, templates representing methylated and unmethylated alleles are distributed across the nanowells according to Poisson statistics; each well then undergoes endpoint PCR, yielding a binary fluorescent signal that reflects the presence or absence of the target molecule. The summation of positive partitions allows direct calculation of methylation ratios with high precision, a feature particularly advantageous for low-abundance epialleles (Hindson et al. 2013; Whale et al. 2016).

Nanoplate-based systems, such as those implemented on the QIAcuity platform (QIAGEN), use fixed, pre-fabricated reaction plates with integrated partitioning. This closed-cartridge architecture minimizes pipetting steps and substantially reduces the risk of contamination, a critical consideration for forensic laboratories. In addition, the workflow remains compatible with bisulfite-treated DNA and short amplicons typical of forensic epigenetic assays. Recent studies have demonstrated that dPCR can sensitively and accurately quantify age-associated methylation levels at loci such as ELOVL2, FHL2, and PDE4C, even when input DNA is highly degraded or present in sub-nanogram quantities (Vidaki and Kayser 2018).

The primary forensic advantage of nanoplate dPCR is its robustness at the lower limits of detection, where qPCR often exhibits stochasticity and reduced reproducibility (Munjakovic et al. 2025). Partitioning ensures that template competition is minimized and that minor methylated fractions, whether age-related or tissue-specific, can be resolved with confidence. However, its limitations include moderate throughput, higher per-sample cost, and dependence on specialized instrumentation (Whale et al. 2016). Even so, nanoplate dPCR increasingly serves as a bridge between high-precision research methodologies and operational forensic workflows.

2.3.4 Droplet Digital PCR

Droplet digital PCR (ddPCR) achieves absolute nucleic acid quantification by emulsifying the reaction mixture into tens of thousands of nanoliter-sized droplets, each functioning as an independent PCR microreactor (Hindson et al. 2013). Following bisulfite conversion of DNA, methylated and unmethylated templates are distributed across droplets according to Poisson statistics. After amplification, fluorescence detection classifies each droplet as positive or negative for the target sequence, allowing direct calculation of copy number and methylation proportion without reliance on standard curves (Hindson et al. 2011).

The high partition number and endpoint detection confer exceptional precision for low-abundance epialleles, making ddPCR well-suited to forensic age estimation as well. Several studies have applied ddPCR to age-associated methylation markers such as ELOVL2, FHL2, and PDE4C (Dias and Manco 2024; Fakhr et al. 2025). In one representative study, a three-locus ddPCR model achieved mean absolute deviations of approximately 4–5 years in blood, demonstrating analytical stability across different input quantities and degrees of DNA degradation (Zbiec-Piekarska et al. 2015a). The ability to use short amplicons (typically <80 bp) further enhances performance with degraded or environmentally exposed forensic samples.

Operationally, ddPCR offers distinct advantages: strong resistance to PCR inhibitors commonly encountered in casework samples, high reproducibility at low template copy numbers, and a workflow that is conceptually similar to qPCR while delivering superior quantitative resolution (Hindson et al. 2013). However, implementation requires investment in dedicated droplet generators and readers, and assay development demands careful optimization to ensure complete bisulfite conversion and proper droplet separation (Hindson et al. 2011). Additionally, the probabilistic nature of droplet partitioning and quantification may require clear explanation when results are presented in forensic or legal contexts. Despite these considerations, ddPCR has emerged as a robust platform for targeted epigenetic quantification providing precise and reproducible methylation measurements that complement conventional STR profiling.

2.3.5 *Amplification Refractory Mutation System PCR*

Amplification Refractory Mutation System PCR (ARMS-PCR), originally developed for allele-specific detection of single-nucleotide variants (Newton et al. 1989), has been adapted in forensic epigenetics to discriminate methylated from unmethylated alleles following bisulfite conversion. Because bisulfite treatment converts unmethylated cytosines to uracil while leaving methylated cytosines unchanged, a single CpG site effectively becomes a sequence polymorphism. ARMS-PCR exploits this by using primers whose 3′ terminal nucleotide is complementary either to the bisulfite-converted (unmethylated) or unconverted (methylated) state, enabling exclusive amplification of one variant under stringent conditions (Herman et al. 1996).

The method yields a binary endpoint—presence or absence of amplification—which can serve as a rapid screening tool for methylation patterns associated with tissue source, physiological state, or environmentally responsive loci (Medrano and de Oliveira 2014). Its conceptual simplicity, compatibility with standard thermocyclers, and short assay time make it appealing for preliminary forensic assessments where rapid triage is needed before deploying more quantitative platforms such as digital PCR or targeted sequencing.

However, ARMS-PCR requires highly specific primer design, as even minor sequence mismatches or suboptimal annealing can compromise allele discrimination (Honardoost et al. 2014). Incomplete bisulfite conversion, sequence polymorphisms adjacent to the CpG site, or degraded templates can result in allele drop-out or reduced specificity, limiting the method's reliability for precise methylation quantification. The assay's qualitative nature also restricts its applicability to situations where the presence of a methylated signal—rather than its proportion—is forensically informative. Despite these limitations, ARMS-PCR illustrates how canonical PCR architectures have been repurposed to interrogate epigenetic modifications, enabling targeted detection of methylation signatures that provide contextual information beyond traditional DNA profiling (Tian et al. 2020).

2.4 Capillary Electrophoresis-Based (SNaPshot)

Capillary electrophoresis (CE) has long been central to forensic DNA analysis, underpinning STR profiling and SNP genotyping with high sensitivity and reproducibility (Butler 2011). Building on this established platform, CE has been adapted for epigenetic applications through the SNaPshot single-base extension (SBE) assay (Dias et al. 2020). By combining bisulfite conversion or miRNA-specific reverse transcription with multiplex SBE and electrophoretic separation, SNaPshot enables simultaneous interrogation of multiple CpG sites or small-RNA markers within a workflow already familiar to forensic laboratories.

In methylation-based assays, genomic DNA is first subjected to bisulfite conversion, which converts unmethylated cytosines to uracil while preserving methylated CpGs. Short amplicons spanning the targeted CpGs are then generated using bisulfite-compatible primers. These PCR products serve as templates for a multiplex SBE reaction in which fluorescently labeled dideoxynucleotides are incorporated opposite the CpG of interest (Mehta et al. 2017). Because bisulfite conversion renders methylated and unmethylated templates sequence-distinct (cytosine vs. thymine), each allele produces a different fluorescent signal. The extension products are separated by CE, and their peak heights in the electropherogram reflect the relative proportion of methylated and unmethylated alleles (Kaminsky and Petronis 2009). The visual output closely resembles traditional STR electropherograms, which facilitates interpretation for analysts accustomed to CE-based workflows.

Several studies have demonstrated that CE-based SNaPshot methylation assays are suitable for forensic applications, most notably age estimation and body-fluid or tissue identification. For age estimation, Jung et al. developed a multiplex SNaPshot assay targeting CpGs in ELOVL2, FHL2, KLF14, C1orf132/MIR29B2C, and TRIM59, achieving mean absolute deviations of approximately 3.5–4.3 years across blood, saliva, and buccal swabs, indicating that minisequencing formats can be applied to casework-type samples (Jung et al. 2019). Dias et al. later replicated and validated this assay in a Portuguese population, including blood from deceased individuals, and reported comparable prediction errors (~4–5 years), supporting its reproducibility across tissues and post-mortem intervals (Dias et al. 2020). Additional refinements—such as cartilage-specific or saliva-specific SNaPshot methylation models—have further shown that the approach can be adapted to different biological matrices (Jung et al. 2025; Xiao et al. 2025).

In parallel, SNaPshot chemistry has been applied to body-fluid identification. Forat et al. developed a methylation-sensitive single-nucleotide primer extension (msSNuPE) assay using the ABI SNaPshot platform to quantify CpGs capable of distinguishing peripheral blood, menstrual blood, semen, saliva, and vaginal secretions, including under mixture and environmental stress conditions (Forat et al. 2016a, b). More recently, Chinese forensic laboratories have constructed methylation-sensitive SNaPshot panels for tissue and fluid assignment—several incorporating machine-learning classifiers—to improve discrimination power and enable interpretation of mixed stains (Fang et al. 2023c; Kim et al. 2024). Collectively, these studies demonstrate that SNaPshot-based CE assays offer a feasible and technically accessible route for implementing methylation marker analysis in forensic laboratories using existing CE infrastructure.

Beyond CpG methylation, SNaPshot chemistry has also been applied to mRNA-based analyses through the detection of coding-region SNPs (cSNPs) embedded within body-fluid–specific transcripts. Rather than relying solely on differential expression, these assays use single-base extension to genotype cSNPs in mRNA markers, enabling both body-fluid identification and donor attribution on standard CE platforms. Wang et al. demonstrated that cSNP genotypes obtained from mRNA using a SNaPshot assay were fully concordant with those derived from genomic DNA, confirming the reliability of RNA-derived SNP typing (Wang et al. 2019a, b). Liu et al. subsequently developed a multiplex PCR–SNaPshot system targeting cSNPs within blood-specific transcripts, achieving high sensitivity in mixture stains and demonstrating strong discrimination power in population testing (Liu et al. 2020). Further work extended this strategy to mRNA markers from semen and vaginal secretions, supporting the broader utility of cSNP-based SNaPshot assays for contributor identification in complex biological stains (Liu et al. 2025; Neis et al. 2024; Zhang et al. 2022).

The practical advantages of CE-based SNaPshot assays are significant: minimal input requirements, compatibility with widely used forensic CE platforms, and robust multiplexing with clear electrophoretic separation of extension products. The method's reliance on familiar fluorescent peak patterns reduces implementation barriers in accredited laboratories. Limitations include the DNA degradation inherent to bisulfite conversion, the need for careful calibration to ensure quantitative reliability, and dye-related variability that can influence peak intensity. When these factors are appropriately controlled, SNaPshot assays provide reproducible, cost-effective measurements suitable for both research and applied forensic settings.

In summary, CE-based SNaPshot assays represent an effective adaptation of routine forensic technology to the emerging field of epigenetic analysis. Whether applied to CpG methylation or miRNA expression, they translate biochemical modifications into highly interpretable electropherograms and provide a practical bridge between discovery-level epigenetic studies and operational forensic casework.

2.5 Sequencing-Based Techniques

There are two general types of sequencing-based techniques to study DNA methylation (Hu et al. 2021):

1. Short-read next-generation sequencing (NGS), also known as second-generation sequencing, represents an advancement over traditional Sanger sequencing. These technologies enable high-throughput sequencing of short (250–800 bp), clonally amplified DNA fragments in parallel. The workflow includes library preparation, sequencing, and data analysis (Tucker et al. 2009). Applied specifically to DNA methylation in forensic sciences, Illumina's NGS technology, Pyrosequencing, and Ion Torrent will be covered in this section.
2. Long-read technologies, also known as third-generation sequencing, can generate sequences exceeding 10 kb directly from native DNA (van Dijk et al. 2018). Recent applications to forensic sciences include the use of Oxford Nanopore Technology (ONT), which will be covered in this section.

2.5.1 Sequencing by Synthesis

"Sequencing-by-synthesis (SBS)" refers to a method in which DNA polymerase directly synthesizes a complementary strand, producing a detectable signal as each nucleotide is incorporated (Heather and Chain 2016). Applying to the analysis of DNA methylation in forensic sciences, pyrosequencing is the gold standard technique used in several studies. However, Illumina-based methodologies have also been applied to forensic methylation works based on the possibility of simultaneous analysis of many DNA methylation markers (Castagnola et al. 2024).

2.5.1.1 Pyrosequencing

This method uses chemiluminescence detection to identify the sequence of interest following bisulfite conversion and PCR amplification. Deoxynucleotide triphosphates (dNTPs) are sequentially dispensed into a chamber containing the DNA template. When a complementary dNTP is incorporated by DNA polymerase, inorganic pyrophosphate (PPi) is released. ATP sulfurylase then converts PPi into ATP, which serves as a substrate for luciferase to oxidize luciferin into oxyluciferin, producing light. The intensity of this light, proportional to the amount of PPi released, is captured by a charge-coupled device (CCD) camera. Because the identity of each dispensed nucleotide is known, the DNA sequence can be determined. The resulting chemiluminescent signal is plotted as a pyrogram, with peak heights corresponding to the quantity of nucleotides incorporated. By comparing the relative heights of cytosine and thymine peaks at specific positions, the methylation status can be inferred (Ghemrawi et al. 2023; Hyman 1988; Nyren and Lundin 1985). It is important to observe that the currently available pyrosequencer on the market—the PyroMark series by Qiagen—operates in a non-massively parallel manner. This contrasts with the original 454 pyrosequencers developed by Roche (discontinued in 2013), which utilized clonal amplification and supported massively parallel sequencing (Ghemrawi et al. 2023). Despite its limitation of targeting only a limited number of genes and methylation sites, this method has been widely adopted in forensic studies due to its simplicity, high reproducibility, ability to detect methylation differences as small as 5%, and suitability for analyzing heterogeneous samples (Kurdyukov and Bullock 2016). Among these works, age estimation in different body fluids and tissues (Alghanim et al. 2017; Bekaert et al. 2015; Zapico et al. 2021; Fang et al. 2023b; Marquez-Ruiz et al. 2020; Soares Bispo Santos Silva et al. 2015; Zbiec-Piekarska et al. 2015b); identification of body fluids (Alghanim et al. 2020; Antunes et al. 2015, 2016; Balamurugan et al. 2014; Ghemrawi et al. 2024; Madi et al. 2012; Silva et al. 2016; Vidaki et al. 2017c); lifestyle (Alghanim et al. 2018).

2.5.1.2 Illumina

Illumina's NGS technology is based on SBS using fluorescently labeled reversible terminators (McCombie et al. 2019). Before sequencing begins, DNA libraries undergo clonal amplification via "bridge amplification" PCR directly on the flow cell, a process controlled by the sequencing instrument. During each sequencing cycle, a single fluorescently labeled reversible terminator-bound dNTP is incorporated into the growing DNA strand by DNA polymerase. The resulting fluorescent signal is captured optically, and the terminator and dye are subsequently cleaved to allow the next nucleotide to be added. Additionally, Illumina platforms support paired-end sequencing, which reads both ends of a DNA fragment, providing high-quality data with deep coverage and a large number of reads (Bentley et al. 2008; Fedurco et al. 2006; Hu et al. 2021; Voelkerding et al. 2009). The combination of reversible terminator chemistry and paired-end sequencing makes Illumina's technology the most accurate base-by-base sequencing method available, with an error rate of just 0.1%, primarily due to substitution errors and rarely insertions or deletions (Heather and Chain 2016; Hu et al. 2021). Moreover, the possibility of gene and sample multiplexing reduces the cost per sample compared to other techniques. Additionally, it is possible to apply to low quality/quantity DNA samples, frequently encountered in forensic scenarios (Aliferi et al. 2022). However, the disadvantages include high equipment and infrastructure costs, as well as the complexity of data analysis, which will require extensive training for laboratory personnel (Barros-Silva et al. 2018). Despite these disadvantages, several studies were conducted for forensic age estimation purposes applying this methodology in different body fluids and tissues (Guan et al. 2021; Heidegger et al. 2022; Naue et al. 2017, 2018; Vidaki et al. 2017a; Wozniak et al. 2021).

2.5.2 Semiconductor Sequencing—Ion Torrent

Ion Torrent was the first platform to implement semiconductor-based sequencing (Goodwin et al. 2016; Levy and Myers 2016). On this platform, clonal amplification is achieved through a bead-based method using Ion Sphere particles within microwells via emulsion PCR. DNA fragments are ligated to adapter sequences and encapsulated in water-in-oil emulsion droplets (micelles), each containing a bead coated with complementary adapters, dNTPs, primers, and DNA polymerase. Each micelle acts as an independent micro-PCR reactor. Sequencing occurs on a semiconductor chip equipped with microwells and a complementary metal-oxide semiconductor (CMOS) pH sensor. During sequencing, unmodified nucleotides are sequentially introduced. When a nucleotide is incorporated, a hydrogen ion is released, which is detected by the CMOS sensor—eliminating the need for optical detection (Hu et al. 2021). A known limitation of this technology is its reduced accuracy in homopolymer regions, where multiple identical bases can lead to nonlinear voltage responses and insertion/deletion errors. However, it maintains a low substitution error rate (<0.1%) (Merriman et al. 2012). In forensic science, most applications of the Ion Torrent platform have focused on STR and SNP profiling (Guo et al. 2016, 2022; Zubakov et al. 2015), with only a limited number of studies exploring its use for epigenetic analysis (Hanggi et al. 2025; Pospiech et al. 2023).

2.5.3 Long-Read Technologies—Nanopore

Nanopore sequencing does not rely on nucleotide labeling or optical detection. This technique emerged from research on DNA translocation through artificial nanopores. Nanopore sequencing operates by converting electrical signals generated as nucleotides pass through a nanopore—specifically, an α-hemolysin pore covalently linked to a cyclodextrin molecule, which serves as the nucleotide binding site. As a DNA strand moves through the pore, the ionic current is modulated, revealing structural characteristics such as diameter, length, and conformation. During sequencing, exonuclease-cleaved nucleotides interact with cyclodextrin, temporarily blocking the ionic current. The duration of this blockage is unique to each base, allowing the DNA sequence to be determined based on the electrical signal pattern (Pareek et al. 2011). The chemistry of this technology enables the continuous sequencing of long DNA fragments, with approximately accuracies ranges from 87% to 98% (Logsdon et al. 2020). A general drawback is its susceptibility to signal-to-noise limitations, which contribute to a higher error rate (ranging from 2% to 15%) compared to short-read technologies. However, these error rates continue to decline as base-calling algorithms improve over time (Hu et al. 2021; McCombie and McPherson 2019; McCombie et al. 2019; Rang et al. 2018). One of the key advantages of this technology is its ability to analyze DNA without prior bisulfite treatment, overcoming a major limitation when working with already degraded forensic samples. Additionally, the MinION platform from ONT is low-cost, highly portable, and compact, making it suitable for field deployment, such as in mass disaster scenarios (Quick et al. 2016). In recent years, nanopore sequencing has emerged as a potential valuable tool in forensic science, with applications expanding to include STR and SNP profiling (de Bruin et al. 2025), as well as DNA methylation analysis for body fluid identification (Beentjes et al. 2025; Yuen et al. 2024), age estimation (Beentjes et al. 2025; de Bruin et al. 2025; Yuen et al. 2024), and monozygotic twin differentiation (Zeng et al. 2025).

2.6 Mass Array-Based Technique—Sequenom EpiTyper

The Sequenom (CA, USA) EpiTyper™ assay is a widely used bisulfite-based method for analyzing regional DNA methylation with high precision and reproducibility. It employs matrix-assisted laser desorption/ionization time-of-flight (MALDI-TOF) mass spectrometry to detect mass differences in RNA cleavage fragments derived from post-bisulfite PCR products (Ehrich et al. 2005; Fouse et al. 2010; Raval et al. 2007). These products are generated using a T7-promoter tagged reverse primer, transcribed in vitro, and then cleaved with RNase A to produce small fragments. When these fragments contain the complement of a CpG site, methylation status can be inferred from the mass difference between guanine and adenine (Chatterjee et al. 2016). This approach enables high-throughput, quantitative methylation analysis at single CpG resolution across hundreds of loci, making it particularly useful for candidate loci screening or as a follow-up to genome-wide profiling (Fouse et al. 2010). In forensic science, although less common, it has been used with age estimation applications in different body fluids and tissues to assess specific genes and methylation sites towards this goal (Freire-Aradas et al. 2016, 2018; Giuliani et al. 2016).

2.7 Conclusions

Epigenetic techniques now offer a structured toolkit for forensic applications, ranging from discovery-oriented platforms to targeted assays suitable for routine analysis. Microarrays, Illumina short-read sequencing, nanopore long-read sequencing, and MassARRAY EpiTyper enable genome-wide or regional screening and have been instrumental in identifying informative CpG loci and refining epigenetic models. In contrast, pyrosequencing, MS-qPCR/MS-HRM, RT-qPCR for miRNAs, digital PCR, ARMS-PCR, and CE-based SNaPshot provide locus-specific, quantitative, or qualitative readouts that can be implemented on existing forensic infrastructure with relatively modest adaptations.

Across these methods, three main forensic applications recur: age estimation, body-fluid and tissue identification, and, more recently, differentiation of

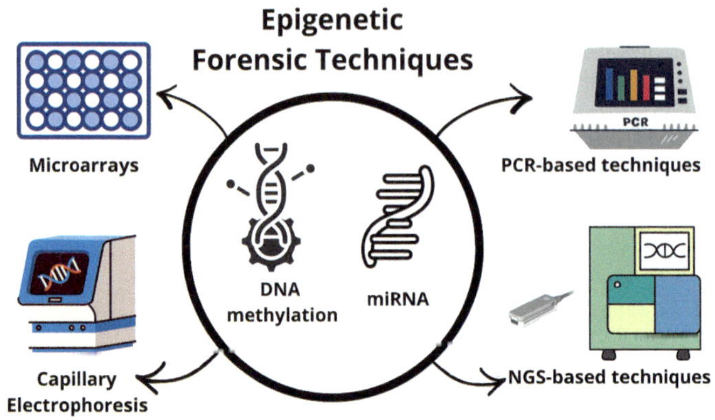

Fig. 2.1 Summary of the main epigenetic techniques described in this chapter

challenging scenarios such as monozygotic twins or complex mixtures. Age-associated methylation markers (notably in genes such as ELOVL2 and FHL2) have been validated on multiple platforms, while methylation and miRNA signatures have improved the specificity of body-fluid attribution. At the same time, each technique carries characteristic constraints, including DNA/RNA input requirements, sensitivity to degradation, bisulfite-induced damage, multiplexing limits, cost, and data-analysis complexity.

Taking together, these approaches outline a practical division of roles: high-throughput platforms for marker discovery and model development, and targeted assays for operational deployment in casework. Continued work on inter-laboratory validation, standardization of workflows, and optimization for low-quantity, degraded samples will be key to expanding the routine use of epigenetic evidence in forensic science.

Figure 2.1 summarizes the techniques described in this chapter.

Compliance with Ethical Standards The authors have no conflicts of interests to declare that are relevant to the content of this chapter.
This chapter is a review of previously published accounts, as such, no animal or human studies were performed.

References

Alghanim H, Antunes J, Silva D, Alho CS, Balamurugan K, McCord B (2017) Detection and evaluation of DNA methylation markers found at SCGN and KLF14 loci to estimate human age. Forensic Sci Int Genet 31:81–88. https://doi.org/10.1016/j.fsigen.2017.07.011

Alghanim H, Wu W, McCord B (2018) DNA methylation assay based on pyrosequencing for determination of smoking status. Electrophoresis 39(21):2806–2814. https://doi.org/10.1002/elps.201800098

Alghanim H, Balamurugan K, McCord B (2020) Development of DNA methylation markers for sperm, saliva and blood identification using pyrosequencing and qPCR/HRM. Anal Biochem 611:113933. https://doi.org/10.1016/j.ab.2020.113933

Aliferi A, Sundaram S, Ballard D, Freire-Aradas A, Phillips C, Lareu MV, Court DS (2022) Combining current knowledge on DNA methylation-based age estimation towards the development of a superior forensic DNA intelligence tool. Forensic Sci Int Genet 57:102637. https://doi.org/10.1016/j.fsigen.2021.102637

Antunes J, Balamurugan K, Duncan G, McCord B (2015) Tissue-specific DNA methylation patterns in forensic samples detected by pyrosequencing(R). Methods Mol Biol 1315:397–409. https://doi.org/10.1007/978-1-4939-2715-9_27

Antunes J, Silva DS, Balamurugan K, Duncan G, Alho CS, McCord B (2016) Forensic discrimination of vaginal epithelia by DNA methylation analysis through pyrosequencing. Electrophoresis 37(21):2751–2758. https://doi.org/10.1002/elps.201600037

Balamurugan K, Bombardi R, Duncan G, McCord B (2014) Identification of spermatozoa by tissue-specific differential DNA methylation using bisulfite modification and pyrosequencing. Electrophoresis 35(21–22):3079–3086. https://doi.org/10.1002/elps.201400175

Barros-Silva D, Marques CJ, Henrique R, Jeronimo C (2018) Profiling DNA methylation based on next-generation sequencing approaches: new insights and clinical applications. Genes (Basel) 9(9):429. https://doi.org/10.3390/genes9090429

Beentjes I, Haagmans MA, de Bruin D, Permana A, Pospiech E, Branicki W, M'Charek AA, van der Gaag KJ, Sijen T, Henneman P (2025) DNA methylation-based forensic framework for age prediction and body fluid identification using nanopore sequencing. Forensic Sci Int Genet 81:103370. https://doi.org/10.1016/j.fsigen.2025.103370

Bekaert B, Kamalandua A, Zapico SC, Van de Voorde W, Decorte R (2015) Improved age determination of blood and teeth samples using a selected set of DNA methylation markers. Epigenetics 10(10):922–930. https://doi.org/10.1080/15592294.2015.1080413

Bentley DR, Balasubramanian S, Swerdlow HP, Smith GP, Milton J, Brown CG, Hall KP, Evers DJ, Barnes CL, Bignell HR, Boutell JM, Bryant J, Carter RJ, Keira Cheetham R, Cox AJ, Ellis DJ, Flatbush MR, Gormley NA, Humphray SJ et al (2008) Accurate whole human genome sequencing using reversible terminator chemistry. Nature 456(7218):53–59. https://doi.org/10.1038/nature07517

Bibikova M, Barnes B, Tsan C, Ho V, Klotzle B, Le JM, Delano D, Zhang L, Schroth GP, Gunderson KL, Fan JB, Shen R (2011) High density DNA methylation array with single CpG site resolution. Genomics 98(4):288–295. https://doi.org/10.1016/j.ygeno.2011.07.007

Bocklandt S, Lin W, Sehl ME, Sanchez FJ, Sinsheimer JS, Horvath S, Vilain E (2011) Epigenetic predictor of age. PLoS One 6(6):e14821. https://doi.org/10.1371/journal.pone.0014821

Butler JM (2011) Advanced topics in forensic DNA typing: methodology. Academic

Castagnola MJ, Medina-Paz F, Zapico SC (2024) Uncovering forensic evidence: a path to age estimation through DNA methylation. Int J Mol Sci 25(9):4917. https://doi.org/10.3390/ijms25094917

Chatterjee A, Rodger EJ, Morison IM, Eccles MR, Stockwell PA (2016) Tools and strategies for analysis of genome-wide and gene-specific DNA methylation patterns. In: Seymour G, Cullinan M, Heng N (eds) Oral biology. Methods in molecular biology, vol 1537. Humana Press, pp 249–277

Christiansen SN, Andersen JD, Kampmann ML, Liu J, Andersen MM, Tfelt-Hansen J, Morling N (2022) Reproducibility of the Infinium methylationEPIC BeadChip assay using low DNA amounts. Epigenetics 17(12):1636–1645. https://doi.org/10.1080/15592294.2022.2051861

Consortium B (2016) Quantitative comparison of DNA methylation assays for biomarker development and clinical applications. Nat Biotechnol 34(7):726–737. https://doi.org/10.1038/nbt.3605

Courts C, Madea B (2011) Specific micro-RNA signatures for the detection of saliva and blood in forensic body-fluid identification. J Forensic Sci 56(6):1464–1470. https://doi.org/10.1111/j.1556-4029.2011.01894.x

de Bruin D, Haagmans MA, van der Gaag KJ, Hoogenboom J, Weiler NEC, Tesi N, Salazar A, Zhang Y, Holstege H, Reinders M, M'Charek AA, Sijen T, Henneman P (2025) Exploring nanopore direct sequencing performance of forensic STRs, SNPs, InDels, and DNA methylation markers in a single assay. Forensic Sci Int Genet 74:103154. https://doi.org/10.1016/j.fsigen.2024.103154

Dias HC, Manco L (2024) Predicting age from blood by droplet digital PCR using a set of three DNA methylation markers. Forensic Sci Int 356:111950. https://doi.org/10.1016/j.forsciint.2024.111950

Dias HC, Cordeiro C, Pereira J, Pinto C, Real FC, Cunha E, Manco L (2020) DNA methylation age estimation in blood samples of living and deceased individuals using a multiplex SNaPshot assay. Forensic Sci Int 311:110267. https://doi.org/10.1016/j.forsciint.2020.110267

Ehrich M, Nelson MR, Stanssens P, Zabeau M, Liloglou T, Xinarianos G, Cantor CR, Field JK, van den Boom D (2005) Quantitative high-throughput analysis of DNA methylation patterns by base-specific cleavage and mass spectrometry. Proc Natl Acad Sci USA 102(44):15785–15790. https://doi.org/10.1073/pnas.0507816102

Fakhr R, Koh UN, Lee SE, Lim SK (2025) A hexaplex droplet digital PCR assay for DNA methylation-based age prediction from blood samples targeting ELOVL2, FHL2, and TRIM59. Forensic Sci Int Genet 81:103396. https://doi.org/10.1016/j.fsigen.2025.103396

Fang C, Zhou P, Li R, Guo J, Qiu H, Zhang J, Li M, Yu C, Meng D, Xu X, Liu X, Guan D, Yan J (2023a) Development of a novel forensic age estimation strategy for aged blood samples by combining piRNA and miRNA markers. Int J Legal Med 137(5):1327–1335. https://doi.org/10.1007/s00414-023-03028-8

Fang Y, Chen M, Cai M, Lei F, Zhu B (2023b) Selection and validation of a novel set of specific differential methylation markers and construction of a random forest prediction model for the accurate tissue origin identifications of body fluids involving young and middle-aged group of Chinese Han population. Int J Legal Med 137(5):1395–1405. https://doi.org/10.1007/s00414-023-03049-3

Fang Y, Chen M, Zhu B (2023c) Construction and evaluation of in-house methylation-sensitive SNaPshot system and three classification prediction models for identifying the tissue origin of body fluid. J Zhejiang Univ Sci B 24(9):839–852. https://doi.org/10.1631/jzus.B2200555. (鉴定体液来源的甲基化敏感性SNaPshot体系及三种预测模型的构建与评估)

Fedurco M, Romieu A, Williams S, Lawrence I, Turcatti G (2006) BTA, a novel reagent for DNA attachment on glass and efficient generation of solid-phase amplified DNA colonies. Nucleic Acids Res 34(3):e22. https://doi.org/10.1093/nar/gnj023

Forat S, Huettel B, Reinhardt R, Fimmers R, Haidl G, Denschlag D, Olek K (2016a) Correction: methylation markers for the identification of body fluids and tissues from forensic trace evidence. PLoS One 11(5):e0156472. https://doi.org/10.1371/journal.pone.0156472

Forat S, Huettel B, Reinhardt R, Fimmers R, Haidl G, Denschlag D, Olek K (2016b) Methylation markers for the identification of body fluids and tissues from forensic trace evidence. PLoS One 11(2):e0147973. https://doi.org/10.1371/journal.pone.0147973

Fouse SD, Nagarajan RO, Costello JF (2010) Genome-scale DNA methylation analysis. Epigenomics 2(1):105–117. https://doi.org/10.2217/epi.09.35

Freire-Aradas A, Phillips C, Mosquera-Miguel A, Giron-Santamaria L, Gomez-Tato A, Casares de Cal M, Alvarez-Dios J, Ansede-Bermejo J, Torres-Espanol M, Schneider PM, Pospiech E, Branicki W, Carracedo A, Lareu MV (2016) Development of a methylation marker set for forensic age estimation using analysis of public methylation data and the Agena Bioscience EpiTYPER system. Forensic Sci Int Genet 24:65–74. https://doi.org/10.1016/j.fsigen.2016.06.005

Freire-Aradas A, Phillips C, Giron-Santamaria L, Mosquera-Miguel A, Gomez-Tato A, Casares de Cal MA, Alvarez-Dios J, Lareu MV (2018) Tracking age-correlated DNA methylation markers in the young. Forensic Sci Int Genet 36:50–59. https://doi.org/10.1016/j.fsigen.2018.06.011

Fujimoto S, Manabe S, Morimoto C, Ozeki M, Hamano Y, Hirai E, Kotani H, Tamaki K (2019) Distinct spectrum of microRNA expression in forensically relevant body fluids and probabilistic discriminant approach. Sci Rep 9(1):14332. https://doi.org/10.1038/s41598-019-50796-8

Ghemrawi M, Tejero NF, Duncan G, McCord B (2023) Pyrosequencing: current forensic methodology and future applications—a review. Electrophoresis 44(1–2):298–312. https://doi.org/10.1002/elps.202200177

Ghemrawi M, Fernandez-Tejero N, Vaquero L, Wanna A, Carmel JH, McCord B (2024) An examination of differences in epigenetic methylation of saliva type samples based on collection method. Electrophoresis 45(9–10):897–905. https://doi.org/10.1002/elps.202300240

Giuliani C, Cilli E, Bacalini MG, Pirazzini C, Sazzini M, Gruppioni G, Franceschi C, Garagnani P, Luiselli D (2016) Inferring chronological age from DNA methylation patterns of human teeth. Am J Phys Anthropol 159(4):585–595. https://doi.org/10.1002/ajpa.22921

Goodwin S, McPherson JD, McCombie WR (2016) Coming of age: ten years of next-generation sequencing technologies. Nat Rev Genet 17(6):333–351. https://doi.org/10.1038/nrg.2016.49

Guan X, Ohuchi T, Hashiyada M, Funayama M (2021) Age-related DNA methylation analysis for forensic age estimation using post-mortem blood samples from Japanese individuals. Leg Med (Tokyo) 53:101917. https://doi.org/10.1016/j.legalmed.2021.101917

Guardado-Estrada M, Cardenas-Monroy CA, Martinez-Rivera V, Cortez F, Pedraza-Lara C, Millan-Catalan O, Perez-Plasencia C (2023) A miRNome analysis at the early postmortem interval. PeerJ 11:e15409. https://doi.org/10.7717/peerj.15409

Guo F, Zhou Y, Song H, Zhao J, Shen H, Zhao B, Liu F, Jiang X (2016) Next generation sequencing of SNPs using the HID-Ion AmpliSeq Identity Panel on the Ion Torrent PGM platform. Forensic Sci Int Genet 25:73–84. https://doi.org/10.1016/j.fsigen.2016.07.021

Guo F, Lang Y, Long G, Liu Z, Jing G, Zhou Y, Zhang B, Yu S (2022) Ion torrent genexus integrated sequencer and ForeNGS analysis software—an automatic NGS-STR workflow from DNA to profile for forensic science. Forensic Sci Int Genet 61:102753. https://doi.org/10.1016/j.fsigen.2022.102753

Hamano Y, Manabe S, Morimoto C, Fujimoto S, Ozeki M, Tamaki K (2016) Forensic age prediction for dead or living samples by use of methylation-sensitive high resolution melting. Leg Med (Tokyo) 21:5–10. https://doi.org/10.1016/j.legalmed.2016.05.001

Hanggi NV, Neubauer J, Marti Y, Banemann R, Kulstein G, Courts C, Gosch A, Hadrys T, Haas C, Dorum G (2025) Assessing transcriptomic signatures of aging: testing an mRNA marker panel for forensic age estimation of blood samples. Forensic Sci Int Genet 78:103282. https://doi.org/10.1016/j.fsigen.2025.103282

Hannum G, Guinney J, Zhao L, Zhang L, Hughes G, Sadda S, Klotzle B, Bibikova M, Fan JB, Gao Y, Deconde R, Chen M, Rajapakse I, Friend S, Ideker T, Zhang K (2013) Genome-wide methylation profiles reveal quantitative views of human aging rates. Mol Cell 49(2):359–367. https://doi.org/10.1016/j.molcel.2012.10.016

Hanson EK, Ballantyne J (2013) Circulating microRNA for the identification of forensically relevant body fluids. Methods Mol Biol 1024:221–234. https://doi.org/10.1007/978-1-62703-453-1_18

Hanson EK, Lubenow H, Ballantyne J (2009) Identification of forensically relevant body fluids using a panel of differentially expressed microRNAs. Anal Biochem 387(2):303–314. https://doi.org/10.1016/j.ab.2009.01.037

Heather JM, Chain B (2016) The sequence of sequencers: the history of sequencing DNA. Genomics 107(1):1–8. https://doi.org/10.1016/j.ygeno.2015.11.003

Heidegger A, Pisarek A, de la Puente M, Niederstatter H, Pospiech E, Wozniak A, Schury N, Unterlander M, Sidstedt M, Junker K, Ventayol Garcia M, Laurent FX, Ulus A, Vannier J, Bastisch I, Hedman J, Sijen T, Branicki W, Xavier C et al (2022) Development and interlaboratory validation of the VISAGE enhanced tool for age estimation from semen using quantitative DNA methylation analysis. Forensic Sci Int Genet 56:102596. https://doi.org/10.1016/j.fsigen.2021.102596

Herman JG, Graff JR, Myohanen S, Nelkin BD, Baylin SB (1996) Methylation-specific PCR: a novel PCR assay for methylation status of CpG islands. Proc Natl Acad Sci USA 93(18):9821–9826. https://doi.org/10.1073/pnas.93.18.9821

Hernandez HG, Tse MY, Pang SC, Arboleda H, Forero DA (2013) Optimizing methodologies for PCR-based DNA methylation analysis. BioTechniques 55(4):181–197. https://doi.org/10.2144/000114087

Hindson BJ, Ness KD, Masquelier DA, Belgrader P, Heredia NJ, Makarewicz AJ, Bright IJ, Lucero MY, Hiddessen AL, Legler TC, Kitano TK, Hodel MR, Petersen JF, Wyatt PW, Steenblock ER, Shah PH, Bousse LJ, Troup CB, Mellen JC et al (2011) High-throughput droplet digital PCR system for absolute quantitation of DNA copy number. Anal Chem 83(22):8604–8610. https://doi.org/10.1021/ac202028g

Hindson CM, Chevillet JR, Briggs HA, Gallichotte EN, Ruf IK, Hindson BJ, Vessella RL, Tewari M (2013) Absolute quantification by droplet digital PCR versus analog real-time PCR. Nat Methods 10(10):1003–1005. https://doi.org/10.1038/nmeth.2633

Honardoost MA, Tabatabaeian H, Akbari M, Salehi M (2014) Investigation of sensitivity, specificity and accuracy of Tetra primer ARMS PCR method in comparison with conventional ARMS PCR, based on sequencing technique outcomes in IVS-II-I genotyping of beta thalassemia patients. Gene 549(1):1–6. https://doi.org/10.1016/j.gene.2014.05.071

Horvath S (2013) DNA methylation age of human tissues and cell types. Genome Biol 14(10):R115. https://doi.org/10.1186/gb-2013-14-10-r115

Hu T, Chitnis N, Monos D, Dinh A (2021) Next-generation sequencing technologies: an overview. Hum Immunol 82(11):801–811. https://doi.org/10.1016/j.humimm.2021.02.012

Hyman ED (1988) A new method of sequencing DNA. Anal Biochem 174(2):423–436. https://doi.org/10.1016/0003-2697(88)90041-3

Illumina (2025) Infinium MethylationEPIC v2.0 kit. https://www.illumina.com/products/by-type/microarray-kits/infinium-methylation-epic.html#accordion-b63ea058e5-item-9db1fc51cc

Jiang L, Zhang K, Wei X, Li J, Wang S, Wang Z, Zhou Y, Zha L, Luo H, Song F (2023) Developing a male-specific age predictive model based on Y-CpGs for forensic analysis. Forensic Sci Int 343:111566. https://doi.org/10.1016/j.forsciint.2023.111566

Jung SE, Lim SM, Hong SR, Lee EH, Shin KJ, Lee HY (2019) DNA methylation of the ELOVL2, FHL2, KLF14, C1orf132/MIR29B2C, and TRIM59 genes for age prediction from blood, saliva, and buccal swab samples. Forensic Sci Int Genet 38:1–8. https://doi.org/10.1016/j.fsigen.2018.09.010

Jung JY, Song YW, Jeong KS, Park HC, So MH, Lee HY (2025) A SNaPshot assay for epigenetic age prediction of costal cartilage. Electrophoresis 46(7–8):413–423. https://doi.org/10.1002/elps.8132

Kaija H, Pakanen L, Porvari K (2020) RNU6B, a frequent reference in miRNA expression studies, differentiates between deaths caused by hypothermia and chronic cardiac ischemia. Int J Legal Med 134(1):159–162. https://doi.org/10.1007/s00414-019-02041-0

Kaminsky Z, Petronis A (2009) Methylation SNaPshot: a method for the quantification of site-specific DNA methylation levels. Methods Mol Biol 507:241–255. https://doi.org/10.1007/978-1-59745-522-0_18

Kim BM, Park SU, Schmelzer L, Yang SB, Lee SD, Kim MY, Naue J, Lee HY (2024) DNA methylation-based organ tissue identification: marker identification, SNaPshot multiplex assay development, and interlaboratory comparison. Forensic Sci Int Genet 71:103052. https://doi.org/10.1016/j.fsigen.2024.103052

Kurdyukov S, Bullock M (2016) DNA methylation analysis: choosing the right method. Biology (Basel) 5(1):3. https://doi.org/10.3390/biology5010003

Lee HY, An JH, Jung SE, Oh YN, Lee EY, Choi A, Yang WI, Shin KJ (2015) Genome-wide methylation profiling and a multiplex construction for the identification of body fluids using epigenetic markers. Forensic Sci Int Genet 17:17–24. https://doi.org/10.1016/j.fsigen.2015.03.002

Lee HY, Hong SR, Lee JE, Hwang IK, Kim NY, Lee JM, Fleckhaus J, Jung SE, Lee YH (2020) Epigenetic age signatures in bones. Forensic Sci Int Genet 46:102261. https://doi.org/10.1016/j.fsigen.2020.102261

Levine ME, Lu AT, Quach A, Chen BH, Assimes TL, Bandinelli S, Hou L, Baccarelli AA, Stewart JD, Li Y, Whitsel EA, Wilson JG, Reiner AP, Aviv A, Lohman K, Liu Y, Ferrucci L, Horvath S (2018) An epigenetic biomarker of aging for lifespan and healthspan. Aging (Albany NY) 10(4):573–591. https://doi.org/10.18632/aging.101414

Levy SE, Myers RM (2016) Advancements in next-generation sequencing. Annu Rev Genomics Hum Genet 17:95–115. https://doi.org/10.1146/annurev-genom-083115-022413

Liu Z, Gao Z, Wang J, Shi J, Liu J, Chen D, Li W, Guo J, Cheng X, Hao T, Li Z, Li Y, Yan J, Zhang G (2020) A method of identifying the blood contributor in mixture stains through detecting blood-specific mRNA polymorphism. Electrophoresis 41(15):1364–1373. https://doi.org/10.1002/elps.202000053

Liu Z, Wang N, Zhang Q, Li R, Liu J, Luo B, Wang Y, Xue L, Wu R, Cong B, Xie B, Sun H (2025) Developmental validation of an mRNA-cSNP profiling panel for body fluids identification and individualization. Int J Legal Med 139(4):1495–1509. https://doi.org/10.1007/s00414-025-03434-0

Logsdon GA, Vollger MR, Eichler EE (2020) Long-read human genome sequencing and its applications. Nat Rev Genet 21(10):597–614. https://doi.org/10.1038/s41576-020-0236-x

Madi T, Balamurugan K, Bombardi R, Duncan G, McCord B (2012) The determination of tissue-specific DNA methylation patterns in forensic biofluids using bisulfite modification and pyrosequencing. Electrophoresis 33(12):1736–1745. https://doi.org/10.1002/elps.201100711

Marquez-Ruiz AB, Gonzalez-Herrera L, Luna JD, Valenzuela A (2020) DNA methylation levels and telomere length in human teeth: usefulness for age estimation. Int J Legal Med 134(2):451–459. https://doi.org/10.1007/s00414-019-02242-7

McCombie WR, McPherson JD (2019) Future promises and concerns of ubiquitous next-generation sequencing. Cold Spring Harb Perspect Med 9(9):a025783. https://doi.org/10.1101/cshperspect.a025783

McCombie WR, McPherson JD, Mardis ER (2019) Next-generation sequencing technologies. Cold Spring Harb Perspect Med 9(11):a036798. https://doi.org/10.1101/cshperspect.a036798

McDonald C, Taylor D, Linacre A (2024) PCR in forensic science: a critical review. Genes (Basel) 15(4):438. https://doi.org/10.3390/genes15040438

Medrano RF, de Oliveira CA (2014) Guidelines for the tetra-primer ARMS-PCR technique development. Mol Biotechnol 56(7):599–608. https://doi.org/10.1007/s12033-014-9734-4

Mehta B, Daniel R, Phillips C, McNevin D (2017) Forensically relevant SNaPshot((R)) assays for human DNA SNP analysis: a review. Int J Legal Med 131(1):21–37. https://doi.org/10.1007/s00414-016-1490-5

Merriman B, Ion Torrent R, Team D, Rothberg JM (2012) Progress in ion torrent semiconductor chip based sequencing. Electrophoresis 33(23):3397–3417. https://doi.org/10.1002/elps.201200424

Morley AA (2014) Digital PCR: a brief history. Biomol Detect Quantif 1(1):1–2. https://doi.org/10.1016/j.bdq.2014.06.001

Munjakovic H, Povsic K, Poljak M, Seme K, Gaspersic R, Skubic L (2025) Digital PCR outperforms quantitative real-time PCR for the detection and quantification of major periodontal pathobionts. J Oral Microbiol 17(1):2537439. https://doi.org/10.1080/20002297.2025.2537439

Naue J, Hoefsloot HCJ, Mook ORF, Rijlaarsdam-Hoekstra L, van der Zwalm MCH, Henneman P, Kloosterman AD, Verschure PJ (2017) Chronological age prediction based on DNA methylation: massive parallel sequencing and random forest regression. Forensic Sci Int Genet 31:19–28. https://doi.org/10.1016/j.fsigen.2017.07.015

Naue J, Sanger T, Hoefsloot HCJ, Lutz-Bonengel S, Kloosterman AD, Verschure PJ (2018) Proof of concept study of age-dependent DNA methylation markers across different tissues by massive parallel sequencing. Forensic Sci Int Genet 36:152–159. https://doi.org/10.1016/j.fsigen.2018.07.007

Neis M, Gross T, Schneider H, Schneider PM, Courts C (2024) Comprehensive body fluid identification and contributor assignment by combining targeted sequencing of mRNA and coding region SNPs. Forensic Sci Int Genet 73:103125. https://doi.org/10.1016/j.fsigen.2024.103125

Newton CR, Graham A, Heptinstall LE, Powell SJ, Summers C, Kalsheker N, Smith JC, Markham AF (1989) Analysis of any point mutation in DNA. The amplification refractory mutation system (ARMS). Nucleic Acids Res 17(7):2503–2516. https://doi.org/10.1093/nar/17.7.2503

Nussbaumer C, Gharehbaghi-Schnell E, Korschineck I (2006) Messenger RNA profiling: a novel method for body fluid identification by real-time PCR. Forensic Sci Int 157(2–3):181–186. https://doi.org/10.1016/j.forsciint.2005.10.009

Nyren P, Lundin A (1985) Enzymatic method for continuous monitoring of inorganic pyrophosphate synthesis. Anal Biochem 151(2):504–509. https://doi.org/10.1016/0003-2697(85)90211-8

Ogata A, Kondo M, Yoshikawa M, Okano M, Tsutsumi T, Aboshi H (2022) Dental age estimation based on DNA methylation using real-time methylation-specific PCR. Forensic Sci Int 340:111445. https://doi.org/10.1016/j.forsciint.2022.111445

Pajares MJ, Palanca-Ballester C, Urtasun R, Alemany-Cosme E, Lahoz A, Sandoval J (2021) Methods for analysis of specific DNA methylation status. Methods 187:3–12. https://doi.org/10.1016/j.ymeth.2020.06.021

Paparazzo E, Geracitano S, Lagani V, Bartolomeo D, Aceto MA, D'Aquila P, Citrigno L, Bellizzi D, Passarino G, Montesanto A (2022) A blood-based molecular clock for biological age estimation. Cells 12(1):32. https://doi.org/10.3390/cells12010032

Pareek CS, Smoczynski R, Tretyn A (2011) Sequencing technologies and genome sequencing. J Appl Genet 52(4):413–435. https://doi.org/10.1007/s13353-011-0057-x

Park JL, Kwon OH, Kim JH, Yoo HS, Lee HC, Woo KM, Kim SY, Lee SH, Kim YS (2014) Identification of body fluid-specific DNA methylation markers for use in forensic science. Forensic Sci Int Genet 13:147–153. https://doi.org/10.1016/j.fsigen.2014.07.011

Pospiech E, Pisarek A, Rudnicka J, Noroozi R, Boron M, Masny A, Wysocka B, Migacz-Gruszka K, Lisman D, Pruszkowska-Przybylska P, Kobus M, Szargut M, Dowejko J, Stanisz K, Zacharczuk J, Zielinski P, Sitek A, Ossowski A, Spolnicka M, Branicki W (2023) Introduction of a multiplex amplicon sequencing assay to quantify DNA methylation in target cytosine markers underlying four selected epigenetic clocks. Clin Epigenetics 15(1):128. https://doi.org/10.1186/s13148-023-01545-2

Quick J, Loman NJ, Duraffour S, Simpson JT, Severi E, Cowley L, Bore JA, Koundouno R, Dudas G, Mikhail A, Ouedraogo N, Afrough B, Bah A, Baum JH, Becker-Ziaja B, Boettcher JP, Cabeza-Cabrerizo M, Camino-Sanchez A, Carter LL et al (2016) Real-time, portable genome sequencing for Ebola surveillance. Nature 530(7589):228–232. https://doi.org/10.1038/nature16996

Rang FJ, Kloosterman WP, de Ridder J (2018) From squiggle to basepair: computational approaches for improving nanopore sequencing read accuracy. Genome Biol 19(1):90. https://doi.org/10.1186/s13059-018-1462-9

Raval A, Tanner SM, Byrd JC, Angerman EB, Perko JD, Chen SS, Hackanson B, Grever MR, Lucas DM, Matkovic JJ, Lin TS, Kipps TJ, Murray F, Weisenburger D, Sanger W, Lynch J, Watson P, Jansen M, Yoshinaga Y et al (2007) Downregulation of death-associated protein kinase 1 (DAPK1) in chronic lymphocytic leukemia. Cell 129(5):879–890. https://doi.org/10.1016/j.cell.2007.03.043

Sabeeha, Hasnain SE (2019) Forensic epigenetic analysis: the path ahead. Med Princ Pract 28(4):301–308. https://doi.org/10.1159/000499496

Sandoval J, Heyn H, Moran S, Serra-Musach J, Pujana MA, Bibikova M, Esteller M (2011) Validation of a DNA methylation microarray for 450,000 CpG sites in the human genome. Epigenetics 6(6):692–702. https://doi.org/10.4161/epi.6.6.16196

Sauer E, Babion I, Madea B, Courts C (2014) An evidence based strategy for normalization of quantitative PCR data from miRNA expression analysis in forensic organ tissue identification. Forensic Sci Int Genet 13:217–223. https://doi.org/10.1016/j.fsigen.2014.08.005

Sauer E, Reinke AK, Courts C (2016) Differentiation of five body fluids from forensic samples by expression analysis of four microRNAs using quantitative PCR. Forensic Sci Int Genet 22:89–99. https://doi.org/10.1016/j.fsigen.2016.01.018

Schmidt M, Maixner F, Hotz G, Pap I, Szikossy I, Palfi G, Zink A, Wagner W (2021) DNA methylation profiling in mummified human remains from the eighteenth-century. Sci Rep 11(1):15493. https://doi.org/10.1038/s41598-021-95021-7

Silva SS, Lopes C, Teixeira AL, Carneiro de Sousa MJ, Medeiros R (2015) Forensic miRNA: potential biomarker for body fluids? Forensic Sci Int Genet 14:1–10. https://doi.org/10.1016/j.fsigen.2014.09.002

Silva D, Antunes J, Balamurugan K, Duncan G, Alho CS, McCord B (2016) Developmental validation studies of epigenetic DNA methylation markers for the detection of blood, semen and saliva samples. Forensic Sci Int Genet 23:55–63. https://doi.org/10.1016/j.fsigen.2016.01.017

Soares Bispo Santos Silva D, Antunes J, Balamurugan K, Duncan G, Sampaio Alho C, McCord B (2015) Evaluation of DNA methylation markers and their potential to predict human aging. Electrophoresis 36(15):1775–1780. https://doi.org/10.1002/elps.201500137

Soliman AM, Das S, Abd Ghafar N, Teoh SL (2018) Role of microRNA in proliferation phase of wound healing. Front Genet 9:38. https://doi.org/10.3389/fgene.2018.00038

Stirzaker C, Taberlay PC, Statham AL, Clark SJ (2014) Mining cancer methylomes: prospects and challenges. Trends Genet 30(2):75–84. https://doi.org/10.1016/j.tig.2013.11.004

Tian H, Bai P, Tan Y, Li Z, Peng D, Xiao X, Zhao H, Zhou Y, Liang W, Zhang L (2020) A new method to detect methylation profiles for forensic body fluid identification combining ARMS-PCR technique and random forest model. Forensic Sci Int Genet 49:102371. https://doi.org/10.1016/j.fsigen.2020.102371

Tucker T, Marra M, Friedman JM (2009) Massively parallel sequencing: the next big thing in genetic medicine. Am J Hum Genet 85(2):142–154. https://doi.org/10.1016/j.ajhg.2009.06.022

van Dijk EL, Jaszczyszyn Y, Naquin D, Thermes C (2018) The third revolution in sequencing technology. Trends Genet 34(9):666–681. https://doi.org/10.1016/j.tig.2018.05.008

Vidaki A, Kayser M (2018) Recent progress, methods and perspectives in forensic epigenetics. Forensic Sci Int Genet 37:180–195. https://doi.org/10.1016/j.fsigen.2018.08.008

Vidaki A, Ballard D, Aliferi A, Miller TH, Barron LP, Syndercombe Court D (2017a) DNA methylation-based forensic age prediction using artificial neural networks and next generation sequencing. Forensic Sci Int Genet 28:225–236. https://doi.org/10.1016/j.fsigen.2017.02.009

Vidaki A, Diez Lopez C, Carnero-Montoro E, Ralf A, Ward K, Spector T, Bell JT, Kayser M (2017b) Epigenetic discrimination of identical twins from blood under the forensic scenario. Forensic Sci Int Genet 31:67–80. https://doi.org/10.1016/j.fsigen.2017.07.014

Vidaki A, Johansson C, Giangasparo F, Denise Syndercombe C (2017c) Differentially methylated embryonal Fyn-associated substrate (EFS) gene as a blood-specific epigenetic marker and its potential application in forensic casework. Forensic Sci Int Genet 29:165–173. https://doi.org/10.1016/j.fsigen.2017.04.010

Vidaki A, Kalamara V, Carnero-Montoro E, Spector TD, Bell JT, Kayser M (2018) Investigating the epigenetic discrimination of identical twins using buccal swabs, saliva, and cigarette butts in the forensic setting. Genes (Basel) 9(5):252. https://doi.org/10.3390/genes9050252

Voelkerding KV, Dames SA, Durtschi JD (2009) Next-generation sequencing: from basic research to diagnostics. Clin Chem 55(4):641–658. https://doi.org/10.1373/clinchem.2008.112789

Wang S, Wang Z, Tao R, He G, Liu J, Li C, Hou Y (2019a) The potential use of Piwi-interacting RNA biomarkers in forensic body fluid identification: a proof-of-principle study. Forensic Sci Int Genet 39:129–135. https://doi.org/10.1016/j.fsigen.2019.01.002

Wang S, Wang Z, Tao R, Song F, Hou Y (2019b) Validating the consistency of cSNPs analysis results between DNA and RNA using SNaPshot method. Forensic Sci Int Genet Suppl Ser 7(1):76–78

Wei S, Hu S, Han N, Wang G, Chen H, Yao Q, Zhao Y, Ye J, Ji A, Sun Q (2023) Screening and evaluation of endogenous reference genes for miRNA expression analysis in forensic body fluid samples. Forensic Sci Int Genet 63:102827. https://doi.org/10.1016/j.fsigen.2023.102827

Weidner CI, Lin Q, Koch CM, Eisele L, Beier F, Ziegler P, Bauerschlag DO, Jockel KH, Erbel R, Muhleisen TW, Zenke M, Brummendorf TH, Wagner W (2014) Aging of blood can be tracked by DNA methylation changes at just three CpG sites. Genome Biol 15(2):R24. https://doi.org/10.1186/gb-2014-15-2-r24

Whale AS, Huggett JF, Tzonev S (2016) Fundamentals of multiplexing with digital PCR. Biomol Detect Quantif 10:15–23. https://doi.org/10.1016/j.bdq.2016.05.002

Wojdacz TK, Dobrovic A (2007) Methylation-sensitive high resolution melting (MS-HRM): a new approach for sensitive and high-throughput assessment of methylation. Nucleic Acids Res 35(6):e41. https://doi.org/10.1093/nar/gkm013

Wojdacz TK, Dobrovic A, Hansen LL (2008) Methylation-sensitive high-resolution melting. Nat Protoc 3(12):1903–1908. https://doi.org/10.1038/nprot.2008.191

Wozniak A, Heidegger A, Piniewska-Rog D, Pospiech E, Xavier C, Pisarek A, Kartasinska E, Boron M, Freire-Aradas A, Wojtas M, de la Puente M, Niederstatter H, Ploski R, Spolnicka M, Kayser M, Phillips C, Parson W, Branicki W, Consortium V (2021) Development of the VISAGE enhanced tool and statistical models for epigenetic age estimation in blood, buccal cells and bones. Aging (Albany NY) 13(5):6459–6484. https://doi.org/10.18632/aging.202783

Xiao B, Zhou Y, Zhang Z, Wang X, Xiang J, Lv Z, Liao M, Luo H, Song F (2025) Combining a novel ensemble model and multiplex methylation SNaPshot assays for saliva age prediction and cross-platform data analysis. BMC Genomics 26(1):546. https://doi.org/10.1186/s12864-025-11713-8

Yuen ZW, Shanmuganandam S, Stanley M, Jiang S, Hein N, Daniel R, McNevin D, Jack C, Eyras E (2024) Profiling age and body fluid DNA methylation markers using nanopore adaptive sampling. Forensic Sci Int Genet 71:103048. https://doi.org/10.1016/j.fsigen.2024.103048

Zapico SC, Gauthier Q, Antevska A, McCord BR (2021) Identifying methylation patterns in dental pulp aging: application to age-at-death estimation in forensic anthropology. Int J Mol Sci 22(7):3717. https://doi.org/10.3390/ijms22073717

Zbiec-Piekarska R, Spolnicka M, Kupiec T, Makowska Z, Spas A, Parys-Proszek A, Kucharczyk K, Ploski R, Branicki W (2015a) Examination of DNA methylation status of the ELOVL2 marker may be useful for human age prediction in forensic science. Forensic Sci Int Genet 14:161–167. https://doi.org/10.1016/j.fsigen.2014.10.002

Zbiec-Piekarska R, Spolnicka M, Kupiec T, Parys-Proszek A, Makowska Z, Paleczka A, Kucharczyk K, Ploski R, Branicki W (2015b) Development of a forensically useful age prediction method based on DNA methylation analysis. Forensic Sci Int Genet 17:173–179. https://doi.org/10.1016/j.fsigen.2015.05.001

Zeng K, Yang KB, Du J, Rakha A, Liu Z, Wang DY, Chen YZ, Wang SW, Sun ML, Wang H, Wang YL, Adnan A, Yao J (2025) Epigenetic differentiation of monozygotic twins in forensic investigations using nanopore sequencing. Forensic Sci Int Genet 81:103344. https://doi.org/10.1016/j.fsigen.2025.103344

Zhang X, Li J, Liu J, Wang J, Liu Z, Liu Y, Zhang G (2022) Identification of the vaginal secretion donor in mixture stains using polymorphic cSNPs on mRNA biomarkers. Forensic Sci Int Genet 59:102703. https://doi.org/10.1016/j.fsigen.2022.102703

Zubakov D, Boersma AW, Choi Y, van Kuijk PF, Wiemer EA, Kayser M (2010) MicroRNA markers for forensic body fluid identification obtained from microarray screening and quantitative RT-PCR confirmation [Research Support, Non-U.S. Gov't]. Int J Legal Med 124(3):217–226. https://doi.org/10.1007/s00414-009-0402-3

Zubakov D, Kokmeijer I, Ralf A, Rajagopalan N, Calandro L, Wootton S, Langit R, Chang C, Lagace R, Kayser M (2015) Towards simultaneous individual and tissue identification: a proof-of-principle study on parallel sequencing of STRs, amelogenin, and mRNAs with the Ion Torrent PGM. Forensic Sci Int Genet 17:122–128. https://doi.org/10.1016/j.fsigen.2015.04.002

Chapter 3
Towards an Epigenetic Identity: Discriminating Monozygotic Twins in Forensics

Mariza Panoutsopoulou and Athina Vidaki

> *Chance events—injuries, infections, infatuations; the haunting trill of that particular nocturne—impinge on one twin and not on the other. Genes are turned on and off in response to these events, as epigenetic marks are gradually layered above genes, etching the genome with its own scars, calluses, and freckles.*
>
> *Siddhartha Mukherjee, The Gene: An Intimate History, 2016*

Abstract Forensic DNA profiling using short tandem repeats (STRs) can individualize humans with great discrimination power, except in the case of monozygotic (MZ) twins. Originating from the same zygote, MZ twins share identical DNA sequences and STR profiles, creating a significant challenge in forensic investigations involving such individuals. While ultra-deep whole-genome sequencing offers a potential solution, its technical, computational, and financial demands limit its widespread use. Alternatively, epigenetic profiling, particularly the analysis of DNA methylation, has emerged as a promising approach for distinguishing MZ twins. This review highlights the potential of epigenetics as a dynamic and individualizing biological layer, drawing on two decades of research largely driven by large-scale twin cohort studies. Furthermore, it summarizes a growing body of forensic research focused on identifying and applying differential DNA methylation markers, alongside emerging approaches involving novel biomarkers, such as non-coding RNAs. Finally, key considerations for the future forensic implementation of epigenetic profiling for MZ twin discrimination are discussed.

M. Panoutsopoulou
Department of Clinical Genetics, Maastricht University Medical Center, Maastricht, The Netherlands

A. Vidaki (✉)
Department of Genetics and Cell Biology, GROW and CARIM Institutes, Maastricht University, Maastricht, The Netherlands
e-mail: athina.vidaki@mumc.nl

S. C. Zapico (ed.), *Forensic Science and Epigenetics*, Epigenetics and Human Health 18, https://doi.org/10.1007/978-3-032-23085-0_3

Keywords Epigenetics · DNA methylation · Monozygotic twins · Identical twins · Human identification

Abbreviations

AI	Artificial Intelligence
circRNA	circular RNA
CpG	Cytosine-phosphate-Guanine
DMC	Differentially Methylated CpG
DML	Differentially Methylated Locus/Loci
DMR	Differentially Methylated Region
DNMT	DNA MethylTransferase
HRM	High-Resolution Melting
lncRNA	long non-coding RNA
LINE	Long Interspersed Nuclear Element
LRS	Long-Read Sequencing
MeDIP-seq	Methylated DNA Immuno-Precipitation sequencing
miRNA	microRNA
mtDNA	mitochondrial DNA
MZ	Monozygotic
ncRNA	non-coding RNA
NGS	Next-Generation Sequencing
ONT	Oxford Nanopore Technologies
piRNA	PIWI-interacting RNAs
PZM	Post-Zygotic Mutations
SAM	S-Adenyl Methionine
siRNA	small-interfering RNA
SMRT	Single-Molecule Real-Time
SNP	Single Nucleotide Polymorphism
STR	Short Tandem Repeat
tDMS	twin-Differentially Methylated Site
tsRNA	tRNA-derived small RNAs
WGS	Whole-Genome Sequencing
5mC	5-methyl Cytosine

3.1 Introduction

Forensic genetics is based on the principle that every individual carries a unique DNA signature. For more than three decades, short tandem repeat (STR) profiling has been the main tool for DNA-based human identification, leading to not only establishing biological relationships worldwide, but also solving crimes (Roewer

2013). STR-based identification is enabled by the analysis of different, allele-specific lengths of short DNA sequences repeating in tandem at specific locations across all chromosomes. Hence, this approach eventually creates a unique "barcode" for every person. However, while DNA is frequently thought to be the ultimate personal identifier in forensics, there is still a group of individuals that cannot be discriminated: the so-called "identical" or monozygotic (MZ) twins (Marqueta-Gracia et al. 2018). Worldwide, four in every 1000 births result in MZ twins, meaning that based on current global birth rates there should be more than 30,000,000 MZ twins alive on earth (Fuchs and D'Alton 2018). In cases where MZ twins are involved in civil or criminal cases, either as victims or perpetrators, their personal identification is currently possible only via their different fingerprints.

MZ twins are uniquely formed with the underlying biology still remaining largely unknown. After the first week of fertilization the initial single zygote undergoes a rapid series of cell divisions and initiates a cascade of cellular events in this early phase of embryonic development (Fuchs and D'Alton 2018). As a result, a blastocyte is formed, which can then split, resulting in the creation of two separate embryos. Based on the timing of embryonic division, three types of MZ twins can be distinguished: monochorionic–monoamniotic MZ twins (1–2%), who share both the chorion and amniotic sac; monochorionic–diamniotic MZ twins (70–75%), who share the chorion but develop in separate amniotic sacs; and finally, dichorionic-diamniotic MZ twins (20–25%), who do share neither chorion nor amnion (Silva et al. 2011) (Fig. 3.1). Given that they originate from the same embryo, MZ twins are considered to be genetically identical, sharing an extremely similar genome, i.e., DNA sequence (Aourangzaib et al. 2024). Due to their nearly identical genomes MZ twins have, therefore, given rise to the so-called "twin paradox" in forensics, as they share identical STR profiles, leading to thousands of unresolved cases around the world without any other type of evidence.

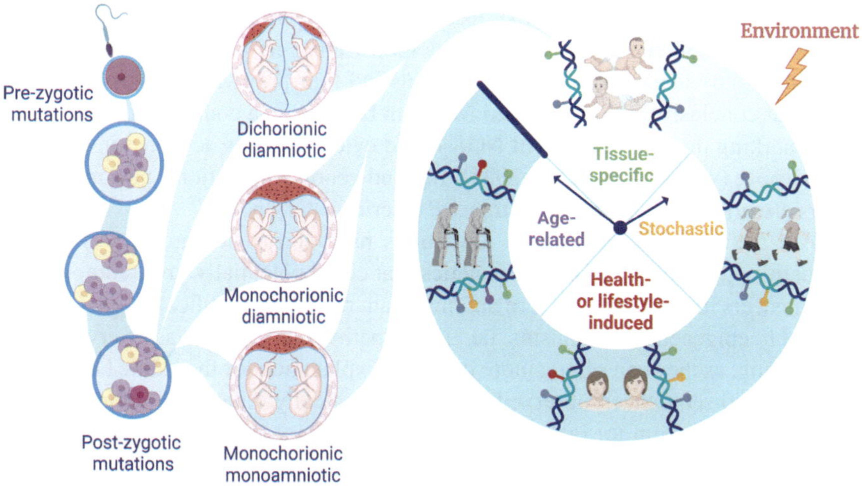

Fig. 3.1 Expected (epi)genetic divergence of MZ twins from conception and throughout life

Yet, MZ twins are distinct individuals, hence not perfectly identical. Stochastic and rare mutations, somatic mosaicism, and epigenetic modifications can introduce subtle differences over time (Fig. 3.1), which are increasingly being used in forensic analyses (Rolf and Krawczak 2021). More specifically, genetic approaches were the first line of investigation in attempts to distinguish MZ twins in the forensic scenario. Early studies focused on de novo mutations—single nucleotide polymorphisms (SNPs), small insertions or deletions, or copy number variations—that could potentially arise during embryonic cell divisions very early in development, usually before the initial embryo splits (Weber-Lehmann et al. 2014). However, initial efforts using standard whole-genome sequencing (WGS) approaches often failed to reveal robust and consistent differences between MZ twins, mainly due to the low sequencing coverage (30X) and small portion of "mutated" cells. Often, depending on the timing, these rare mutations are typically present in only a subset of cells and are, hence, inherited in different proportions in the cells of each MZ twin, generating minor genetic differences between them (Ouwens et al. 2018).

With the advent of more advanced technologies, such as ultra-deep next-generation sequencing (NGS) combined with strict bioinformatic filtering, these rare, potentially discriminating SNPs could be identified more robustly, both in somatic (buccal) and germline (sperm) cells (Weber-Lehmann et al. 2014). In parallel, post-zygotic mutations (PZMs)—genetic alterations that occur after embryo splitting and fertilization—are also possible although even rarer (Acuna-Hidalgo et al. 2016). While twin pair-specific genetic differences can now be detected with high-resolution sequencing, their interpretation complexity and extreme rarity make this approach risky, complex, and potentially unreliable in forensics. For example, a large MZ twin study revealed that on average, MZ twins show about 5.2 early developmental mutations, but only in roughly 15% of pairs one of the twins carries a considerable number of unique PZMs (Jonsson et al. 2021). Moreover, the high cost associated with these techniques further limits their practical application in forensic labs (Weber-Lehmann et al. 2014). Recently, whole-genome sequencing has been successfully applied in real casework, demonstrating its potential in practice. A landmark study identified one post-twinning and four pre-twinning somatic mutations that conclusively differentiated MZ twin brothers in a complex sexual-assault case, marking the first instance of NGS-based evidence being accepted in court for this purpose (van der Gaag et al. 2025). Beyond genomic mutations, new molecular layers are now being explored to improve discriminatory power.

Over the last few years, these limitations have shifted research focus toward alternative approaches in forensic genetics that could potentially give (complementary) answers to this relatively rare but important personal identification problem. In particular, epigenetic mechanisms, i.e., DNA methylation, oppose as a very promising solution, as they can offer a more dynamic explanation for the subtle but present phenotypic discordances observed between genetically "identical" twins. In this book chapter, we aim to: (a) provide a basic but comprehensive definition of epigenetics; (b) review the progress of twin epigenetics studies of the last 20 years, mainly driven by understanding their inter-individual variability and phenotypic discordance; (c) highlight recent forensic epigenetics studies focused on uncovering differential and potentially individualizing methylation patterns; (d) inspire future integrative multi-omics research by presenting a few pilot epigenetics studies beyond methylation; and finally, (e) discuss potential considerations for their future application in forensic casework.

3.2 Epigenetics as an Individual and Dynamic Layer

The term "epigenetics" was first introduced in the early 1940s by Conrad Waddington, who described it as the field of biology that investigates how genes and their products interact with each other to produce a phenotype (Dupont et al. 2009; Allis and Jenuwein 2016). Today, epigenetics refers to a set of chemical modifications and mechanisms that do not alter the DNA sequence directly but play a pivotal role in gene expression and regulation. Epigenetic patterns are (re)set very early in embryonic development to guide the process of cellular differentiation, eventually giving rise to a wide range of distinct cell types with divergent epigenetic identities (Leeb and Wutz 2012). Therefore, every cell type has a unique epigenetic signature as different genes need to be "on" and "off" for proper function. One way to visualize this concept is to picture the genome as a book with the DNA sequencing acting like words and epigenetic modifications acting like highlighter marks emphasizing certain passages while leaving others untouched.

Human epigenetic variation is shaped via multiple mechanisms at different scales, including histone modifications and variants, DNA or RNA base modifications, non-coding RNAs (ncRNAs), 3D-looping, and phase separation (Fig. 3.2). First, histones are core proteins (H2A, H2B, H3, H4) that resemble beads, around which DNA is tightly wrapped to form nucleosomes, enabling nearly two meters of DNA to be compacted into the tiny space of the cell nucleus (Kimura 2013). These beads can often carry post-translational modifications like acetylation, methylation, and phosphorylation affecting the accessibility and regulation of genes. These modifications can further regulate chromatin structure by directly altering nucleosome interactions and by recruiting enzymatic complexes that influence transcription, DNA repair, replication, and recombination (Bannister and Kouzarides 2011). Additionally, recent studies have turned attention to regulatory ncRNAs, including microRNAs (miRNAs), PIWI-interacting RNAs (piRNAs), small interfering RNAs (siRNAs), long non-coding RNAs (lncRNAs), tRNA-derived small RNAs (tsRNAs), and circular RNAs (circRNAs) highlighting them as key regulators of gene expression that do not encode proteins (Chen and Kim 2024). Beyond their classical post-transcriptional roles, they directly participate in epigenetic control by recruiting specialized enzymes and histone-modifying complexes to specific genomic loci, modulating chromatin structure, and influencing the transcriptome. Importantly, these ncRNA-mediated pathways often interact with other epigenetic regulators, such as histone modifications (Wei et al. 2017).

Building on the regulation provided by histone modifications and ncRNAs, DNA methylation represents another foundational epigenetic layer and is currently the

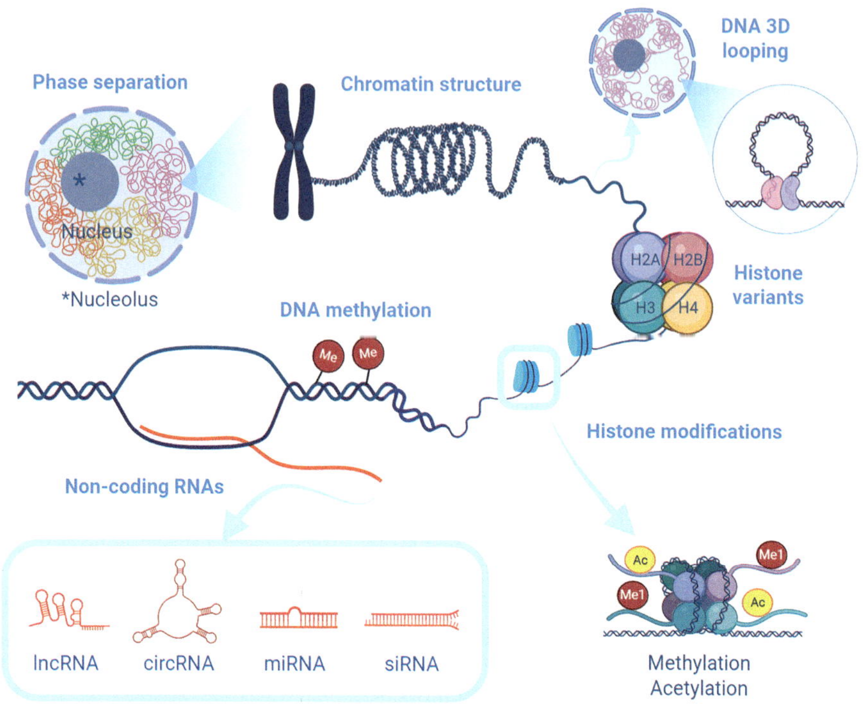

Fig. 3.2 Key epigenetic mechanisms shaping chromatin structure and gene expression. *circRNA* circular RNA, *lncRNA* long non-coding RNA, *miRNA* microRNA, *siRNA* small-interfering RNA

most widely studied epigenetic mechanism in both fundamental and forensic studies. In humans, the vast majority of DNA methylation is found in cytosines followed by guanines—the so-called CpG sites—and involves the direct chemical modification of the double strand by transferring a methyl group from S-adenyl methionine (SAM) to the fifth carbon of the cytosine residue to form 5-methylcytosine (5mC) (Moore et al. 2013). This process is catalysed by a family of enzymes called DNA methyltransferases (DNMTs). In particular, DNMT1 ensures that DNA methylation patterns are maintained after DNA replication, while DNMT3a and DNMT3b establish new de novo methylation patterns especially during embryonic development. This process plays a crucial role in silencing transposable elements, controlling tissue-specific gene expression, mediating genomic imprinting, and inactivating the second X chromosome in females (Moore et al. 2013).

The human genome contains 28–30 million CpG sites, the majority of which are methylated under normal/healthy conditions, contributing to genome stability and gene expression regulation. While most CpG islands located in gene promoters remain unmethylated to allow transcription, intergenic and repetitive regions are heavily methylated, protecting the genome from unwanted transcriptional noise and transposon activation (Ziller et al. 2013). Despite this general pattern, approximately 20% of CpG sites exhibit variable methylation among individuals, reflecting the influence of cell type, sex, age, and environmental exposures (Gu et al. 2016). Unlike DNA sequence, methylation is potentially reversible, modulated by enzymatic activity in response to physiological or environmental factors. However, many methylation marks remain remarkably stable over time, particularly in CpG-rich regions, enabling the reconstruction of an individual's biological history (Fraga et al. 2005).

So far, many technologies are being employed to accurately capture DNA methylation patterns and their fluctuations. Traditionally, bisulfite sequencing has served as the golden standard for decades. This technique is based on chemical conversion using sodium bisulfite that transforms unmethylated cytosines into uracils, while leaving methylated cytosines unchanged. This nucleotide conversion allows researchers to distinguish between cytosines of different methylation status at single-base resolution when followed by either whole-genome or amplicon sequencing (Madi et al. 2012). The bisulfite-based pyrosequencing technology that offers cheap, high-throughput, and real-time sequencing of short DNA fragments (up to 100 bp) has been widely used in forensics, particularly for the identification of tissues and body fluids (Vidaki et al. 2017a). Moreover, as high-throughput analysis became essential, microarray-based methods such as the Illumina HumanMethylation450 and MethylationEPIC BeadChip arrays quickly gained popularity (Leung et al. 2024). These bisufite- and fluorescence-based technologies simultaneously interrogate hundreds of thousands of CpG sites (450,000 and 850,000, respectively), offering cost-effective and large-scale methylation profiling that has been widely used in population, ageing, and twin studies. In forensic contexts, several additional targeted methylation assays have been employed. qPCR-based methylation assays, often relying on bisulfite conversion or methylation-sensitive restriction enzymes, offer rapid and sensitive quantification of specific CpG sites, making them well-suited for low-input forensic samples. High-resolution melting (HRM) analysis distinguishes methylation states by detecting bisulfite-induced sequence differences through DNA melting profiles, enabling fast and cost-efficient screening (Stewart et al. 2015). Targeted NGS further allows deep sequencing of predefined loci, providing high accuracy and resolution at selected CpG markers while remaining scalable and affordable for forensic applications. Recently, long-read sequencing (LRS) technologies such as nanopore sequencing from Oxford Nanopore Technologies (ONT) and single-molecule real-time (SMRT) sequencing from PacBio have transformed methylation detection by enabling direct analysis of native DNA without bisulfite conversion (Angeloni et al. 2022). Comparative analyses show that LRS approaches can reach accuracy comparable to golden standard bisulfite sequencing, while also preserving haplotype and allele-specific information (Sigurpalsdottir et al. 2024).

To sum up, advances in methylation detection technologies have transformed our understanding of the epigenome from a static layer to a dynamic record of biological and environmental influences. These tools now allow us to explore subtle molecular variations that could potentially define individual identity, even among those with nearly identical genomes. This technological progress laid the foundation for the past two decades of research on MZ twins, an invaluable model for studying how epigenetic patterns diverge over time and how such differences can inform forensic investigations.

3.3 20 Years of Epigenetic Profiling of Monozygotic Twins

MZ twins have long represented a unique biological model for dissecting the complex interplay between nature (genetics) and nurture (environment). Their near-identical genomes eliminate the effects of genetic variability on discordant phenotyping, allowing the exploration of how life experiences, environmental exposures, and stochastic biological processes shape molecular individuality (van Dongen et al. 2012). As a result, epigenomic studies in MZ twins have provided a powerful framework for understanding how DNA methylation changes naturally over time, independently of the underlying DNA sequence (Fraga et al. 2005). Early studies established that even among genetically identical individuals, methylation marks are neither static nor perfectly synchronized, with many CpG sites showing substantial individual-specific methylation differences (Zhang et al. 2015). This phenomenon, referred to as epigenetic drift, reflects the influence of ageing-related dynamics, environmental exposure and stochastic processes on the epigenome in a tissue-specific manner (Teschendorff et al. 2013). Recent large-scale studies have further demonstrated that MZ twins carry a persistent and distinctive methylation signature that originates during early embryonic development and remains stable across tissues and over time. These epigenetic imprints, enriched in regions associated with developmental regulation, suggest that the process of zygotic splitting itself leaves a lasting molecular trace detectable well into adulthood (van Dongen et al. 2021).

First, MZ twin epigenetic research has increasingly dived into the wealth of information that individual DNA methylation profiling can reveal about the rate of biological ageing. To capture these complex dynamics, advanced mathematical models and machine learning approaches have been developed to construct the so-called "epigenetic ageing clocks", biological age predictors based on differential DNA methylation patterns at specific CpG sites across the genome (Teschendorff and Horvath 2025). These clocks, including well-established models such as the Horvath, Hannum, PhenoAge, GrimAge, and DunedinPACE clocks, have become indispensable tools in the individual epigenomics field (Hong et al. 2025). In particular, longitudinal twin studies have provided valuable insights into how these epigenetic clocks behave across the lifespan. While the heritability of epigenetic age acceleration is relatively stable over short time frames, it gradually decreases as individuals grow older, implying that environmental and lifestyle factors increasingly influence methylation patterns with age (Miao et al. 2024). Additionally, a large-scale analysis of adult twin pairs confirmed that the genetic contribution to epigenetic ageing declines with ageing, highlighting that the longer MZ twins live (apart), the more obvious their methylation divergence becomes (Hong et al. 2025). The dynamic nature of epigenetic ageing is further supported by long-term follow-up studies showing that the predictive power of certain clocks, such as GrimAge, extends beyond simple chronological ageing to reflect health-related outcomes, including mortality risk. Specifically, in an 18-year follow-up of elderly female MZ twins, GrimAge acceleration emerged as a robust predictor of mortality (Föhr et al. 2021). Furthermore, X-chromosome methylation research in MZ twins showed that age-dependent variability differs between males and females, highlighting sex-specific epigenetic dynamics that may influence ageing trajectories and disease susceptibility (Tan et al. 2024). Apart from ageing itself, epigenetic profiling in MZ twins has also been used to explore links between DNA methylation and general health decline. For example, an epigenome-wide analysis of frailty across two large European twin cohorts identified specific CpG sites associated with reduced physical performance and resilience, suggesting that methylation changes may reflect cumulative physiological deterioration rather than disease-specific mechanisms (Mak et al. 2024).

Second, beyond ageing, several studies have focused on how DNA methylation variability relates to broader "healthy" physiological traits. For example, epigenome-wide analyses in Chinese MZ twins identified CpGs associated with functional traits such as handgrip strength, suggesting that epigenetic regulation extends beyond age to reflect biological resilience and physical performance (Luo et al. 2024).

Third, given that environmental influence affects the methylome, attention was drawn early to the extend in which an individual's life choices and experiences have an impactful outcome on individual DNA methylation patterns, particularly by influencing the rate of epigenetic ageing. MZ twin studies computing a healthy lifestyle score reveal that individuals who abstain from tobacco smoking, limit alcohol intake, maintain regular physical activity, and follow diets rich in fruits and vegetables consistently display slower epigenetic ageing compared to their less healthy co-twins, with physical activity and nutrition having especially strong effects on methylation clocks (Peng et al. 2021). Additionally, large-scale associations between DNA methylation and alcohol consumption highlight specific methylation changes linked to regular alcohol (ab)use, which is associated with accelerated ageing, as seen in genome-wide methylation analyses of twin pairs (Lu et al. 2021). Moreover, suboptimal dietary patterns, including high intake of processed foods and low consumption of plant-based nutrients, correlate with faster biological ageing, whereas vegan diets are associated with more favourable methylation profiles and a slower ageing trajectory (Ravi et al. 2025; Dwaraka et al. 2024). Furthermore, beyond lifestyle choices, twin studies emphasize that significant life events, from early-life difficulties to stressful experiences in adulthood, can leave enduring marks on the epigenome, leading to serious health adversities long after the event itself. For example, analyses of MZ twins demonstrate that human habits, such as smoking, are critical determinants of DNA methylation patterns later in life, affecting late-life health and the pace of biological ageing (Li et al. 2022). Exposure to negative life events is associated with longitudinal accelerated epigenetic ageing: MZ twins who have experienced negative and traumatic life events, like sexual crimes, losing a loved one or financial adversities, display increased rates of age-associated methylation changes compared to their co-twin (Gonggrijp et al. 2025).

Finally, epigenomic profiling of MZ twins has played a key role in uncovering how disease states influence DNA methylation. MZ twin pairs discordant for cancer have provided one of the clearest examples of (ageing) disease-related epigenetic divergence. When comparing affected and unaffected co-twins, measurable differences in methylation-based age prediction suggest that oncogenic processes and their systemic effects alter the epigenetic stability (Bode et al. 2022). Similarly, metabolic syndrome has emerged as another health-related phenotype linked to methylation variation. In a MZ twin study discordant for metabolic syndrome, affected individuals showed significantly accelerated methylation ageing across several clocks, implying that systemic metabolic stress imprints molecular signatures on the genome (Föhr et al. 2024). Extending this framework to neurological conditions, WGS of MZ twins discordant for amyotrophic lateral sclerosis revealed variability in PZMs, indicating that even subtle somatic mutations arising after fertilization may interact with epigenetic mechanisms to influence disease manifestation (Tazelaar et al. 2023). On the other hand, moving beyond physical health, the influence of mental health on epigenetic profiles has also been explored through co-twin control designs. MZ twins discordant for depression displayed consistent differences in methylation-based measures of biological state, suggesting that chronic psychological stress and altered neuroendocrine signalling may leave substantial epigenetic marks (Liu et al. 2022).

To sum up, over the past two decades, studies on MZ twins have fundamentally advanced our understanding of how environmental, behavioural, and stochastic factors shape the human epigenome beyond the DNA sequence. By disentangling nature from nurture at an unprecedented resolution, twin-based epigenetic profiling has revealed that methylation patterns are dynamic markers of both ageing and health, reflecting the biological consequences of lifestyle habits, life experiences, and disease states. As technology progresses from microarray-based to single-cell and LRS, future research will be able to capture even finer layers of epigenetic individuality, including cell type-specific and allele-specific signatures. Integrating these multi-omic, longitudinal, and environmental datasets promises to refine our understanding of molecular ageing, resilience, and disease susceptibility, paving the way for personalized health monitoring, improved forensic applications, and deeper insights into what truly makes identical twins unique.

3.4 Differential DNA Methylation of Monozygotic Twins in Forensics

Since the early 2010s, epigenetic profiling has been increasingly applied in forensic genetics, offering potential answers to some of the field's most persistent challenges. Particularly, following the recognition of epigenetic marks as powerful cellular biomarkers, forensic scientists have begun utilizing DNA methylation patterns to identify the tissue or body fluid origin of DNA evidence (Frumkin et al. 2011). Tissue-specific DNA methylation markers have shown excellent performance in determining the cellular source of forensic-type samples (e.g., blood, saliva, semen, vaginal fluid, menstrual fluid), while also remaining stable in degraded (DNA) material, a potential advantage over RNA-based methods (Vidaki and Kayser 2018). Another major advancement in forensic epigenetics came from age estimation models built on DNA methylation levels at just a handful of CpG sites that strongly correlate with and can quite accurately predict (±2–3 years) a person's chronological age. These models have opened up the horizon of narrowing down suspect pools based on registered population databases, making this research topic one of the most popular in the last few years (Gerra et al. 2024). Beyond ageing there are other ongoing efforts to employ differential DNA methylation for the prediction of additional phenotypic characteristics or lifestyle traits, such as smoking (Vidaki et al. 2023) or alcohol drinking (Ambroa-Conde et al. 2024). Such approaches could, in the future, extend forensic DNA phenotyping to provide valuable investigative leads, potentially aiding in the identification of individuals when conventional genetic profiling is inconclusive (Vidaki et al. 2013). However, current phenotypic characteristic methylation assays are still constrained by methodological and biological limitations, including bisulfite conversion-related DNA degradation, PCR amplification bias, limited CpG coverage, and confounding effects driven by age, cell type composition, and (sub)population group. Interestingly, when testing a blood-based age estimation method in 53 MZ twin pairs, researchers demonstrated high predictive accuracy but also observed a mean age difference of 2.5 years among co-twins (Vidaki et al. 2017b). Similarly, in a more recent forensic study testing a model for legal age estimation using a large cohort of 18-year-old twins showed that almost 30% of twin pairs had "opposite" results: one being predicted as minor (<18 years) and one as adult (>18 years) (Boullón-Cassau et al. 2025). This demonstrates that ageing rates can be very personal, but no study so far has proposed this approach for MZ twin differentiating. This is likely not only due to DNA methylation's dynamic nature but also due to potential time differences in reference sample collection between twins, often observed in such relevant cases.

Moving towards the complex topic of differentiating and identifying between MZ twins, it comes with no surprise that researchers proposed epigenetic profiling as a potential solution. For this approach to work, there is a need to identify methylation marks that are not only MZ twin-differentiating but also robust and stable longitudinally. An initial study applied bisulfite conversion followed by High-Resolution Melt (HRM) analysis on buccal swabs from 5 MZ twin pairs, targeting *Alu-E2F3* and *Alu-SP* loci (Stewart et al. 2015). Significant methylation differences were detected in all pairs for *Alu-E2F3* and four of five pairs for *Alu-SP*, supporting HRM's potential. A complementary strategy employed the analysis of specific CpG regions through PCR-HRM, a rapid and cost-effective technique capable of detecting small methylation differences without the need of sequencing. Distinct methylation profiles were successfully identified in several twin pairs, further supporting the notion that individual-specific methylation patterns can emerge despite genetic identity. However, the small sample size, limited number of loci examined, use of a single tissue type, and observed variability among twin pairs highlight the need for expanded marker panels and longitudinal validation to achieve reliable and reproducible results (Marqueta-Gracia et al. 2018).

Among the earliest and most influential studies exploring the potential of DNA methylation for discriminating MZ twins were those based on Illumina's 27K and 450K BeadChip platforms. These array-based technologies enabled high-throughput, genome-wide interrogation of methylation variation across thousands of CpG sites. Using Illumina 27K array, one of the first large-scale studies analysed whole-blood DNA from 22 MZ twin pairs and identified 92 CpG sites showing significant methylation differences between co twins, providing the first genome scale evidence that methylation patterns can serve as potential individualizing biomarkers even among genetically identical individuals (Li et al. 2013). However, these findings derive from a pilot study and require validation in larger cohorts, as well as careful assessment of the temporal stability and tissue specificity of the identified CpG sites before their application in routine forensic casework. Subsequent research employing the higher-resolution 450K array examined 10 pairs of adult twins and observed that 0.087–1.53% of CpG sites exhibited differential methylation between co-twins, revealing that methylation discordance is widespread but occurs at relatively low frequencies across the genome (Zhang et al. 2015). Further investigations performed genome-wide screening of blood DNA from 10 MZ twin pairs and identified 19–111 twin-differential methylation sites (tDMSs) per pair (at the level of 30% methylation difference), which were subsequently validated using methylation-specific qPCR, highlighting the feasibility of applying methylation analysis in realistic forensic scenarios (Vidaki et al. 2017c). This methodological framework was later extended to compare different forensic biological sources. Buccal swabs, saliva, and even cigarette butts have been analysed to assess whether tissue-specific DNA methylation profiles can provide sufficient discriminatory power for forensic purposes (Vidaki et al. 2018). Comparative analyses revealed that distinct tissues exhibit unique methylation signatures, and that inter-twin differences are not uniformly expressed across all sample types. Buccal and saliva samples, in particular, demonstrated measurable variation at selected CpG sites, highlighting the influence of cellular composition and environmental exposure on epigenetic profiles recovered from forensic materials. At the same time this also limits the applicability of methylation markers across forensic substrates and necessitates tissue-aware marker selection and validation.

The introduction of the Illumina MethylationEPIC BeadChip (850K) represented a major advancement in genome-wide methylation profiling, enabling broader coverage of CpG sites and higher resolution in detecting methylation variability across tissues and developmental stages. Large-scale twin studies were based on this platform. One comprehensive investigation identified 333 CpG sites exhibiting comparable methylation variability between MZ co-twins and unrelated individuals, suggesting the presence of general epigenetic dissimilarity independent of genetic background (Planterose Jiménez et al. 2021). These variable sites were enriched at regions known for stochastic methylation in developing neurons, and were further replicated in adipose tissue, underscoring their biological robustness. Another study employed the EPIC array to analyse blood samples from 47 MZ twin pairs, identifying recurrent differentially methylated CpG sites (DMCs) predominantly enriched in immune-related genes, which were subsequently validated using public datasets from the Gene Expression Omnibus (Kim et al. 2023). Later investigations extended this comparison to blood and buccal DNA derived from MZ triplets, showing that both tissue origin and external factors contribute significantly to methylation divergence (El-Hossary et al. 2024). These findings underscored that while epigenetic differences between genetically identical individuals can indeed be detected, their extend and consistency depend heavily on the biological material analysed. Therefore, the integration of multiple tissue types and the identification of tissue-specific methylation markers are essential steps toward achieving reliable and reproducible forensic discrimination of MZ individuals.

Beyond array-based platforms, sequencing approaches have provided a complementary strategy for detecting methylation differences between MZ twins at both global and locus-specific scales. Using methylated DNA immunoprecipitation sequencing (MeDIP-seq), a genome-wide scan of whole blood DNA from 4 pairs of healthy MZ twins identified 38 differentially methylated regions (DMRs) consistently differing within each pair (Du et al. 2015). These regions were primarily located in CpG islands, including promoter-associated and intergenic sites, and were functionally linked to genes involved in cell proliferation, differentiation, and G-protein signalling, highlighting biologically relevant methylation variability between twins. Another line of investigation used pyrosequencing and focused on the methylation status of long interspersed nuclear elements (LINE-1), which comprise a large fraction of the human genome (Xu et al. 2015). Comparative analyses of LINE-1 methylation between MZ twins revealed slight but reproducible differences, indicating that even highly conserved repetitive sequences may undergo epigenetic divergence over time. These results suggested that global methylation markers could serve as useful preliminary indicators for the differentiation of genetically identical individuals in forensic applications, but the tissue- and individual-specific methylation variability could limit the broader applicability of these findings. Large genome-wide studies investigating the methylation patterns of various repetitive elements including (sub)types of transposable elements in MZ twin samples are lacking.

The emergence of LRS technologies, particularly those developed by ONT, has marked a new era in forensic epigenetics by enabling direct detection of DNA methylation without the need for bisulfite conversion and by providing single-molecule precision. A recent study applied ONT nanopore sequencing to genome-wide methylation profiling of six MZ twin pairs, identifying 3820 differentially methylated loci (DMLs) enriched in metabolic and neural pathways (Zeng et al. 2025). The method achieved >99.5% alignment efficiency and an average read length exceeding 13 kb, demonstrating the robustness and resolution of long-read, PCR-free methylation profiling in distinguishing between genetically identical individuals.

Altogether, these studies summarized in Table 3.1 illustrate the progress of DNA methylation–based approaches in addressing one of forensic genetics' most

Table 3.1 Summary of DNA methylation studies for MZ twin differentiation in forensics

Study	MZ twin pairs	Tissue	Technology	Main outcome
Li et al. (2013)	22	Whole blood	Illumina 27K	92 tDMSs
Du et al. (2015)	4	Whole blood	MeDIP-seq	38 DMRs
Stewart et al. (2015)	5	Buccal cells	PCR-HRM	Alu-E2F3 CpGs
Xu et al. (2015)	119	Whole blood, buccal cells	Pyrosequencing	LINE-1 CpGs
Zhang et al. (2015)	10	Whole blood	Illumina 450K	0.087–1.530% array CpGs
Vidaki et al. (2017a, b, c)	53	Whole blood, saliva	Illumina 27K/450K	16 age CpGs
Vidaki et al. (2017a, b, c)	10	Whole blood	Illumina 450K	19–111 tDMSs
Marqueta-Gracia et al. (2018)	18	Saliva	PCR-HRM	3 tDMSs
Vidaki et al. (2018)	1	Buccal cells, saliva, cigarette butts	Illumina 450K	25 buccal/6 saliva tDMSs
Planterose Jiménez et al. (2021)	1008	Whole blood, adipose tissue	Illumina 450K	333 tDMSs
Kim et al. (2023)	47	Whole blood	Illumina EPIC	Immune-related DMCs
El-Hossary et al. (2024)	1	Whole blood, buccal cells	Illumina EPIC	24 blood/11 buccal DMSs
Boullón-Cassau et al. (2025)	426	Whole blood	Illumina 450K	391 age CpGs
Zeng et al. (2025)	6	Whole blood	ONT	3820 DMRs

CpGs Cytosine-phosphate-Guanine, *DMCs* Differentially Methylated CpG, *DMR* Differentially Methylated Region, *HRM* High-Resolution Melting, *LINE-1* Long Interspersed Nuclear Element 1, *MeDIP-seq* Methylated Immuno-Precipitation sequencing, *ONT* Oxford Nanopore Technologies, *tDMS* twin-Differentiating Methylation Site

persistent challenges, the discrimination of MZ twins. Yet, despite significant technological advances and the increasing resolution of methylation profiling, this layer of the epigenome represents only part of a much broader regulatory landscape.

3.5 Beyond DNA Methylation: Towards Multi-omic Characterization

Human individuality is a complex topic that may affect multiple molecular layers beyond DNA sequence and methylation. To achieve a more complete understanding of possible MZ twin divergence, and to eventually enhance discrimination power in future forensic casework, research is shifting towards other omics layers, which could be potentially integrated as a single multi-omic approach in the future (Fig. 3.3). This approach could combine insights from (epi)genomics,

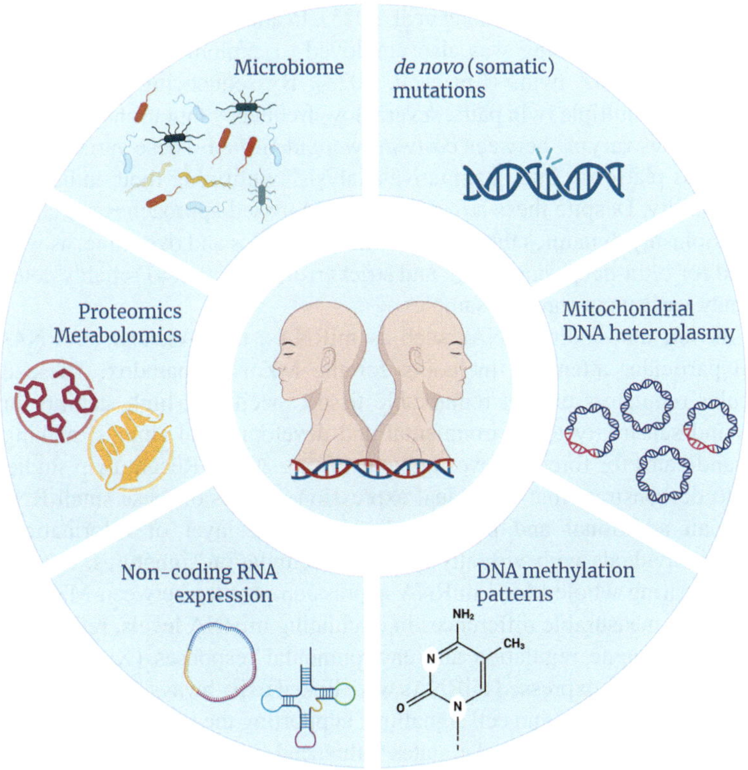

Fig. 3.3 Potential differences between MZ twins across multiple biological layers

mito-genomics, transcriptomics, metabolomics, and others to move beyond the static view of the genome and offer a more comprehensive molecular signature (Hagenbeek et al. 2022).

Recent technological advances have introduced novel possibilities for distinguishing MZ twins beyond nuclear DNA profiling. Long-read SMRT sequencing has enabled the generation of highly accurate and complete mitochondrial DNA (mtDNA) haplotypes, providing an alternative source of genetic variation for forensic investigation (Chen et al. 2020). Analysis of mtDNA from MZ twins revealed subtle but consistent differences in heteroplasmy levels, minor variants in mtDNA sequences that can differ between individuals despite shared nuclear genomes. In this study, the ONT-enabled detection of heteroplasmic differences allowed for the reliable differentiation of genetically identical twins, complementing with an additional layer of individual specificity beyond DNA and methylation methods.

Complementary studies using a probe capture enrichment strategy followed by NGS demonstrated that even a small number of heteroplasmic variants, down to fractions of 1% in frequency, can serve as individual-specific markers (Zhong et al. 2023). This targeted approach increased sequencing depth across the mitogenome, enabling the detection of rare variants that are often missed by conventional whole-genome sequencing (Zhong et al. 2023). In another investigation, Ion Torrent semiconductor sequencing was also employed to explore rare differences in the mitogenomes of MZ twins (Liu et al. 2023). By sequencing complete mtDNA genomes from multiple twin pairs, several low-frequency mutations and heteroplasmic imbalances varying between co-twins were identified. These variants were validated across platforms and quantitative analysis, confirming their authenticity and reproducibility. Despite these advances, mtDNA-based approaches are constrained by heteroplasmy dynamics that may vary across tissues and over time, as well as by the need for ultra-deep sequencing and strict error correction to reliably detect low-frequency variants in forensic samples.

Moreover, non-coding RNAs such as miRNAs, tsRNAs, and circRNAs have gained particular attention, including for the MZ twin paradox. These diverse molecular regulators exhibit remarkable tissue specificity, high stability in body fluids, and sensitivity to environmental and developmental signals, making them ideal candidates for forensic investigations (Wong 2015). Recent twin studies have begun to demonstrate that individual expression profiles of these small RNAs can provide an additional and possibly discriminating layer of information, even between individuals with virtually identical methylation signatures. For instance, when comparing whole-blood miRNA expression profiles between MZ twins, co-twins exhibit measurable differences in circulating miRNA levels, reflecting subtle divergences in gene regulation and environmental responses (Xiao et al. 2019). Many differentially expressed miRNAs were linked to pathways involved in immune regulation, metabolism, and cell signalling, supporting the idea that miRNA expression captures dynamic molecular states influenced by both lifestyle and stochastic biological variation. These findings were validated by another twin-based prospective study of MZ and DZ pairs discordant for non-alcoholic fatty liver disease. In this study, specific serum microRNA signatures are associated with disease status, independent of shared genetics and environmental exposures (Zarrinpar et al. 2016). Furthermore, using small RNA sequencing, researchers profiled tsRNA expression in peripheral blood of MZ twins and revealed distinct expression signatures between co-twins (Tian et al. 2025). These tsRNA profiles showed higher inter-individual variability compared to miRNAs, suggesting that even short regulatory RNAs can reflect subtle molecular individuality. Similarly, studies on circRNAs, a class of molecularly closed RNA molecules with exceptional stability, have uncovered consistent expression differences between MZ twins. Through RNA sequencing in blood, significant divergences were found in circRNA abundance and isoform diversity, particularly in transcripts related to immune function and cellular metabolism (Wang et al. 2024). Due to their resistance to degradation and cell-type specificity, circRNAs could complement methylation markers by providing a transcriptome-level fingerprint, unique to each co-twin.

Expanding this view, broader reviews of non-coding RNA biology in forensics have highlighted how multiple small RNA species, including miRNAs, piRNAs, lncRNAs, can be applied in casework. Their ability to reflect both genetic background and environmental exposure makes them valuable in tissue source identification, age prediction, and postmortem interval estimation (Song et al. 2024).

Integrating these RNA-based markers with existing DNA methylation data offers a multi-dimensional approach to resolving complex forensic challenges, particularly those involving MZ twins where sequence-based techniques are not always efficient and applicable. Future studies should aim to include larger and more diverse populations, account for both technical and biological variability, and standardise experimental protocols. Such improvements are essential for enhancing the reproducibility and reliability of results and could lead to translating these molecular techniques into routine forensic investigations (Tian et al. 2025). Moreover, RNA-based approaches are sensitive to tissue composition, temporary fluctuations, and post-collection degradation, raising concerns about longitudinal stability and interpretability when the reference and trace samples are obtained at different time points or under varying conditions.

Beyond these approaches, recent advances in proteomics and metabolomics have opened new dimensions for differentiating MZ twins by capturing downstream molecular effects of gene regulation, environment, and disease. Proteomic analyses offer a snapshot of functional biology at the protein expression and modification level. In a landmark twin study, high-resolution mass spectrometry was applied to plasma samples from MZ twins discordant for ischemic stroke, uncovering distinct proteomic signatures between affected and unaffected co-twins (Vadgama et al. 2019). Differentially expressed proteins were enriched in pathways related to inflammation, coagulation, lipid metabolism, and endothelial function, highlighting the capacity of proteomic profiling to reflect disease-specific molecular alterations. Metabolomic profiling using liquid chromatography-mass spectrometry has also demonstrated possibility to discriminate MZ twins based solely on their metabolic fingerprints. A recent study analysing peripheral blood samples from twin pairs revealed significant differences in metabolite composition and abundance, identifying a panel of small-molecule biomarkers capable of efficiently distinguishing between co-twins (Zeng et al. 2024). The differential metabolites were primarily associated with amino acid, lipid, and energy metabolism, reflecting both genetic regulation and external influences such as diet, stress, and environmental exposure. However, proteomic and metabolomic profiles are particularly susceptible to short-term physiological variation, medication use, diet, and circadian clock, which may complicate reproducibility and limit their suitability without extensive contextual and longitudinal validation.

Taken all together, these multi-omic approaches highlight that true individual identity extends far beyond early induced but static DNA sequence variation. By integrating information from the epigenome, mitogenome, transcriptome, proteome, and metabolome, researchers are uncovering increasingly refined molecular "fingerprints" that define each person's biological individuality. For MZ twins, these advances reveal that even genetically identical individuals possess measurable biochemical distinctions shaped by time, environment, and stochastic events.

3.6 Considerations for Implementing in Future Forensic Casework

Despite remarkable progress over the past decade, the application of epigenetic profiling in forensic genetics remains largely experimental, though its potential for future casework is becoming increasingly evident. Several practical, biological, technical, and ethical challenges still stand between research and casework implementation. To bridge this gap, further research is required across multiple levels: from understanding the biology of epigenetic regulation, to establishing robust lab protocols, and to defining clear legal and ethical frameworks for the interpretation of these data, particularly for the forensic MZ twin discrimination question.

From a practical and casework-driven perspective, additional critical challenges emerge for successful implementation. The main consideration is the temporal stability of epigenetic signatures, as reference samples are often collected months or even years after the crime has been committed. Given that DNA methylation patterns are known to change over time in response to ageing, environmental exposures, lifestyle factors, and stochastic processes, it remains unclear whether twin—differentiating methylation signals that are identified at one time point remain sufficiently stable for later comparison. This uncertainty also applies in scenarios where reference samples from co-twins are obtained at different time points, for example, when one twin provides a sample substantially later than the other, or when one twin cannot be compelled by law to donate a reference sample at all. From a biological standpoint, such situations necessitate the identification of CpG markers with demonstrated longitudinal stability or the construction of age- and time-sensitive modelling strategies that can consider the temporal divergence.

Moreover, epigenetic approaches face conceptual limitations in cases where both twins may have contributed to the crime, as differential methylation alone cannot disentangle mixed contributions from genetically identical individuals, highlighting the need to explore complementary molecular layers, such as somatic mutations, transcriptomic signatures, or microbiome profiles, where possible and/or applicable. Additional factors, including sex, age, and developmental stage also need careful consideration, as these variables influence baseline methylation levels and rates of epigenetic change, potentially affecting both discrimination power and interpretability. Addressing these issues will require dedicated validation studies that explicitly simulate realistic forensic scenarios, including delayed sampling, partial reference availability, mixed traces, and diverse demographic backgrounds, to ensure that the proposed methods are scientifically robust for real-world forensic casework.

From a biological perspective, one of the greatest challenges arises from the tissue- and cell-type specificity of epigenetic marks. Methylation and other epigenetic signatures can vary between different cell (type) populations, even within the same individual, reflecting distinct transcriptional patterns and biological functions. As a result, a marker that performs well in white blood cells may yield inconsistent or misleading results when applied to DNA extracted from saliva, sperm, or epithelial cells. This, naturally, limits its reliability for distinguishing MZ twins in forensic casework when reference and trace samples often originate from different tissues. This variability undermines reproducibility and limits the universal applicability of current methylation-based assays including for the "twin paradox" (Vidaki and Kayser 2018). Given that reference samples are usually collected via buccal swabs, it is particularly challenging to replicate any tDMSs in forensic trace samples often recovered as blood stains in physical assault or semen stains in sexual assault cases. As a result, it is important that tissue-matched reference sampling may be required in exceptional MZ twin investigations. Moreover, reliable interpretation of epigenetic differences needs prior identification of the tissue or body fluid source of the forensic trace, which, although feasible with current molecular approaches, inevitably consumes additional DNA sample that is often limited in trace samples. Additionally, the presence of mixed traces, i.e., including DNA from the victim, a common scenario in forensics, introduces further complexity in distinguishing true variation from cellular heterogeneity.

An additional layer of complexity arises from lifestyle-associated epigenetic variation. On the one hand, such lifestyle differences could enhance the discriminative power for MZ twins; on the other hand, they introduce important interpretational challenges. More specifically, while differences in smoking status, diet, or environmental exposures could amplify epigenetic divergence between co-twins, such effects are very dynamic, raising concerns regarding temporal stability. Future research should, therefore, aim to identify cross-tissue, stable epigenetic markers, or alternatively, develop computational normalization strategies that account for the tissue origin of each sample. In parallel, the number of CpG markers required to achieve reliable twin differentiation represents both a biological and practical constraint, as small marker panels may lack sufficient resolution and be more prone to assay variation. In contrary, large-scale panels increase technical complexity, costs, and validation demands in forensic casework. Setting an optimal balance between marker number, discriminatory power, and operational feasibility will be essential for translating such epigenetic approaches into routine forensic practice. Only through such thorough investigation and standardization can the field move toward biologically meaningful and reproducible forensic epigenetic signatures (Gerra et al. 2024).

These biological challenges are closely intertwined with a range of technical limitations that restrict the adaptation of epigenetic profiling into routine forensic workflows. Most biological traces encountered at crime scenes contain very small quantities of highly degraded DNA, often below the current threshold required for bisulfite conversion or NGS. Bisulfite treatment, while still the gold standard for methylation detection, is notoriously harsh, causing DNA fragmentation and loss of material that can reach up to 70% of the input sample (Xiao et al. 2019). Addressing this issue will likely require a combination of strategies, including the development of more sensitive methylation detection methods, the adoption of alternative, less DNA-damaging conversion techniques, and continued optimization of low-input and degraded DNA-suitable workflows. In parallel, DNA input quantity and quality should be accounted for during assay validation and result interpretation, acknowledging their influence on methylation estimates.

Emerging technologies such as ONT-based LRS provide direct methylation detection without chemical conversion and may alleviate some of these issues; yet their current sensitivity does not meet forensic standards. Alternatively, digital PCR offers a promising route for low-input material due to its high sensitivity and quantitative precision; however, its marker multiplex-capability is very limited. This might be particularly problematic for the MZ twin differentiation question, which may rely on a large number of tDMSs. Additionally, differences in DNA extraction methods, amplification protocols, and sequencing platforms lead to substantial inter-laboratory variation, compromising the comparability of results. Because replicates are rarely available in forensic contexts, these discrepancies are difficult to experimentally control and account for during data interpretation. The establishment of standardized procedures for sample handling, processing, data analysis, and interpretation is, therefore, a critical step toward reliability.

In the context of MZ twin differentiation, these issues are further complicated by the common use of different technologies between tDMS discovery and trace analysis, where sensitivity is often prioritized over consistency. This becomes particularly relevant when laboratories seek to reproduce twin-differentiating markers reported in the literature using alternative methods, as technical variation may mask the true biological differences between co-twins. In addition, population compatibility must be considered, as markers identified in one population may not perform similarly in others, highlighting the need for population-specific validation prior to forensic application. Beyond these considerations, additional technical challenges arise when extending epigenetic twin differentiation towards more complex or realistic forensic scenarios. The integration of multi-omics layers, such as combining DNA methylation with somatic variation, transcriptomic signals, or microbial profiles, may enhance discriminatory power but also substantially increases analytical complexity, data requirements, and costs. These challenges are further amplified when dealing with highly fragmented or environmentally compromised samples, such as bone DNA in missing-person investigations involving MZ twins. Importantly, many proposed tDMSs exhibit small effect sizes, raising concerns about their reliable detection in low-input casework samples and the risk of false positives or negatives. Addressing this issue will require validation of candidate markers across multiple technologies, careful definition of analytical thresholds, and realistic assessment of measurement error.

Beyond the laboratory, the integration of epigenetic tools into forensic casework raises important ethical, interpretative, and legal considerations. Unlike traditional DNA profiling, which relies on static genetic information, epigenetic data reflect dynamic biological states shaped by age, environment, stress, and lifestyle factors. This highlights the need to distinguish between dynamic, phenotype-associated methylation changes and more stable but still discriminatory epigenetic differences when selecting markers for MZ twin differentiation. While this dynamic nature makes epigenetic data rich sources of information, it raises particularly sensitive questions, where attempts to distinguish MZ individuals may reveal age, lifestyle, or health-related information beyond identity itself. To address this, the application of methylation-based MZ twin discrimination may need to be restricted to exceptional or high-profile forensic cases, with strict limitations on marker selection and interpretation (Vidaki et al. 2017c). From a legal perspective, this is important because traditional DNA profiling cannot distinguish MZ twins, and the use of epigenetic evidence in such cases will require careful evaluation by courts regarding of this new type of evidence (Turrina et al. 2021). Building large, population-based reference databases, ensuring standardization of analytical procedures, and integrating artificial intelligence (AI)-driven analytical tools to support interpretation might be necessary steps toward achieving this goal. Such databases would be beneficial for defining baseline variability and population-specific behaviour of the tDMSs, enabling the accurate interpretation of observed differences. Ultimately, the successful implementation of epigenetics for the forensic discrimination of MZ twins will depend on coordinated efforts between scientists, policymakers, and legal authorities to ensure that these powerful molecular tools are applied responsibly, reproducibly, and with full consideration of their societal implications.

3.7 Conclusion

MZ twins embody one of the most fascinating paradoxes in human biology: two individuals sharing an identical genetic code, yet diverging in countless molecular, physiological, and phenotypic ways over time. Their existence has long challenged the foundations of forensic genetics, which traditionally relied on immutable DNA sequence variation for identification. However, the growing body of research on epigenetic mechanisms has reshaped this perspective, revealing that the genome is not a static entity but a dynamic landscape that records the interplay between genes, environment, and life experience. DNA methylation, histone modifications, and non-coding RNA activity together create unique molecular signatures that can, in principle, distinguish even genetically identical individuals. In this context, the study of epigenetic differences between MZ twins has not only advanced our understanding of human individuality but also redefined the frontiers of forensic science. Although substantial work remains before these insights can be reliably translated into routine forensic casework, they have already illuminated a starting point toward molecular-level discrimination that was once thought impossible. At the same time, the dynamic nature of epigenetic variation demands careful consideration of marker stability and interpretability, particularly for use in high-profile forensic cases involving MZ twins. Each advance, from genome-wide methylation profiling to the integration of multi-omic data, brings us closer to recognizing individuality as a multi-layered phenomenon rather than a purely genetic one. Ultimately, the successful implementation of these approaches will depend not only on technological innovation but also on rigorous validation, standardization, as well as ethical and legal governance. MZ twins may be genetically indistinguishable, but they are epigenetically unique. Within the subtle DNA modifications lies the record of their distinct lives; proof that even in genetic identity, individuality persists.

Acknowledgments The study was funded by the Dutch Organization for Scientific Research (NWO) [grant number VI.Vidi.223.130] awarded to A.V.

Competing Interests The authors have no conflicts of interests to declare that are relevant to the content of this chapter.

Author Contributions A.V. conceived the idea of the book chapter and created its structure. M.P. researched the literature and prepared the display items, under the supervision of A.V. Both authors wrote and edited the draft of the manuscript including its final version.

References

Acuna-Hidalgo R, Veltman JA, Hoischen A (2016) New insights into the generation and role of de novo mutations in health and disease. Genome Biol 17(1):241

Allis CD, Jenuwein T (2016) The molecular hallmarks of epigenetic control. Nat Rev Genet 17(8):487–500

Ambroa-Conde A, Casares de Cal MA, Gómez-Tato A, Robinson O, Mosquera-Miguel A, de la Puente M et al (2024) Inference of tobacco and alcohol consumption habits from DNA methylation analysis of blood. Forensic Sci Int Genet 70:103022

Angeloni A, Ferguson J, Bogdanovic O (2022) Nanopore sequencing and data analysis for base-resolution genome-wide 5-methylcytosine profiling. Methods Mol Biol 2458:75–94

Aourangzaib M, Chandra M, Maham R, Naz A, Malathi H, Qadeer S et al (2024) Solving the twin paradox-forensic strategies to identify the identical twins. Forensic Sci Int 363:112205

Bannister AJ, Kouzarides T (2011) Regulation of chromatin by histone modifications. Cell Res 21(3):381–395

Bode HF, Heikkinen A, Lundgren S, Kaprio J, Ollikainen M (2022) Differences in DNA methylation-based age prediction within twin pairs discordant for cancer. Twin Res Hum Genet 25(4–5):171–179

Boullón-Cassau M, Ambroa-Conde A, Casares de Cal MA, Gómez-Tato A, Mosquera-Miguel A, Ruiz-Ramírez J et al (2025) Exploring legal age estimation using DNA methylation. Forensic Sci Int Genet 74:103142

Chen LL, Kim VN (2024) Small and long non-coding RNAs: past, present, and future. Cell 187(23):6451–6485

Chen L, Wang J, Tan L, Lu C, Fu G, Fu L et al (2020) Highly accurate mtGenome haplotypes from long-read SMRT sequencing can distinguish between monozygotic twins. Forensic Sci Int Genet 47:102306

Du Q, Zhu G, Fu G, Zhang X, Fu L, Li S et al (2015) A genome-wide scan of DNA methylation markers for distinguishing monozygotic twins. Twin Res Hum Genet 18(6):670–679

Dupont C, Armant DR, Brenner CA (2009) Epigenetics: definition, mechanisms and clinical perspective. Semin Reprod Med 27(5):351–357

Dwaraka VB, Aronica L, Carreras-Gallo N, Robinson JL, Hennings T, Carter MM et al (2024) Unveiling the epigenetic impact of vegan vs. omnivorous diets on aging: insights from the Twins Nutrition Study (TwiNS). BMC Med 22(1):301

El-Hossary NM, El-Desouky MA, Sabry GM, Omar MF, Ali MY, Elzayat MG et al (2024) A new insight of blood vs. buccal DNA methylation in the forensic identification of monozygotic triplets. Forensic Sci Int 364:112247

Föhr T, Waller K, Viljanen A, Sanchez R, Ollikainen M, Rantanen T et al (2021) Does the epigenetic clock GrimAge predict mortality independent of genetic influences: an 18 year follow-up study in older female twin pairs. Clin Epigenetics 13(1):128

Föhr T, Hendrix A, Kankaanpää A, Laakkonen EK, Kujala U, Pietiläinen KH et al (2024) Metabolic syndrome and epigenetic aging: a twin study. Int J Obes 48(6):778–787

Fraga MF, Ballestar E, Paz MF, Ropero S, Setien F, Ballestar ML et al (2005) Epigenetic differences arise during the lifetime of monozygotic twins. Proc Natl Acad Sci USA 102(30):10604–10609
Frumkin D, Wasserstrom A, Budowle B, Davidson A (2011) DNA methylation-based forensic tissue identification. Forensic Sci Int Genet 5(5):517–524
Fuchs KM, D'Alton ME (2018) 159—monochorionic monoamniotic twin gestations. In: Copel JA, D'Alton ME, Feltovich H, Gratacós E, Krakow D, Odibo AO et al (eds) Obstetric imaging: fetal diagnosis and care, 2nd edn. Elsevier, pp 642–5.e1
Gerra MC, Dallabona C, Cecchi R (2024) Epigenetic analyses in forensic medicine: future and challenges. Int J Legal Med 138(3):701–719
Gonggrijp BMA, van de Weijer SGA, Bijleveld C, Boomsma DI, van Dongen J (2025) Negative life events and epigenetic ageing: a study in the Netherlands twin register. Behav Genet 55(3):215–230
Gu J, Stevens M, Xing X, Li D, Zhang B, Payton JE et al (2016) Mapping of variable DNA methylation across multiple cell types defines a dynamic regulatory landscape of the human. Genome G3 (Bethesda) 6(4):973–986
Hagenbeek FA, van Dongen J, Pool R, Boomsma DI (2022) Chapter 32—twins and omics: the role of twin studies in multi-omics. In: Tarnoki A, Tarnoki D, Harris J, Segal N (eds) Twin research for everyone. Academic, pp 547–584
Hong X, Cao H, Cao W, Lv J, Yu C, Huang T et al (2025) Trends of genetic contributions on epigenetic clocks and related methylation sites with aging: a population-based adult twin study. Aging Cell 24(3):e14403
Jonsson H, Magnusdottir E, Eggertsson HP, Stefansson OA, Arnadottir GA, Eiriksson O et al (2021) Differences between germline genomes of monozygotic twins. Nat Genet 53(1):27–34
Kim JY, Lee HY, Lee SY, Kim SY, Park JL, Lee SD (2023) DNA methylome profiling of blood to identify individuals in a pair of monozygotic twins. Genes Genomics 45(10):1273–1279
Kimura H (2013) Histone modifications for human epigenome analysis. J Hum Genet 58(7):439–445
Leeb M, Wutz A (2012) Establishment of epigenetic patterns in development. Chromosoma 121(3):251–262
Leung ML, Abdullaev Z, Santana-Santos L, Skaugen JM, Moore S, Ji J (2024) Microarray-based DNA methylation profiling: validation considerations for clinical testing. J Mol Diagn 26(6):447–455
Li C, Zhao S, Zhang N, Zhang S, Hou Y (2013) Differences of DNA methylation profiles between monozygotic twins' blood samples. Mol Biol Rep 40(9):5275–5280
Li S, Ye Z, Mather KA, Nguyen TL, Dite GS, Armstrong NJ et al (2022) Early life affects late-life health through determining DNA methylation across the lifespan: a twin study. EBioMedicine 77:103927
Liu C, Wang Z, Hui Q, Goldberg J, Smith NL, Shah AJ et al (2022) Association between depression and epigenetic age acceleration: a co-twin control study. Depress Anxiety 39(12):741–750
Liu J, Zhang S, Wen Y, Su Y, Jiang L, Li S et al (2023) Exploring rare differences in mitochondrial genome between MZ twins using Ion Torrent semiconductor sequencing. Forensic Sci Int 348:111708
Lu M, Xueying Q, Hexiang P, Wenjing G, Hägg S, Weihua C et al (2021) Genome-wide associations between alcohol consumption and blood DNA methylation: evidence from twin study. Epigenomics 13(12):939–951
Luo J, Wang W, Li J, Duan H, Xu C, Tian X et al (2024) Epigenome-wide association study identifies DNA methylation loci associated with handgrip strength in Chinese monozygotic twins. Front Cell Dev Biol 12:1378680
Madi T, Balamurugan K, Bombardi R, Duncan G, McCord B (2012) The determination of tissue-specific DNA methylation patterns in forensic biofluids using bisulfite modification and pyrosequencing. Electrophoresis 33(12):1736–1745
Mak JKL, Skovgaard AC, Nygaard M, Kananen L, Reynolds CA, Wang Y et al (2024) Epigenome-wide analysis of frailty: results from two European twin cohorts. Aging Cell 23(6):e14135

Marqueta-Gracia JJ, Álvarez-Álvarez M, Baeta M, Palencia-Madrid L, Prieto-Fernández E, Ordoñana JR et al (2018) Differentially methylated CpG regions analyzed by PCR-high resolution melting for monozygotic twin pair discrimination. Forensic Sci Int Genet 37:e1–e5
Miao K, Liu S, Cao W, Lv J, Yu C, Huang T et al (2024) Five years of change in adult twins: longitudinal changes of genetic and environmental influence on epigenetic clocks. BMC Med 22(1):289
Moore LD, Le T, Fan G (2013) DNA methylation and its basic function. Neuropsychopharmacology 38(1):23–38
Ouwens KG, Jansen R, Tolhuis B, Slagboom PE, Penninx B, Boomsma DI (2018) A characterization of postzygotic mutations identified in monozygotic twins. Hum Mutat 39(10):1393–1401
Peng H, Gao W, Cao W, Lv J, Yu C, Wu T et al (2021) Combined healthy lifestyle score and risk of epigenetic aging: a discordant monozygotic twin study. Aging (Albany NY) 13(10):14039–14052
Planterose Jiménez B, Liu F, Caliebe A, Montiel González D, Bell JT, Kayser M et al (2021) Equivalent DNA methylation variation between monozygotic co-twins and unrelated individuals reveals universal epigenetic inter-individual dissimilarity. Genome Biol 22(1):18
Ravi S, Kankaanpää A, Bogl LH, Heikkinen A, Pietiläinen KH, Kaprio J et al (2025) Suboptimal dietary patterns are associated with accelerated biological aging in young adulthood: a study with twins. Clin Nutr 45:10–21
Roewer L (2013) DNA fingerprinting in forensics: past, present, future. Investig Genet 4(1):22
Rolf B, Krawczak M (2021) The germlines of male monozygotic (MZ) twins: very similar, but not identical. Forensic Sci Int Genet 50:102408
Sigurpalsdottir BD, Stefansson OA, Holley G, Beyter D, Zink F, Hardarson M et al (2024) A comparison of methods for detecting DNA methylation from long-read sequencing of human genomes. Genome Biol 25(1):69
Silva S, Martins Y, Matias A, Blickstein I (2011) Why are monozygotic twins different? J Perinat Med 39(2):195–202
Song B, Qian J, Fu J (2024) Research progress and potential application of microRNA and other non-coding RNAs in forensic medicine. Int J Legal Med 138(2):329–350
Stewart L, Evans N, Bexon KJ, van der Meer DJ, Williams GA (2015) Differentiating between monozygotic twins through DNA methylation-specific high-resolution melt curve analysis. Anal Biochem 476:36–39
Tan Q, Alo H, Nygaard M, Sørensen M, Saleh A, Mengel-From J et al (2024) Age-dependent DNA methylation variability on the X-chromosome in male and female twins. Epigenomes 8(4):43
Tazelaar GHP, Hop PJ, Seelen M, van Vugt J, van Rheenen W, Kool L et al (2023) Whole genome sequencing analysis reveals post-zygotic mutation variability in monozygotic twins discordant for amyotrophic lateral sclerosis. Neurobiol Aging 122:76–87
Teschendorff AE, Horvath S (2025) Epigenetic ageing clocks: statistical methods and emerging computational challenges. Nat Rev Genet 26(5):350–368
Teschendorff AE, West J, Beck S (2013) Age-associated epigenetic drift: implications, and a case of epigenetic thrift? Hum Mol Genet 22(R1):R7–r15
Tian M, Liu X, Wang D, Wang Y, Wang S, Wei J et al (2025) Differences of tsRNA expression profiles efficiently discriminate monozygotic twins in peripheral blood. Forensic Sci Int Genet 77:103242
Turrina S, Bortoletto E, Giannini G, De Leo D (2021) Monozygotic twins: identical or distinguishable for science and law? Med Sci Law 61(1_suppl):62–66
Vadgama N, Lamont D, Hardy J, Nasir J, Lovering RC (2019) Distinct proteomic profiles in monozygotic twins discordant for ischaemic stroke. Mol Cell Biochem 456(1–2):157–165
van der Gaag KJ, van Marion V, van den Berg RR, Weiler NEC, Hoogenboom J, Kal A et al (2025) Identifying a monozygotic twin brother as a donor of DNA in minimal, mixed forensic stains—a case example. Forensic Sci Int Genet 78:103292
van Dongen J, Slagboom PE, Draisma HH, Martin NG, Boomsma DI (2012) The continuing value of twin studies in the omics era. Nat Rev Genet 13(9):640–653

van Dongen J, Gordon SD, McRae AF, Odintsova VV, Mbarek H, Breeze CE et al (2021) Identical twins carry a persistent epigenetic signature of early genome programming. Nat Commun 12(1):5618

Vidaki A, Kayser M (2018) Recent progress, methods and perspectives in forensic epigenetics. Forensic Sci Int Genet 37:180–195

Vidaki A, Daniel B, Court DS (2013) Forensic DNA methylation profiling—potential opportunities and challenges. Forensic Sci Int Genet 7(5):499–507

Vidaki A, Johansson C, Giangasparo F (2017a) Differentially methylated embryonal Fyn-associated substrate (EFS) gene as a blood-specific epigenetic marker and its potential application in forensic casework. Forensic Sci Int Genet 29:165–173

Vidaki A, Ballard D, Aliferi A, Miller TH, Barron LP, Syndercombe Court D (2017b) DNA methylation-based forensic age prediction using artificial neural networks and next generation sequencing. Forensic Sci Int Genet 28:225–236

Vidaki A, Díez López C, Carnero-Montoro E, Ralf A, Ward K, Spector T et al (2017c) Epigenetic discrimination of identical twins from blood under the forensic scenario. Forensic Sci Int Genet 31:67–80

Vidaki A, Kalamara V, Carnero-Montoro E, Spector TD, Bell JT, Kayser M (2018) Investigating the epigenetic discrimination of identical twins using buccal swabs, saliva, and cigarette butts in the forensic setting. Genes (Basel) 9(5):252

Vidaki A, Planterose Jiménez B, Poggiali B, Kalamara V, van der Gaag KJ, Maas SCE et al (2023) Targeted DNA methylation analysis and prediction of smoking habits in blood based on massively parallel sequencing. Forensic Sci Int Genet 65:102878

Wang J, Fu G, Wang Q, Ma G, Wang Z, Lu C et al (2024) Differences of circular RNA expression profiles between monozygotic twins' blood, with the forensic application in bloodstain and saliva. Forensic Sci Int Genet 69:103001

Weber-Lehmann J, Schilling E, Gradl G, Richter DC, Wiehler J, Rolf B (2014) Finding the needle in the haystack: differentiating "identical" twins in paternity testing and forensics by ultra-deep next generation sequencing. Forensic Sci Int Genet 9:42–46

Wei JW, Huang K, Yang C, Kang CS (2017) Non-coding RNAs as regulators in epigenetics (review). Oncol Rep 37(1):3–9

Wong DT (2015) Salivary extracellular noncoding RNA: emerging biomarkers for molecular diagnostics. Clin Ther 37(3):540–551

Xiao C, Pan C, Liu E, He H, Liu C, Huang Y et al (2019) Differences of microRNA expression profiles between monozygotic twins' blood samples. Forensic Sci Int Genet 41:152–158

Xu J, Fu G, Yan L, Craig JM, Zhang X, Fu L et al (2015) LINE-1 DNA methylation: a potential forensic marker for discriminating monozygotic twins. Forensic Sci Int Genet 19:136–145

Zarrinpar A, Gupta S, Maurya MR, Subramaniam S, Loomba R (2016) Serum microRNAs explain discordance of non-alcoholic fatty liver disease in monozygotic and dizygotic twins: a prospective study. Gut 65(9):1546–1554

Zeng K, Du J, Chen YZ, Wang DY, Sun ML, Li YZ et al (2024) Metabolomics efficiently discriminates monozygotic twins in peripheral blood. Int J Legal Med 138(6):2249–2258

Zeng K, Yang KB, Du J, Rakha A, Liu Z, Wang DY et al (2025) Epigenetic differentiation of monozygotic twins in forensic investigations using nanopore sequencing. Forensic Sci Int Genet 81:103344

Zhang N, Zhao S, Zhang SH, Chen J, Lu D, Shen M et al (2015) Intra-monozygotic twin pair discordance and longitudinal variation of whole-genome scale DNA methylation in adults. PLoS One 10(8):e0135022

Zhong Y, Zeng K, Adnan A, Li YZ, Hou XK, Pan Y et al (2023) Discrimination of monozygotic twins using mtDNA heteroplasmy through probe capture enrichment and massively parallel sequencing. Int J Legal Med 137(5):1337–1345

Ziller MJ, Gu H, Müller F, Donaghey J, Tsai LT, Kohlbacher O et al (2013) Charting a dynamic DNA methylation landscape of the human genome. Nature 500(7463):477–481

Chapter 4
Epigenetic Identification of Body Fluids

Georgiana Gibson-Daw, Deborah S. B. S. Silva, and Bruce R. McCord

Abstract Body fluid identification at crime scenes is essential for reconstructing the events of the crime, providing key evidence to assist in determining the perpetrator, and, in some cases, the victim. Currently, a variety of presumptive and confirmatory tests exist to detect the main body fluids, blood, semen, and saliva. However, this task can be difficult when encountering trace levels, mixtures, or uncommon body fluids such as vaginal fluid or menstrual blood. A number of recent studies have explored the possibility of using DNA methylation to distinguish between body fluids, based on tissue-specific differentially methylated regions (tDMRs). Additional markers, such as non-coding RNAs, have also been utilized based on their stability and tissue-specificity. Many of these studies point to the possibility of developing assays to simultaneously detect various body fluids at the same time. This chapter will review the current state-of-the-art of the applications of epigenetics to body fluid identification, particularly DNA methylation and non-coding RNAs, including advantages and limitations, and their future applicability to forensic casework.

Keywords Body fluid identification · Epigenetics · DNA methylation · Non-coding RNA · MicroRNA

G. Gibson-Daw
Department of Physical and Biological Sciences, Western New England University, Springfield, MA, USA
e-mail: georgiana.gibson-daw@wne.edu

D. S. B. S. Silva (✉)
Chemistry Department, Hofstra University, Hempstead, NY, USA
e-mail: Deborah.Silva@hofstra.edu

B. R. McCord
Department of Chemistry and Biochemistry, Florida International University, Miami, FL, USA
e-mail: mccordb@fiu.edu

S. C. Zapico (ed.), *Forensic Science and Epigenetics*, Epigenetics and Human Health 18, https://doi.org/10.1007/978-3-032-23085-0_4

Abbreviations

AI	Artificial Intelligence
AP	Acid Phosphatase
BSP	Bisulfite Sequencing PCR
cDNA	Complementary DNA
CpG	Cytosine–Phosphate–Guanine Dinucleotide
C_t	Cycle Threshold
ddPCR	Droplet Digital PCR
DNA	Deoxyribonucleic Acid
DNMT	DNA methyltransferases
DSI	DNA Source Identifier
HRM	High-Resolution Melting
LMG	Leucomalachite Green
lncRNA	Long Non-Coding RNA
MBD	Methyl-CpG Binding Domains
MBD-seq	Methyl-CpG Binding Domains Sequencing
MeDIP-chip	Methylated DNA Immunoprecipitation Microarray
MeDIP-Seq	Methylated DNA Immunoprecipitation Sequencing
miRNA	MicroRNA
mRNA	Messenger RNA
MRE-Seq	Methylation-Sensitive Restriction Enzyme Sequencing
MSP	Methylation-Specific PCR
MSRE	Methylation-Sensitive Restriction Enzyme
MSRE-PCR	Methylation-Sensitive Restriction Enzyme PCR
msSNuPE	Methylation-Sensitive Single Nucleotide Primer Extension
ncRNA	Non-Coding RNA
NGS	Next-Generation Sequencing
PCR	Polymerase Chain Reaction
piRNA	Piwi-Interacting RNA
PCA	Principal Component Analysis
PSA	Prostate-Specific Antigen
qPCR	Quantitative PCR
RISC	RNA-Induced Silencing Complex
RNA	Ribonucleic Acid
RNA S-FISH	RNA Suspension Fluorescence In Situ Hybridization
RRBS	Reduced Representation Bisulfite Sequencing
rRNA	Ribosomal RNA
RT-qPCR	Reverse Transcription Quantitative PCR
SBE	Single-Base Extension
siRNA	Small Interfering RNA
SNaPshot	Single-Nucleotide Primer Extension Assay
STR	Short Tandem Repeat
SVM	Support Vector Machine
SWGDAM	Scientific Working Group on DNA Analysis Methods

tDMR	Tissue-Specific Differentially Methylated Region
TET	Ten-Eleven Translocation
Tm	Melting Temperature
UV	Ultraviolet
WGBS	Whole-Genome Bisulfite Sequencing

4.1 Introduction

Forensic science has always been driven by the imperative to transform trace biological evidence into reliable, interpretable information. Since the advent of DNA profiling in the 1980s, short tandem repeat (STR) analysis has revolutionized the field, providing investigators with the ability to match individuals to crime scene evidence with extraordinary statistical power (Butler 2011). Yet, as powerful as STR analysis is for identifying "who" contributed a biological trace, it tells us little about the biological context of that material. Was the stain blood, saliva, semen, or vaginal fluid? What was the biological age of the contributor? What environmental exposures had they suffered? Or lifestyle choices had they made? These questions remain central to forensic casework because the probative value of DNA evidence is not only in its ability to associate an individual with a crime scene but also in its capacity to contextualize the biological event itself. Such information can further assist triers of fact by helping to reconstruct events, corroborate testimonies, or establish the relevance of a sample to a case.

These are areas where classical genetics falls short and where epigenetics emerges as a powerful complementary tool to the traditional STR analysis. Epigenetics is the study of heritable changes in gene expression that occur without alterations to the underlying DNA sequence, often through mechanisms like DNA methylation, histone modification, and non-coding RNAs (Bird 2007). While being able to answer all of these questions is important, here we will focus on body fluid identification, as this is an important courtroom issue as it can help differentiate between an innocent touch of the hand in a crowd and more serious forms of sexual or aggravated assault.

Traditional methods for body fluid identification—presumptive chemical assays and enzymatic tests—have been used as the backbone of forensic biology for decades (An et al. 2012). They are often quick and inexpensive, but also have limitations such as low specificity, inability to reliably differentiate between closely related fluids (e.g., vaginal versus menstrual secretions), and stability when exposed to heat, and humidity. Moreover, many traditional tests are consumptive, destroying a portion of the sample and limiting the amount of material available for DNA profiling. While confirmatory tests based on immunoassays exist, their sensitivity may not be sufficient when compared to that of DNA amplified by the PCR reaction. This leaves a gap in sensitivity and specificity which can only be filled by more advanced molecular methods.

As a result, researchers have begun to investigate new methods for analysis such as mRNA, proteomic, and epigenetic methods. These have the potential to be more sensitive and specific than previous methods. The molecular methods exploit tissue-specific gene and protein expression to discriminate among different body fluids. Ghemrawi et al. (2024) recently evaluated several emerging technologies for body fluid identification. The study found that traditional serologic methods involving immunoassays, while fit for purpose, lacked sensitivity and specificity when confronted with mixtures or vaginal fluids. In contrast, emerging molecular approaches (mRNA, DNA methylation, and targeted proteomics) demonstrated higher specificity, with mRNA currently being the most advanced. Challenges mainly involved samples exposed to inhibitory substances like humic acid. Results for DNA methylation indicated similar accuracy, but specificity was lower due to lack of defined standards for mixture deconvolution, leaving menstrual blood more difficult to characterize. Shotgun proteomics showed reduced specificity due to broad protein detection, while a targeted mass spectrometry method produced no false positives but was unable to identify vaginal fluid or distinguish between peripheral and menstrual blood. Overall, the study demonstrated that these molecular techniques frequently outperform traditional immunoassays, particularly for complex or low-volume samples.

Another emerging approach is microbial profiling, which seeks to differentiate body fluids based on their microbial communities, typically analyzed through housekeeping genes such as 16S rRNA. Studies have shown that fluids such as saliva, vaginal fluid, fecal matter, semen, peripheral and menstrual blood, as well as skin and nasal secretions, each exhibit distinct and relatively stable microbiome patterns. Although promising, microbial profiling is still in its early stages and is not yet widely used in forensic laboratories due to the lack of established standards. Creating standardized protocols would strengthen its acceptance in criminal investigations. A significant challenge, however, is the dynamic nature of the human microbiome, which varies across body sites and can shift over an individual's lifetime, posing challenges for consistent interpretation (Dass et al. 2023).

Although these emerging molecular technologies require further research and validation, published findings are highly encouraging. Omic-based techniques are at varying stages of development, supported by intensive research and databasing efforts. Once fully validated and legally tested, their potential specificity may significantly transform the field of body fluid identification (Kayser 2026).

Against this backdrop, epigenetics has emerged as a particularly transformative tool. Broadly defined, epigenetics refers to heritable changes in gene expression that are not caused by alterations to the underlying DNA sequence (Felsenfeld 2014). Two mechanisms, in particular, have attracted significant attention: DNA methylation and non-coding RNA regulation. DNA methylation involves the addition of a methyl group to cytosine bases, typically in CpG dinucleotides, leading to transcriptional repression or activation depending on genomic context. Non-coding RNAs—including microRNAs (miRNAs), small interfering RNAs (siRNAs), piwi-interacting RNAs (piRNAs), and long non-coding RNAs (lncRNAs)—regulate gene expression at the transcriptional and post-transcriptional level, often in tissue-specific patterns (Esteller 2011).

From a forensic perspective, these mechanisms offer something that neither STR profiling nor traditional biochemical assays can provide: tissue-specific, stable, and quantifiable molecular signatures. While every nucleated cell in the human body contains the same DNA sequence, epigenetic modifications vary dramatically between cell types. For example, red blood cell precursors exhibit unique methylation profiles at hemoglobin gene promoters, sperm cells express distinctive miRNA repertoires, and salivary gland epithelial cells display characteristic epigenetic landscapes around amylase genes (Vidaki and Kayser 2018). These patterns can be exploited to determine the tissue origin of forensic samples with a level of precision and reliability previously unattainable.

The stability of epigenetic markers is particularly valuable in forensic contexts. Proteins and enzymes targeted by conventional assays can degrade under environmental stress, such as heat, humidity, or ultraviolet radiation. RNA, while generally more fragile than DNA, is surprisingly stable in the form of microRNAs due to their small size and association with protein complexes (Zubakov et al. 2010). DNA methylation patterns, by contrast, are chemically robust and can often be detected even in highly degraded or aged samples (Frumkin et al. 2011). This resilience makes epigenetic assays ideally suited to the often challenging conditions of forensic evidence.

Furthermore, epigenetic analysis has the advantage of non-consumptive synergy with STR profiling. Because both assays rely on DNA extraction, forensic laboratories can apply methylation analysis and STR typing to the same sample. Unlike protein- or enzyme-based methods, which can require separation prior to analysis, epigenetic markers can be detected simultaneously with forensic DNA, thus maximizing the probative value of even the smallest and most degraded traces. This compatibility is a crucial consideration in casework, where evidence is often limited and must be preserved as much as possible.

The forensic utility of epigenetics has been demonstrated across a range of studies. Frumkin et al. (2011) were among the first to show that tissue-specific DNA methylation patterns could reliably distinguish between blood and semen. Madi et al. (2012) later demonstrated that bisulfite modification could be used instead of enzymatic detection, increasing its utility. Ultimately, epigenetic DNA methylation markers were developed for a wide variety of different body fluids including blood, semen, saliva, vaginal epithelia, menstrual blood, and skin.

The introduction of epigenetics into forensic science, however, is not without challenges. Validation is essential. Forensic laboratories operate under rigorous quality assurance standards, and any new method must demonstrate sensitivity, specificity, reproducibility, and robustness across a wide range of sample types and conditions (Budowle et al. 2008; ANSI 2020). Inter-individual variability in epigenetic patterns, as well as the influence of environmental factors such as smoking or diet, must be carefully accounted for to avoid misclassification (Sabeeha 2019). Furthermore, the legal system demands not only scientific validity but also admissibility under evidentiary standards such as Daubert or Frye (Kaye 2001). The path from promising laboratory research to courtroom-ready application is therefore both complex and rigorous.

Beyond validation, there are also ethical considerations. Epigenetic information can reveal far more than tissue type. DNA methylation patterns and non-coding RNA profiles have been linked to age, environmental exposures, disease states, and even psychosocial stress (Feil and Fraga 2012). While such information might be of scientific interest, its forensic use raises serious questions about privacy, proportionality, and the potential for misuse. Guidelines must be developed to ensure that epigenetic assays are applied narrowly and ethically, focusing on body fluid identification rather than health or lifestyle inference.

This chapter will explore the principles and applications of forensic epigenetics in depth. We begin by reviewing the traditional methods for body fluid identification, highlighting their strengths and limitations. These methods include the use of tissue-specific markers and multiplexing approaches, which provide crucial insights into the biological origin of samples. We then examine DNA methylation in detail, focusing on various methodologies such as bisulfite modification, enzymatic treatment, and newer approaches that are advancing the precision of analysis. These techniques, when paired with advanced detection approaches like HRM (High Resolution Melting), digital PCR, pyrosequencing, and next-generation sequencing (NGS), offer powerful tools for identifying and characterizing epigenetic changes in forensic samples. Next, we turn to non-coding RNAs, discussing their biogenesis, detection, and growing utility in forensic contexts, particularly in conjunction with innovative technologies. The chapter then addresses integrative workflows that combine epigenetic markers, as well as emerging technologies that are reshaping the field, alongside the validation frameworks necessary to translate research into practice. Finally, we consider the legal and ethical dimensions of forensic epigenetics, before concluding with an outlook on future directions.

4.2 Traditional Methods for Body Fluid Identification

The identification of biological fluids has long been a cornerstone of forensic biology. Long before DNA profiling transformed forensic science, serologists and biochemists relied on chemical, enzymatic, and immunological assays to determine whether a stain was blood, semen, saliva, or another body fluid. While these methods laid the groundwork for modern forensic practice, they also illustrate the challenges inherent in interpreting biological evidence: degradation, inhibition, and stochastic effects. This section examines the major categories of traditional methods used in forensic body fluid identification, detailing their scientific principles, operational utility, and limitations.

4.2.1 Presumptive Chemical Tests

Presumptive tests are rapid, inexpensive assays designed to suggest the possible presence of a given body fluid. They are typically used as the first line of investigation at the crime scene or in the laboratory. The scientific basis of these tests lies in detecting chemical or enzymatic activity associated with the target fluid.

4.2.1.1 Blood

The most widely used presumptive tests for blood are based on the peroxidase-like activity of hemoglobin. Compounds such as phenolphthalein (Kastle-Meyer test) and leucomalachite green (LMG test) are oxidized in the presence of hydrogen peroxide and hemoglobin, producing a characteristic color change (Gaensslen 1983; Fonseca et al. 2019). More modern luminol-based assays rely on the chemiluminescent reaction between luminol and hemoglobin, producing a blue glow visible under dark conditions.

These tests are highly sensitive with luminol, for example, being capable of detecting bloodstains diluted up to $1{:}10^6$ (Barni et al. 2007). However, sensitivity comes at the cost of specificity. Plant peroxidases, some chemical oxidants, and even metals can yield false positives. Additionally, luminol application is destructive, potentially diluting or degrading DNA. While invaluable for crime scene searching, presumptive tests cannot confirm the identity of blood without further analysis.

4.2.1.2 Semen

Presumptive semen tests typically detect acid phosphatase (AP), an enzyme present in high concentrations in seminal fluid. The Brentamine Fast Blue B test is the most commonly used, producing a purple color in the presence of AP (Raju 1964; Harbison and Fleming 2016). The assay is quick and inexpensive, but suffers from limited specificity, as vaginal secretions, feces, and some plant materials may also yield positive results.

4.2.1.3 Saliva

For saliva, the most common presumptive assay is the starch-iodine test, which detects the presence of α-amylase, a digestive enzyme abundant in salivary secretions. Modern assays often employ radial diffusion plates, where amylase activity produces clear zones against a stained starch background (Myers and Adkins 2008). However, amylase is not specific to saliva; it is also found in pancreatic secretions and trace amounts in other body fluids, raising the potential for misinterpretation.

4.2.1.4 Vaginal Fluid and Others

Few reliable presumptive tests exist for vaginal fluid. Some approaches have targeted glycogenated epithelial cells using Lugol's iodine staining, but the method lacks both sensitivity and specificity (Hausmann et al. 1994). Sweat and urine have also been subject to presumptive assays, though their forensic utility is generally limited due to variability and cross-reactivity.

4.2.2 *Confirmatory Tests*

Confirmatory tests aim to establish with high certainty the presence of a specific body fluid. They often target unique biomolecules not shared with other fluids, providing greater specificity than presumptive assays.

4.2.2.1 Blood

Confirmatory blood tests rely on the detection of hemoglobin or its derivatives. The Takayama and Teichmann crystal assays produce characteristic hemochromogen crystals when blood is treated with specific reagents. While highly specific, these tests are less sensitive than presumptive assays and require intact heme, making them less effective on degraded samples (Saferstein 2011). Immunochromatographic assays targeting human hemoglobin have also been developed, offering high specificity and compatibility with forensic workflows (Hochmeister et al. 1999; Cuttaia et al. 2024). The downside of these assays is the amount of sample required and the fact that the test consumes the sample.

4.2.2.2 Semen

Semen confirmatory testing traditionally relies on microscopic identification of spermatozoa using phase contrast or fluorescent microscopy. Techniques such as Christmas Tree staining (nuclear fast red and picroindigocarmine) remain staples in forensic laboratories (Allery et al. 2001) where the presence of sperm heads and tails is considered definitive evidence of semen. These microscopic techniques are time-consuming, dependent on analyst expertise, and limited by the degradation of cellular structures while also confronting problems with azoospermic individuals, vasectomized males, and degraded samples that may lack intact spermatozoa, limiting this method's utility (Golomingi et al. 2022). Prostate-specific antigen (PSA), also known as p30, is a protein abundant in seminal plasma and forms the basis of immunological confirmatory assays. While highly specific to seminal fluid, trace levels of PSA may occasionally be detected in female urine or breast milk, necessitating careful interpretation (Sensabaugh 1978; Sensabaugh and Blake 1990).

4.2.2.3 Saliva

Confirmatory saliva assays typically target salivary-specific proteins, though few have achieved widespread forensic adoption. Immunoassays directed against α-amylase can enhance specificity, but the overlap with pancreatic isoforms continues to complicate interpretation (Hedman et al. 2011; Bunescu et al. 2021).

4.2.2.4 Vaginal Fluid

Confirmatory testing for vaginal fluid remains one of the most challenging areas. Some methods attempt to identify vaginal mucin proteins or other biomarkers, but reproducibility and specificity remain limited. Microscopic analysis of glycogenated epithelial cells has been used historically, though the diagnostic power of this approach is weak (Randall and Riis 1985; Hausmann et al. 1994).

4.2.3 Limitations of Traditional Methods

Despite their long-standing use, traditional methods for body fluid identification suffer from several significant limitations:

1. Lack of Specificity: Many assays rely on enzymes or proteins that are not exclusive to a given body fluid, increasing the risk of false positives. Acid phosphatase, for instance, occurs in vaginal secretions and plant material; amylase is present in pancreatic tissue and other body fluids.
2. Degradation Sensitivity: Proteins and enzymes degrade rapidly under environmental conditions. Samples exposed to heat, humidity, or ultraviolet light can yield negative results even when the fluid was originally present.
3. Consumptive Testing: Many assays require substantial portions of the sample, reducing the material available for DNA profiling. This tension is particularly acute in trace or degraded evidence.
4. Mixed Samples: Traditional assays are poorly suited to mixtures. For example, in sexual assault cases, the presence of vaginal fluid may mask or obscure the detection of semen, complicating interpretation.
5. Legal and Evidentiary Challenges: Because of these limitations, results from these tests are often viewed cautiously in court. While they can be probative, their susceptibility to false positives and negatives reduces their reliability as stand-alone evidence.

4.2.4 Transition to Epigenetic Approaches

The shortcomings of traditional methods create a compelling rationale for exploring new avenues. What forensic scientists require is a system that combines specificity, stability, and compatibility with DNA profiling. Epigenetic markers, particularly DNA methylation signatures and non-coding RNAs, meet these criteria. They are inherently tissue-specific, chemically robust, and analyzable using the same DNA extracts employed for STR typing. Moreover, their quantitative nature offers a way forward in mixed-sample interpretation, a domain where traditional methods have largely failed.

4.3 DNA Methylation in Forensic Science

The discovery of DNA methylation as a stable, tissue-specific epigenetic modification has fundamentally reshaped our understanding of gene regulation. Within the forensic sciences, the recognition that methylation signatures are not only preserved in biological stains but also discriminatory across tissues has opened new avenues for body fluid identification. Unlike traditional enzymatic or protein-based assays, DNA methylation analysis integrates seamlessly with forensic genetic workflows, leveraging the same DNA extracts used for STR typing. This section explores the molecular mechanisms underlying DNA methylation, the tissue-specificity that enables its forensic utility, the laboratory methods available for its analysis, and the challenges associated with its forensic implementation.

DNA methylation refers to the covalent addition of a methyl group ($-CH_3$) to the 5th carbon of cytosine residues, predominantly within CpG dinucleotides (Bird 2002). These CpG sites frequently cluster in genomic regions termed CpG islands, which are often located in gene promoters. Methylation of promoter-associated CpG islands typically suppresses gene transcription by either directly blocking transcription factor binding or by recruiting methyl-binding proteins that compact chromatin into inaccessible heterochromatin (Jones 2012).

The establishment and maintenance of DNA methylation patterns are mediated by a family of DNA methyltransferases (DNMTs). DNMT3A and DNMT3B establish de novo methylation patterns during development, while DNMT1 maintains these patterns following DNA replication. Demethylation can occur either passively, through replication without DNMT1 activity, or actively via the ten-eleven translocation (TET) family of enzymes, which oxidize 5-methylcytosine to 5-hydroxymethylcytosine (Tahiliani et al. 2009).

In the context of forensic science, the importance of these mechanisms lies not in the enzymes themselves but in the resulting methylation landscapes. Because different tissues adopt specific transcriptional programs, their DNA methylation signatures are correspondingly distinct, forming the basis for tissue-specific forensic markers.

4.3.1 *Tissue-Specific Methylation Signatures*

Each human cell type contains the same genetic sequence, but methylation patterns differ across tissues, reflecting their specialized functions. For example, promoters of hemoglobin genes are demethylated in erythroid cells but hypermethylated in other tissues, enabling selective gene expression (Shearstone et al. 2011). Similarly, genes involved in spermatogenesis are hypomethylated in spermatozoa but methylated in somatic cells. Forensically, this means that CpG sites can serve as biomarkers for body fluids. Several studies have identified methylation markers with high specificity for blood, semen, saliva, menstrual blood, and vaginal secretions. For instance, the gene locus *ZC3H12D* has been shown to be hypomethylated in semen, while *BCAS4* methylation levels are distinctive in saliva (Madi et al. 2012).

The strength of DNA methylation markers lies in their dual attributes: high tissue specificity and remarkable chemical stability. Unlike proteins or RNA transcripts, methylation modifications persist in dried stains exposed to environmental stress, making them ideal for forensic casework.

4.3.2 *Forensic Applications of DNA Methylation*

4.3.2.1 Body Fluid Identification

The most mature forensic application of DNA methylation is the identification of body fluids. Panels of tissue-specific CpG sites have been developed to differentiate blood, semen, saliva, menstrual blood, and vaginal fluid with high accuracy. Such assays can be multiplexed, reducing sample consumption and improving efficiency. In a recent comparative assessment by Ghemrawi et al. (2024), DNA methylation analysis was applied to a blind set of 58 forensic-like samples using a conservative decision tree developed by the researchers. The assay correctly identified the body fluid in 39 samples and detected at least one component in an additional 11 mixture samples, producing accurate results for 52 of the 58 samples tested. These successes included both single-source and mixed samples across various substrates, as well as degraded specimens. Two samples containing nontarget fluids (urine and breast milk) were appropriately classified as "not targeted," and no cross-reactivity was observed, demonstrating strong marker specificity. Two samples yielded inconclusive results due to conservative interpretation thresholds, and two false negatives were likely attributable to low or degraded DNA. Overall, the study highlights three key points: (1) DNA methylation offers high specificity with no false positives among targeted fluids; (2) conservative thresholds can lead to inconclusive outcomes that may be improved with refined interpretation guidelines; and (3) false negatives are primarily associated with low-template or degraded DNA.

4.3.2.2 Age Estimation

Beyond tissue identification, methylation patterns change predictably with age at specific CpG sites, forming the basis of so-called "epigenetic clocks" (Horvath 2013). Forensic scientists have adapted these markers to estimate the chronological age of donors from biological stains, with mean absolute errors as low as 3–5 years (Weidner et al. 2014). While not the primary focus of this chapter, the overlap between tissue identification and age estimation illustrates the versatility of methylation analysis.

4.3.2.3 Lifestyle and Environmental Exposures

Emerging studies suggest that DNA methylation signatures may capture environmental exposures relevant to forensic investigations, such as smoking status or drug use (Zeilinger et al. 2013). While still experimental, the potential to glean behavioral or lifestyle information from biological stains raises both opportunities and ethical questions.

4.3.3 Challenges and Limitations

Despite its promise, methylation-based forensic analysis faces several obstacles.

1. DNA Quantity and Quality: Forensic stains are often degraded and low in quantity. Bisulfite treatment exacerbates DNA loss, though newer direct methylation detection methods may help mitigate this problem.
2. Mixture Interpretation: Disentangling methylation signals from mixed fluids remains technically challenging. Statistical deconvolution approaches are promising but require validation in real-world casework.
3. Population Variation: While many methylation markers are consistent across populations, some inter-individual variation exists. Forensic validation must account for genetic ancestry and environmental influences.
4. Standardization: Protocols for bisulfite conversion, sequencing, and interpretation vary widely. Standardization across laboratories is essential for admissibility and comparability.

DNA methylation offers a powerful and stable biomarker for forensic applications. Its tissue-specific signatures provide a solution to many limitations of traditional serological assays, offering specificity, stability, and compatibility with existing DNA workflows. While technical challenges remain, particularly in mixture interpretation and sample degradation, the progress of the past decade suggests that methylation-based assays are on the cusp of routine forensic implementation. As technologies such as nanopore sequencing mature, the integration of DNA methylation into forensic practice will likely accelerate, reshaping the way biological evidence is interpreted.

4.4 Methodologies for Analyzing DNA Methylation

Over the past decades, methodologies for detecting and analyzing DNA methylation have progressed from low-resolution, global assays to highly sensitive, genome-wide approaches. The choice of method depends on the research objective, the required resolution, and the amount of available sample. Common strategies for interrogating DNA methylation at specific regions include bisulfite conversion, which differentiates methylated from unmethylated cytosines; restriction enzyme–based approaches, which exploit methylation-sensitive endonucleases; and affinity enrichment techniques, which selectively isolate methylated DNA fragments.

4.4.1 Bisulfite Conversion-Based Methods

The method of bisulfite conversion relies on the treatment of the DNA sample with sodium bisulfite, which converts unmethylated cytosine to uracil but leaves methylated cytosine unchanged. During the treatment, DNA is first denatured to expose the cytosine bases. When sodium bisulfite is added to the DNA solution, non-methylated cytosines will undergo deamination, converting into uracil, while methylated cytosines are resistant to this conversion, remaining as cytosines. The converted DNA is then amplified using PCR, and during amplification and subsequent sequencing, the uracils are treated as thymines. This process enables single-base resolution since methylated and unmethylated DNA can be distinguished (Li and Tollefsbol 2011; Pajares et al. 2021).

Different techniques for DNA methylation analysis rely on the use of bisulfite-converted DNA. Generally, these techniques can be divided into locus-specific assays or genome-wide assays. These are examples of assays that target specific loci:

- Methylation-Specific PCR (MSP): Following bisulfite conversion, samples are amplified in two separate PCR reactions using sequence-specific primer sets: one designed for methylated DNA and the other for unmethylated DNA. The methylated primer set anneals to sequences retaining methylated cytosines, while the unmethylated primer set anneals to bisulfite-converted sequences where cytosines have been converted to uracils. Analysis of the PCR products yields three possible outcomes: (1) amplification with methylated primers indicates methylated DNA, (2) amplification with unmethylated primers indicates unmethylated DNA, and (3) amplification with both primer sets indicates a mixture of methylated and unmethylated DNA (Derks et al. 2004).

- Bisulfite Sequencing PCR (BSP): After bisulfite treatment and PCR amplification of the target region, the amplified products are cloned into bacterial cells and subsequently sequenced using conventional methods. This allows determination of the methylation status at each individual CpG site (Frommer et al. 1992; Reed et al. 2010).
- Digital PCR: Following bisulfite conversion, DNA in solution is divided into thousands of droplets or nanowells, each serving as an individual PCR reaction chamber. Using sequence-specific primers and fluorescent probes, methylated and unmethylated DNA sequences are amplified and detected in separate droplets or wells. After amplification, fluorescence signals are measured allowing quantification of methylated versus unmethylated DNA (Van Wesenbeeck et al. 2018; Zhou et al. 2024).
- Pyrosequencing: Bisulfite-treated DNA is first amplified at the target region and then analyzed by pyrosequencing, a sequencing-by-synthesis method. During this process, nucleotides are sequentially incorporated into the growing DNA strand, releasing pyrophosphate as a byproduct. This byproduct is converted into a light signal, and the intensity of luminescence corresponds to the number of nucleotides incorporated. By comparing sequence data and luminescence patterns, pyrosequencing enables quantitative determination of methylated and unmethylated cytosines at specific sites (Tost and Gut 2007; Reed et al. 2010).
- High-Resolution Melting (HRM): After bisulfite conversion and targeted PCR amplification, samples are analyzed by real-time PCR coupled with melting curve analysis. Because methylated and unmethylated DNA differ in sequence, they exhibit distinct melting profiles and melting temperatures (Tm). Unmethylated cytosines, converted to uracils, lower the Tm compared to methylated cytosines. Using intercalating dyes and real-time PCR instruments, differences in melting curves allow estimation of methylation levels within specific regions. HRM software further compares melting profiles of unknown samples against fully methylated and unmethylated reference DNA to assess methylation status (Wojdacz and Dobrovic 2007; Hussmann and Hansen 2018).
- SNaPshot: After bisulfite conversion and PCR amplification, samples undergo dideoxy single-base extension (SBE) followed by capillary electrophoresis. The SBE reaction targets a specific CpG site within the region of interest, where a fluorescently labeled nucleotide is incorporated into the SBE primer. The color of the resulting peak indicates which nucleotide has been added, while the peak intensity reflects the methylation level at that CpG (Castagnola et al. 2024).
- MethyLight: Following bisulfite treatment, DNA is amplified using primers specific to bisulfite-converted sequences in combination with methylation-specific probes. Fluorescent probes are designed to selectively bind either methylated or unmethylated DNA. During real-time PCR, fluorescence signals from these probes are monitored, enabling quantification of the proportion of methylated DNA at the target locus (Olkhov-Mitsel et al. 2014; Campan et al. 2018).

Other techniques are genome-wide assays. These are examples of assays that investigate DNA methylation patterns across the genome:

- Whole-Genome Bisulfite Sequencing (WGBS): Considered the gold standard for genome-wide DNA methylation analysis, WGBS involves bisulfite treatment of DNA followed by sequencing with next-generation sequencing (NGS). The resulting reads are processed through bioinformatics pipelines and aligned to a reference genome, enabling determination of the methylation status at single-cytosine resolution across the entire genome (Lin et al. 2019; Gong et al. 2022).
- DNA Methylation Microarrays: This method enables the simultaneous detection and quantification of methylation across multiple loci. Genomic DNA is first treated with bisulfite, then hybridized to a microarray chip containing probes specific for methylated (M) or unmethylated (U) CpG sites. Fluorescent labels are incorporated during single-base extension, and the resulting fluorescence signals are analyzed. The relative intensities of M and U probes are then used to determine the methylation status at each locus (Bibikova et al. 2011; Lehne et al. 2015).

4.4.2 Restriction Enzyme-Based Methods

The methylation-sensitive restriction enzyme (MSRE) method relies on restriction endonucleases that recognize and cleave DNA at specific sequences. These enzymes are sensitive to cytosine methylation: they cut when the cytosine within the recognition site is unmethylated but fail to cleave if it is methylated. This approach offers an alternative to bisulfite conversion, which requires harsh chemical treatment that can damage DNA. Nonetheless, MSRE-based analysis is inherently constrained by the requirement that specific restriction sites be present within the genomic region of interest (Beikircher et al. 2018; Šestáková et al. 2019).

The MSRE strategy can be coupled with other approaches such as quantitative PCR (qPCR) for locus-specific methylation analysis. In this approach, genomic DNA is first subjected to digestion with methylation-sensitive restriction enzymes, followed by amplification of the target region using sequence-specific primers. Parallel reactions are performed with undigested control DNA, and the resulting cycle threshold (C_t) values are compared. The difference in amplification efficiency between digested and undigested samples is then used to calculate the relative proportion of methylated alleles (Šestáková et al. 2019). It has also been proposed the use of ddPCR (MSRE-ddPCR) but this concept still needs further validation studies (Gattuso et al. 2024). MSRE analysis can also be combined with sequencing for detailed methylation profiling. After digestion, selected genomic regions can be amplified using fluorescent-labeled primers and analyzed by capillary electrophoresis to determine methylation status (Frumkin et al. 2011). For higher-resolution or genome-wide studies, the digested DNA fragments can be converted into sequencing libraries and analyzed using next-generation sequencing (NGS) in a method known as methylation-sensitive restriction enzyme sequencing (MRE-Seq), allowing comprehensive detection and mapping of methylation across the genome or at specific loci (Mayne et al. 2018). Another approach is the use of reduced representation bisulfite sequencing (RRBS), which is an alternative to whole-genome bisulfite sequencing that combines restriction enzyme digestion with bisulfite treatment to focus on CpG-rich regions, including CpG islands and promoters. Following enzyme digestion and size selection, DNA is bisulfite-converted and sequenced using NGS. By enriching GC-rich regions, RRBS enables high-resolution methylation profiling while reducing the overall sequencing effort (Nakabayashi et al. 2023).

4.4.3 Affinity Enrichment-Based Methods

Affinity enrichment-based methods isolate DNA methylation using proteins or antibodies that specifically bind to methylated cytosines. The enriched fragments are then analyzed by NGS or microarrays. While affinity enrichment provides genome-wide coverage, it is biased toward regions with a high density of CpG sites (Li and Tollefsbol 2021). One example is the methylated DNA immunoprecipitation sequencing (MeDIP-Seq), which isolates methylated DNA fragments through antibody-based immunoprecipitation, followed by purification, library preparation, and high-throughput sequencing. Variants of this approach include MeDIP-chip, which uses DNA microarrays for analysis instead of sequencing, and MBD-seq, where methylated DNA is captured by proteins containing methyl-CpG binding domains (MBDs) prior to sequencing (Taiwo et al. 2012; Hsu et al. 2020; Pajares et al. 2021).

4.5 DNA Methylation and Body Fluid Identification

Frumkin and collaborators (2011) were among the first researchers to successfully demonstrate the use of DNA methylation for body fluid identification. The authors analyzed several human tissues, including blood, saliva, semen, skin, urine, menstrual blood, and vaginal secretions, using a methylation-sensitive restriction enzyme PCR (MSRE-PCR) approach. After screening 205 CpG islands, the researchers identified 38 genomic loci with differential methylation profiles, enabling distinct tissue-specific patterns. Methylation ratios between pairs of loci such as L91762/L68346 and L76138/L26688 allowed discrimination between body fluids, with semen showing low methylation at L91762 and high methylation at L68346, while skin and blood exhibited opposite patterns. The assay was effective even in mixed samples, reliably detecting semen in mixtures with saliva and urine. Building on Frumkin's work, Wasserstrom et al. (2013) developed and validated a highly specific and automatable assay for the forensic identification of semen using

DNA methylation profiling. The DNA source identifier (DSI)-semen assay uses a methylation-sensitive restriction enzyme (HhaI) followed by PCR amplification and capillary electrophoresis to detect semen-specific methylation signatures at five genomic loci, with three additional control loci ensuring the integrity of the reaction. The assay was designed as a single-tube biochemical procedure followed by automated software analysis, which outputs a semen/non-semen identification along with a statistical confidence level. The DSI-semen assay was tested on 135 samples comprising semen, blood, saliva, menstrual blood, urine, and vaginal swabs. Results showed 100% correct identification of semen and non-semen samples for all successfully amplified cases, achieving confidence levels greater than 0.9999. Failures in six samples (approximately 4%) were attributed to incomplete digestion, PCR failure, or inconclusive results, not misidentification. Importantly, in 33 real forensic casework samples provided by the Israeli Police, results from the DSI-semen assay were completely concordant with those obtained by microscopic examination, except in one case where the DSI assay detected semen in a sample where microscopy did not, likely due to low sperm count or degradation, demonstrating that the assay can be more sensitive than the current gold standard. The authors further validated the method using mock casework samples, showing successful differentiation of semen and vaginal material even when only small amounts of semen were present.

The research by Madi et al. (2012) expanded the search for tissue-specific methylation signatures using bisulfite conversion and pyrosequencing across loci such as ZC3H12D, FGF7, C20orf117, and BCAS4. Their findings reinforced earlier conclusions that ZC3H12D and FGF7 reliably identified semen through hypomethylation, C20orf117 was highly methylated in blood, and BCAS4 emerged as a potential saliva-specific marker. Follow-up experiments confirmed that methylation patterns remained stable even in aged biological stains up to 20 years old, demonstrating the long-term durability of methylation signatures. Further validation by Silva et al. (2016) extended these findings through a comprehensive validation of ZC3H12D, BCAS4, and cg06379435, previously identified as potential tissue-specific DNA methylation marker for blood (Park et al. 2014). Using bisulfite modification, PCR amplification, and pyrosequencing, they measured methylation levels at multiple CpG sites within each locus to generate distinct methylation profiles corresponding to each body fluid type. The study followed SWGDAM guidelines to perform developmental validation across several critical forensic parameters, including specificity, sensitivity, species specificity, stability, reproducibility, and the assay's performance in mixed and degraded samples. Results demonstrated that each marker produced a consistent and statistically distinct methylation pattern for its target body fluid, allowing clear differentiation from non-target tissues such as vaginal, nasal, and menstrual samples. The markers showed high human specificity, with only partial cross-reactivity observed in non-human primates, and no amplification in more distantly related species. Sensitivity tests revealed reliable detection using as little as 0.1–10 ng of DNA, depending on the marker, confirming their suitability for forensic samples with limited DNA quantities. The assays maintained their discriminatory power even under conditions simulating degradation and inhibition, including exposure to heat and common PCR inhibitors like hematin and humic acid. Additionally, mixture studies indicated that methylation values changed predictably with the proportions of different body fluids, demonstrating that the method could detect mixed biological samples. Inter-laboratory reproducibility testing, performed at two independent institutions, produced statistically concordant results, further supporting the robustness of the assays.

Subsequent research shifted toward analyzing tissue-specific differentially methylated regions (tDMRs) using bisulfite modification and next-generation sequencing methods. Lee et al. (2012) investigated five tDMRs (DACT1, USP49, HOXA4, PFN3, and PRMT2) in blood, semen, saliva, vaginal fluid, and menstrual blood. The genes DACT1 and USP49 exhibited semen-specific hypomethylation, while HOXA4, PFN3, and PRMT2 showed distinct methylation profiles across body fluids, with PFN3 useful for identifying vaginal fluid and PRMT2 for differentiating semen from other tissues. These markers were later validated and expanded upon in a study by An et al. (2013), who developed methylation SNaPshot and MSRE-PCR multiplex assays that allowed simultaneous detection of multiple tDMRs with high sensitivity, requiring as little as 250–500 pg of DNA, and consistent methylation profiles in aged or mixed forensic samples. Their results confirmed that semen exhibits complete unmethylation at DACT1 and USP49, while blood and saliva remain hypermethylated, providing a robust basis for semen identification even in complex mixtures, such as post-coital samples from sexual assault simulations.

The study by Lee et al. (2015) used the Illumina HumanMethylation450 BeadChip array to conduct genome-wide methylation profiling on 42 samples of blood, saliva, semen, vaginal fluid, and menstrual blood collected from 100 donors. From this analysis, 64 candidate CpG sites were initially identified as body fluid–specific markers based on significant methylation differences among the fluids. Further validation using targeted bisulfite sequencing led to the selection of eight CpG markers suitable for constructing a multiplex methylation SNaPshot assay. This assay allowed simultaneous analysis of methylation at all eight sites and successfully distinguished between blood, saliva, semen, and body fluids originating from female reproductive organs in a single reaction. The methylation profiles were distinct for each fluid type, and the system was able to detect characteristic patterns even in aged or mixed samples, such as post-coital vaginal and penile swabs. Importantly, semen from vasectomized males lacked semen-specific hypermethylation signals, confirming the assay's specificity for sperm-derived DNA. The assay also demonstrated strong sensitivity, requiring only 0.5 nanograms of bisulfite-converted DNA, and showed no cross-reactivity with skin samples. Although the study highlighted some limitations, such as mild cross-reactivity of blood markers with menstrual blood and reliance on a single saliva-specific marker, it represented a major advance in forensic science by demonstrating that epigenetic profiling could accurately and non-destructively identify biological fluids.

Forat et al. (2016) developed and validated a DNA methylation–based method to identify forensic body fluids with high accuracy and stability. Using genome-wide methylation profiling with Illumina 27K and 450K BeadChips, the researchers identified nine CpG markers capable of distinguishing blood, semen, saliva, vaginal fluid, and menstrual blood. These were analyzed using a methylation-sensitive single-nucleotide primer extension (msSNuPE) assay, which proved reliable even with trace or degraded samples. Tests on over 100 samples per fluid showed clear differentiation, high sensitivity (down to ~100 pg DNA), and resistance to environmental effects such as humidity and long-term storage. The study also examined influences from cancer and genetic variants, finding that only cervical carcinoma affected vaginal fluid markers, while minor sequence variants had limited impact. Comparison with next-generation sequencing and bisulfite methods confirmed the assay's accuracy.

Holtkötter et al. (2017) validated DNA methylation markers for identifying forensic body fluids and developed a four-marker multiplex SNaPshot assay targeting menstrual blood (VF2), saliva (SA1), semen (SE2), and venous blood (BL1). Testing samples from 43 donors, the assay reliably distinguished each fluid through clear methylation patterns visible in electropherograms and was effective even in mixed or mock crime scene samples. It also differentiated menstrual from venous blood, a critical factor in sexual assault cases.

More recently, Zhao et al. (2025) developed an AI-based strategy to identify and validate DNA methylation markers (CpGs) for forensic body fluid identification. Using DNA methylation data from public GEO databases, the team combined machine learning algorithms (elastic net, random forest, mutual information, and information gain) with epigenetic pattern analysis to screen body fluid-specific CpGs from over 350,000 candidates. Fourteen CpGs were validated via pyrosequencing on blood, saliva, semen, vaginal secretions, and menstrual blood samples, showing strong tissue specificity. Machine learning models built on these markers achieved 100% classification accuracy, with a simplified six-CpG model performing equally well.

Collectively, these studies established that DNA methylation profiling can reliably identify the biological origin of forensic samples, even when the DNA is old, degraded, or part of a mixture. Early enzyme-based approaches like Frumkin's laid the groundwork, while later bisulfite and multiplex-based assays enhanced sensitivity, stability, and practicality for real-world forensic applications. Modern methylation assays now offer automation, high sensitivity, and compatibility with standard forensic DNA workflows, paving the way for a new generation of epigenetic-based body fluid identification systems that are both highly specific and operationally feasible in forensic laboratories.

4.6 Non-coding RNAs in Forensics

The sequencing of the human genome at the turn of the twenty-first century was initially celebrated for identifying approximately 20,000 protein-coding genes. Yet this coding fraction accounts for less than 2% of the human genome. The remaining non-coding regions, once dismissed as "junk DNA," are now recognized as transcriptionally active landscapes producing a wide variety of non-coding RNAs (ncRNAs). These molecules regulate gene expression at transcriptional and post-transcriptional levels, shaping virtually every biological process (Esteller 2011).

Non-coding RNAs (ncRNAs) are a diverse class of RNA molecules that are transcribed from the genome but not translated into proteins. Once considered transcriptional noise, ncRNAs are now recognized as crucial regulators of gene expression and genome stability (Esteller 2011; Quinn and Chang 2016). These RNAs perform a wide array of cellular functions through mechanisms such as chromatin remodeling, transcriptional interference, RNA processing, and mRNA stability. They are broadly categorized into small ncRNAs, such as microRNAs (miRNAs) and small interfering RNAs (siRNAs), and long non-coding RNAs (lncRNAs), which are typically longer than 200 nucleotides (Ponting et al. 2009). Despite lacking coding potential, ncRNAs exert complex regulatory functions that are often highly specific to cell type, developmental stage, or environmental conditions.

In forensic science, the discovery that ncRNAs exhibit both tissue specificity and remarkable stability has generated substantial interest. Unlike messenger RNA (mRNA), which is notoriously labile, certain ncRNAs—particularly microRNAs (miRNAs)—remain intact in dried biological stains for extended periods, even under environmental stress. This section examines the biogenesis, characteristics, and forensic applications of non-coding RNAs, focusing on microRNAs, long non-coding RNAs (lncRNAs), and piwi-interacting RNAs (piRNAs).

4.6.1 *Classes of Non-coding RNAs*

Non-coding RNAs encompass a broad and diverse set of molecules. While many exist, three categories are particularly relevant to forensic applications:

4.6.1.1 MicroRNAs (miRNAs)

MiRNAs are short, ~22 nucleotide-long RNAs derived from endogenous hairpin precursors. They regulate gene expression primarily by binding to complementary sequences in the 3′ untranslated regions of target mRNAs, leading to translational repression or mRNA degradation (Bartel 2009). Their small size and association with RNA-induced silencing complexes (RISCs) confer exceptional stability, making them highly attractive for forensic science.

4.6.1.2 Long Non-coding RNAs (lncRNAs)

LncRNAs are transcripts longer than 200 nucleotides with limited protein-coding potential. They modulate gene expression through diverse mechanisms, including chromatin remodeling, transcriptional interference, and post-transcriptional regulation (Ponting et al. 2009). While less studied than miRNAs in forensics, lncRNAs are increasingly recognized as tissue-specific markers with potential applications in body fluid identification (Pritchard et al. 2012).

4.6.1.3 Piwi-Interacting RNAs (piRNAs)

PiRNAs are slightly longer than miRNAs (24–31 nucleotides) and interact with PIWI proteins to silence transposable elements and regulate germline integrity (Siomi et al. 2011). Their expression is largely restricted to reproductive tissues, suggesting forensic utility in semen identification. However, research into piRNAs as forensic biomarkers remains relatively nascent compared to miRNAs.

4.6.2 Stability of Non-coding RNAs in Forensic Contexts

One of the principal advantages of ncRNAs, especially miRNAs, is their stability in forensic samples. While mRNA degrades rapidly due to endogenous RNases and environmental factors, miRNAs are protected by their association with Argonaute proteins in the RISC complex and by encapsulation within extracellular vesicles (Mitchell et al. 2008).

Studies have shown that miRNAs can be detected in biological stains stored for years at room temperature, subjected to UV exposure, or deposited on diverse substrates such as fabric and wood (Hanson et al. 2009). This robustness is critical for forensic casework, where evidence is often exposed to uncontrolled environmental conditions.

4.6.3 Tissue-Specific Expression of ncRNAs

Tissue-specific expression profiles make ncRNAs particularly valuable for body fluid identification. MiRNA profiling studies have identified markers uniquely enriched in blood, semen, saliva, vaginal fluid, and menstrual blood. For instance:

- Blood: miR-451 and miR-16 are consistently abundant in erythrocytes (Kirschner et al. 2011).
- Semen: miR-10b and miR-891a are enriched in spermatozoa (Hanson et al. 2009).
- Saliva: miR-205 is overrepresented in epithelial-rich oral secretions (Park et al. 2009).
- Menstrual Blood: miR-412 has been associated with endometrial tissue (Zubakov et al. 2010).

The short sequence length of miRNAs also enables analysis from highly degraded samples, where longer RNAs or proteins are no longer intact.

4.6.4 Forensic Applications of ncRNAs

4.6.4.1 Body Fluid Identification

The primary forensic application of ncRNAs is in the identification of body fluids. Panels of miRNAs have been validated to distinguish between common forensic fluids with high accuracy. For example, Zubakov et al. (2010) demonstrated >90% classification accuracy using a four-miRNA panel. These assays can be multiplexed, requiring minimal sample and preserving DNA for parallel STR analysis.

4.6.4.2 Mixed Sample Analysis

One of the persistent challenges in forensic biology is interpreting mixed stains. Because ncRNAs are tissue-specific, their relative abundances can provide clues to mixture composition. For example, the simultaneous detection of blood-specific miR-451 and semen-specific miR-891a suggests a mixture of blood and semen (Hanson et al. 2009). However, quantification in mixtures remains complex and may require advanced bioinformatics approaches.

4.6.4.3 Age and Tissue State Estimation

Beyond fluid identification, ncRNAs may provide insight into tissue state. Some lncRNAs and miRNAs exhibit differential expression depending on tissue maturity, menstrual cycle stage, or pathological states (Pritchard et al. 2012). Forensic exploitation of these signatures could extend beyond simple fluid identification to contextual information about the donor.

4.6.5 Challenges in Forensic Implementation

Despite their promise, several challenges remain in adopting ncRNAs for routine forensic practice.

1. Standardization: Protocols for RNA extraction, reverse transcription, and normalization vary widely across laboratories. Without standardization, interlaboratory reproducibility is limited.
2. Normalization and Controls: Identifying appropriate internal controls for forensic RNA analysis remains an unresolved issue. Unlike mRNA assays, which often rely on housekeeping genes, small RNAs vary substantially across tissues and conditions.

3. Mixture Complexity: While tissue-specific markers exist, quantifying contributions in mixed stains remains technically difficult. More sophisticated statistical and computational tools are needed.
4. Legal Admissibility: For any forensic method to be admissible in court, it must meet standards of scientific validity and general acceptance (Daubert/Frye). NcRNA assays are still in the validation phase, limiting their courtroom applicability.

Non-coding RNAs represent a promising class of forensic biomarkers. Their small size, stability, and tissue-specific expression profiles overcome many limitations of traditional serological assays. While methodological and interpretive challenges remain, particularly with normalization and mixtures, ongoing research suggests that miRNAs and other ncRNAs will soon play an important role in forensic workflows. Integration with DNA methylation assays offers an especially powerful strategy, leveraging the complementary strengths of each biomarker class.

4.7 Methodologies for Analyzing Non-coding RNAs

A wide range of tools are available for RNA detection and quantification. Broadly, these methods can be divided into those that use direct oligonucleotide hybridization without RNA amplification and those that incorporate amplification following RNA extraction. Approaches that do not include amplification generally require a larger starting amount of total RNA (Kong et al. 2009). The most widely used techniques for RNA analysis are based on three main technologies: probe hybridization, RNA sequencing, and reverse transcription quantitative polymerase chain reaction (RT-qPCR).

Northern blotting is a probe hybridization-based process that begins with the extraction of RNA, which is then separated by size through agarose gel electrophoresis. The separated RNA fragments are transferred onto a solid support membrane, followed by incubation with a labeled probe. After hybridization, the unbound probe is washed away, and the membrane is analyzed to visualize the labeled probe, often using radioactive or fluorescent detection methods. The resulting image displays bands at positions where the probe has bound, revealing both the presence and size of the target RNA molecules in the sample (Streit et al. 2009; Beckman et al. 2010). Another technique is the bead-based method approach which integrates amplification and hybridization steps, with analysis performed by flow cytometry. miRNA-specific capture probes are immobilized on uniquely fluorescent-coded beads. Following enrichment of total RNA for short RNAs, adapters are ligated to both RNA ends, and reverse transcription using adapter-specific primers generates a cDNA library. The resulting cDNAs are then amplified by PCR with biotinylated primers, enabling detection through streptavidin–phycoerythrin fluorescence. Finally, the amplified cDNAs hybridize with the bead library, and flow cytometry detects bead fluorescence to identify each miRNA and measures streptavidin–phycoerythrin intensity to quantify its abundance (Kong et al. 2009). RNA suspension fluorescence in situ hybridization (RNA S-FISH) is a technique that employs fluorescently labeled probes to detect and quantify specific RNA molecules within individual cells maintained in suspension. Cells are first fixed and permeabilized to enable probe entry. Fluorescent nucleic acid probes are then introduced to hybridize specifically with the target RNA sequences, rendering them visible. The suspended cells are subsequently analyzed using an imaging flow cytometer. The resulting fluorescence images are processed using specialized software to measure RNA signal intensity across large populations of individual cells (Williams et al. 2014; Young et al. 2020). Quantitative PCR (qPCR) is a widely used method for quantifying ncRNAs, particularly when combined with reverse transcription (RT-qPCR). This approach is especially effective for long ncRNAs, while short ncRNAs such as microRNAs (miRNAs) require an additional step to extend the RNA molecule prior to reverse transcription. In general, RT-qPCR involves three main steps: RNA isolation, cDNA synthesis, and real-time PCR amplification to detect and quantify the target ncRNA (Kong et al. 2009; Gordanpour et al. 2012). High-throughput techniques, such as microarray analysis, can be used to detect and measure the expression levels of miRNAs. Following RNA isolation, an enrichment step is essential because these molecules are small and present in low abundance. The extracted RNA is then hybridized onto an array containing capture probes, usually composed of oligonucleotides or complementary DNA (cDNA) fragments. The resulting data are analyzed to determine differential miRNA expression, predict target genes, and perform functional annotation (Yin et al. 2008). And finally, RNA sequencing of non-coding RNA (ncRNA) begins with preparing RNA samples by removing highly abundant RNAs such as ribosomal RNA (rRNA), followed by high-throughput sequencing using methods such as Next Generation Sequencing (NGS). The process includes library preparation, sequencing, and bioinformatic analysis. Library preparation strategies vary based on ncRNA size. For instance, poly-A enrichment may be used for certain long non-coding RNAs (lncRNAs), ribosomal RNA depletion for others, and size selection for small RNAs. Afterward, the RNAs are fragmented, reverse-transcribed into complementary DNA (cDNA), and subjected to end repair, adaptor ligation, and size selection before sequencing. The resulting data is analyzed to identify and quantify ncRNAs (Sun and Chen 2020).

Although RNA analyses are not yet routinely used in crime laboratories, the study of non-coding RNAs has the potential to become an important tool in forensic investigations. In particular, miRNA profiling offers a sensitive and efficient approach that can complement existing forensic techniques once key challenges, such as the lack of standardized protocols across laboratories, are addressed.

4.8 Non-coding RNAs and Body Fluid Identification

Zubakov et al. (2010) were among the first to test whether miRNAs could reliably distinguish body fluids in forensic work. Screening 718 human miRNAs, they found clear, fluid-specific expression patterns for blood, menstrual blood, semen, saliva, and vaginal secretions. But when key candidates were validated by quantitative RT-PCR, only venous blood and semen produced consistent markers. Blood was identified by miR-20a, miR-106a, miR-185, and miR-144; semen by miR-135a, miR-10a, miR-507, miR-943, and especially miR-891a, which proved highly semen-specific despite misleading microarray signals. Markers for saliva, menstrual blood, and vaginal secretions failed validation. The confirmed blood and semen markers were stable for at least a year and detectable from as little as 0.1 pg of RNA. Although some miRNAs shared homology with non-human species, human-specific controls support their forensic use.

In 2012, Wang et al. (2012) aimed to improve the accuracy of miRNA-based body fluid identification by applying a more rigorous analysis model that incorporates each assay's PCR efficiency rather than assuming identical efficiencies. They tested miR16, miR205, and miR658 across blood, semen, saliva, vaginal secretions, menstrual blood, and buccal swabs. With their efficiency-calibrated approach, only miR16 reliably distinguished venous blood, showing consistently higher expression than in all other fluids. miR658 was unstable and often undetectable, while miR205 lacked specificity, being highly expressed across multiple epithelial fluids. The method showed strong sensitivity, detecting miR16 from as little as 50 pg of RNA.

Wang et al. (2013) screened 754 human miRNAs across blood, semen, menstrual blood, vaginal secretions, and saliva to identify fluid-specific candidates. Seven markers emerged, and qPCR validation confirmed five: miR16 and miR486 for blood, miR888 and miR891a for semen, and miR214 for menstrual blood. The proposed vaginal-secretion (miR124a) and saliva (miR138–2) markers did not validate. A key contribution of the study was testing four reference genes (RNU44, RNU48, U6, U6b); only U6 showed stable expression across all fluids, making it the most reliable normalizer for forensic miRNA assays, according to the authors. Each validated marker showed strong over-expression in its target fluid, was detectable from as little as 10 pg RNA, and retained stable relative patterns after 1 month of aging, supporting its suitability for trace and environmentally exposed forensic samples.

In 2015, Tong and co-authors tested whether two miRNAs, miR10b and miR135b, could reliably identify semen stains. Using qPCR across semen, vaginal fluid, menstrual blood, and peripheral blood, they found both miRNAs were expressed at much higher levels in semen than in other fluids. The method was sensitive to 10 ng of RNA and showed strong stability: most stains stored for up to a year at room temperature remained detectable with consistent expression levels. Although the markers occasionally produced false positives in menstrual blood or vaginal swabs, the overall results support their usefulness for identifying semen, even in aged or degraded forensic samples.

Sauer et al. (2016) screened 36 candidate miRNAs and validated them through rigorous qPCR testing, including efficiency checks, reference gene assessment, and statistical analysis. Clustering showed clear separation of most body fluids, though menstrual blood was variable. Four miRNAs proved reliably informative: miR-891a-5p as a semen-specific marker; miR-144-3p to distinguish blood from non-blood fluids; and miR-203a-3p and miR-124a-3p for separating venous vs. menstrual blood and saliva vs. vaginal secretions, respectively. Using these markers, the authors built a stepwise classification algorithm that accurately identified 9 of 10 blinded samples and performed well on stains decades old, demonstrating strong miRNA stability. Tests on mixed samples confirmed reliable detection of semen and blood, though distinguishing saliva from vaginal fluid remained difficult due to overlapping expression profiles.

The study by Sirker et al. (2017) tested 19 miRNAs for identifying six body fluids and tissues. After validating stable reference genes (miR26b, miR92, miR484, and miR144), they reassessed 15 proposed forensic markers. Only six showed the expected patterns, and just two, miR451 for blood and miR10b for semen, displayed clear specificity. Many markers overlapped across similar fluids, making reliable classification difficult. ROC (receiver operating characteristic) analysis showed only five miRNAs (miR10b, miR203, miR374, miR451, miR943) could perfectly distinguish at least one pair of fluids.

Fujimoto et al. (2019) showed that miRNA profiles can distinguish venous blood, saliva, semen, and vaginal secretions using a probabilistic model based on RT-qPCR data. After testing 15 candidates, 10 miRNAs met quality criteria and revealed clear patterns: miR-144-3p and miR-451a-5p identified blood, while miR-888-5p and miR-891a-5p reliably marked semen. Saliva and vaginal secretions were harder to separate because their markers overlapped, so the authors built a multivariate model using seven informative miRNAs. It correctly classified 13 of 14 blinded samples, with only one vaginal secretion sample ambiguously clustering with saliva. The study also confirmed that skin contamination, azoospermia, and menstrual-cycle variation did not compromise marker performance. Overall, miRNAs proved reliable for identifying blood and semen, and saliva vs. vaginal secretions could be distinguished using the probabilistic approach.

More recently in 2024, Altmeyer and co-authors investigated a selected set of miRNA markers to distinguish peripheral blood, menstrual blood, saliva, semen, and vaginal secretion, using standardized RT-qPCR methods. The results revealed strong tissue-specific signals for several markers: miR-144-3p and miR-451a were highly expressed in both venous and menstrual blood; miR-214-3p was specific to menstrual blood; miR-888-5p and miR-891a-5p were exclusive to semen; and miR-193b-3p and miR-1260b were strongly expressed in vaginal secretion. The authors performed principal component analysis (PCA) and unsupervised hierarchical clustering, which showed clear separation of the five body fluids with no overlap between clusters, indicating that the combination of markers produces distinct fluid-specific expression signatures.

In the study by Li et al. (2025), a full miRNA-based workflow was developed for body fluid identification using RT-qPCR data and a multi-class SVM (support vector machine) model. After evaluating six reference genes across five body fluids, they identified miR-484 and miR-191-5p as the most stable for normalization. They then analyzed 28 proposed markers, confirming strong specificity for semen markers (miR-891a-5p, miR-888-5p) but finding substantial overlap for saliva and vaginal secretion, and limited ability to distinguish peripheral from menstrual blood. Many published markers failed to replicate consistently. Using SVM feature weighting, the authors selected 12 informative miRNAs for classification. The resulting model achieved 98% accuracy on 60 test samples. In real forensic samples, 8 of 13 single-source stains were correctly identified despite age and substrate variation, though some predictions had low confidence. Mixed-sample testing showed the model could often detect at least one major component, particularly semen, but frequently misclassified vaginal secretion as saliva.

The above studies consistently showed that miRNAs persist in aged or exposed stains, making them suitable for degraded forensic samples. However, because miRNAs show relative rather than absolute fluid specificity, reliable use requires multi-marker panels and statistical models. Despite their promise, routine forensic use still depends on further standardization, validation, and larger datasets, especially for complex samples.

4.9 Concluding Remarks

Non-coding RNAs, especially miRNAs, and DNA methylation have both emerged as powerful molecular approaches for forensic body fluid identification, and together they offer a more complete and reliable strategy than either method alone. Research has shown that miRNAs are well-suited for forensic analysis due to their structure and relative abundance, allowing detection in aged or environmentally compromised samples. They exhibit distinct expression patterns across fluids such as blood, semen, saliva, and vaginal secretions. However, most miRNAs are typically overexpressed in certain body fluids but not exclusively present in them, and overlapping expression in fluids with similar cellular composition can complicate interpretation. Despite these challenges, miRNA-based classification becomes highly effective when multiple markers are analyzed simultaneously using machine-learning models, which can capture subtle combinational patterns across many miRNAs.

DNA methylation, on the other hand, provides epigenetic signatures that tend to be more tissue-specific and more robustly preserved across environmental conditions. Methylation markers have shown strong discriminatory power for fluids like peripheral blood, semen, and vaginal secretions, and they can sometimes resolve distinctions that miRNAs struggle with, such as differentiating menstrual from peripheral blood. Although methylation analysis requires more complex laboratory processing, it offers specificity that complements the characteristics of miRNAs, making their combined use a promising approach.

Looking into the future and the emergence of new technologies, artificial intelligence (AI) could play an important role in advancing epigenetic body fluid identification, particularly when forensic laboratories need to interpret mixed biological samples. Epigenetic markers such as DNA methylation patterns can generate complex datasets, and when multiple body fluids are present in a sample, their methylation signals overlap in ways that challenge traditional interpretation methods. AI, especially machine-learning models, can be well-suited to analyze complex data and can learn to distinguish the methylation signatures of different body fluids.

Advances in sequencing technologies are increasing both the volume and resolution of methylation data, heightening the need for robust computational analysis tools.

Ultimately, the future of body fluid identification is likely to rely on the continued integration of emerging technologies aimed at improving the precision and objectivity of data interpretation. As validation studies continue and analytical tools further develop, these methods may gradually be incorporated into forensic casework, pending rigorous evaluation and standardization.

Compliance with Ethical Standards Conflict of Interest: Georgiana Gibson-Daw declares that she has no conflict of interest. Deborah S.B.S. Silva declares that she has no conflict of interest. Bruce McCord declares that he has no conflict of interest.

References

Allery JP, Telman N, Mieusset R, Blanc A, Rough D (2001) Cytological detection of spermatozoa: comparison of three staining methods. J Forensic Sci 46(2):349–351

Altmeyer L, Baumer K, Hall D (2024) Differentiation of five forensically relevant body fluids using a small set of microRNA markers. Electrophoresis 45(19–20):1785–1795. https://doi.org/10.1002/elps.202400089

American National Standards Institute/Academy Standards Board (ANSI/ASB) (2020) Standard for the developmental and internal validation of forensic serological methods, 1st edn. ANSI/ASB Standard 077. https://www.aafs.org/sites/default/files/media/documents/077_Std_e1.pdf

An JH, Shin KJ, Yang WI, Lee HY (2012) Body fluid identification in forensics. BMB Rep 45:545–553. https://doi.org/10.5483/bmbrep.2012.45.10.206

An JH, Choi A, Shin KJ, Yang WI, Lee HY (2013) DNA methylation-specific multiplex assays for body fluid identification. Int J Legal Med 127(1):35–43. https://doi.org/10.1007/s00414-012-0719-1

Barni F, Lewis SW, Berti A, Miskelly GM, Lago G (2007) Forensic application of the luminol reaction as a presumptive test for latent blood detection. Talanta 72(3):896–913. https://doi.org/10.1016/j.talanta.2006.12.045

Bartel DP (2009) MicroRNAs: target recognition and regulatory functions. Cell 136(2):215–233. https://doi.org/10.1016/j.cell.2009.01.002

Beckmann BM, Grünweller A, Weber MH, Hartmann RK (2010) Northern blot detection of endogenous small RNAs (approximately 14 nt) in bacterial total RNA extracts. Nucleic Acids Res 38(14):e147. https://doi.org/10.1093/nar/gkq437

Beikircher G, Pulverer W, Hofner M, Noehammer C, Weinhaeusel A (2018) Multiplexed and sensitive DNA methylation testing using methylation-sensitive restriction enzymes (MSRE-qPCR). In: Methods in molecular biology, vol 1708, pp 407–424. https://doi.org/10.1007/978-1-4939-7481-8_21

Bibikova M, Barnes B, Tsan C, Ho V, Klotzle B, Le JM, Delano D, Zhang L, Schroth GP, Gunderson KL, Fan J, Shenet R (2011) High density DNA methylation array with single CpG site resolution. Genomics 98:288–295. https://doi.org/10.1016/j.ygeno.2011.07.007

Bird A (2002) DNA methylation patterns and epigenetic memory. Genes Dev 16(1):6–21. https://doi.org/10.1101/gad.947102

Bird A (2007) Perceptions of epigenetics. Nature 447:396–398. https://doi.org/10.1038/nature05913

Budowle B, Schutzer SE, Morse SA, Martinez KF, Chakraborty R, Marrone BL, Messenger SL, Murch RS, Jackson PJ, Williamson P, Harmon R, Velsko SP (2008) Criteria for validation of methods in microbial forensics. Appl Environ Microbiol 74(18):5599–5607. https://doi.org/10.1128/AEM.00966-08

Bunescu S, Stoica BA, Peptanariu D, Foia L (2021) A sensitive method for saliva detection in forensics using salivary amylase coupled with Amplex Red oxidation. J Exp Mol Biol 21(2):41–46

Butler JM (2011) Advanced topics in forensic DNA typing: methodology. Academic

Campan M, Weisenberger DJ, Trinh B, Laird PW (2018) MethyLight and digital MethyLight. In: Methods in molecular biology, vol 1708, pp 497–513. https://doi.org/10.1007/978-1-4939-7481-8_25

Castagnola MJ, Medina-Paz F, Zapico SC (2024) Uncovering forensic evidence: a path to age estimation through DNA methylation. Int J Mol Sci 25(9):4917. https://doi.org/10.3390/ijms25094917

Cuttaia C, Di Stefano B, Sorçaburu Ciglieri S, Vetrini R, Previderè C, Fattorini P (2024) Immunochromatographic detection of human blood: a forensic review. Separations 11(3):66. https://doi.org/10.3390/separations11030066

Dass M, Singh Y, Ghai M (2023) A review on microbial species for forensic body fluid identification in healthy and diseased humans. Curr Microbiol 80(9):299. https://doi.org/10.1007/s00284-023-03413-x

Derks S, Lentjes MH, Hellebrekers DM, de Bruïne AP, Herman JG, van Engeland M (2004) Methylation-specific PCR unraveled. Cell Oncol 26(5–6):291–299. https://doi.org/10.1155/2004/370301

Esteller M (2011) Non-coding RNAs in human disease. Nat Rev Genet 12(12):861–874. https://doi.org/10.1038/nrg3074

Feil R, Fraga MF (2012) Epigenetics and the environment: emerging patterns and implications. Nat Rev Genet 13(2):97–109. https://doi.org/10.1038/nrg3142

Felsenfeld G (2014) A brief history of epigenetics. Cold Spring Harb Perspect Biol 6(1):a018200. https://doi.org/10.1101/cshperspect.a018200

Fonseca RIB, Ricci EL, Spinosa HS, Bernardi MM, de Abreu GR, Waziry PAF et al (2019) Actual trends in the use of the Kastle–Meyer test: applications in different species and verification of the limit of detection of sensitivity and vestigiality. J Dairy Vet Anim Res 8:166–180. https://doi.org/10.15406/jdvar.2019.08.00261

Forat S, Huettel B, Reinhardt R, Fimmers R, Haidl G, Denschlag D, Olek K (2016) Methylation markers for the identification of body fluids and tissues from forensic trace evidence. PLoS One 11(5):e0156472. https://doi.org/10.1371/journal.pone.0156472

Frommer M, McDonald LE, Millar DS, Collis CM, Watt F, Grigg GW, Molloy PL, Paul CL (1992) A genomic sequencing protocol that yields a positive display of 5-methylcytosine residues in individual DNA strands. Proc Natl Acad Sci USA 89(5):1827–1831. https://doi.org/10.1073/pnas.89.5.1827

Frumkin D, Wasserstrom A, Budowle B, Davidson A (2011) DNA methylation-based forensic tissue identification. Forensic Sci Int Genet 5(5):517–524. https://doi.org/10.1016/j.fsigen.2010.12.001

Fujimoto S, Manabe S, Morimoto C, Ozeki M, Hamano Y, Hirai E, Kotani H, Tamaki K (2019) Distinct spectrum of microRNA expression in forensically relevant body fluids and probabilistic discriminant approach. Sci Rep 9:14332. https://doi.org/10.1038/s41598-019-50796-8

Gaensslen RE (1983) Sourcebook in forensic serology, immunology, and biochemistry. United States Department of Justice

Gattuso G, Lavoro A, Caltabiano R, Madonna G, Capone M, Ascierto PA, Falzone L, Libra M, Candido S (2024) Methylation-sensitive restriction enzyme droplet digital PCR assay for the one step highly sensitive analysis of DNA methylation hotspots. Int J Legal Med 53(5):42. https://doi.org/10.3892/ijmm.2024.5366

Ghemrawi M, Ballantyne J, McCord B, Parker G (2024) Comparative assessment of emerging technologies for body fluid identification, NCJ No. 308841. National Institute of Justice. https://nij.ojp.gov/library/publications/comparative-assessment-emerging-technologies-body-fluid-identification

Golomingi R, Haas C, Dobay A, Kottner S, Ebert L (2022) Sperm hunting on optical microscope slides for forensic use: the challenge of azoospermia and vasectomy. Forensic Chem 30:Article 100401. https://doi.org/10.1016/j.fsigen.2021.102602

Gong T, Borgard H, Zhang Z, Chen S, Gao Z, Deng Y (2022) Analysis and performance assessment of the whole genome bisulfite sequencing data workflow: currently available tools and a practical guide to advance DNA methylation studies. Small Methods 6(3):e2101251. https://doi.org/10.1002/smtd.202101251

Gordanpour A, Nam RK, Sugar L, Bacopulos S, Seth A (2012) MicroRNA detection in prostate tumors by quantitative real-time PCR (qPCR). JoVE 63:*e3874*. https://doi.org/10.3791/3874

Hanson EK, Ballantyne J, Callaghan T (2009) Identification of forensically relevant body fluids using a panel of differentially expressed microRNAs. Anal Biochem 387(2):303–314. https://doi.org/10.1016/j.ab.2009.01.021

Harbison S, Fleming R (2016) Forensic body fluid identification: state of the art. Res Rep Forensic Med Sci 6:11–23. https://doi.org/10.2147/RRFMS.S57994

Hausmann R, Pregler C, Schellmann B (1994) The value of the Lugol's iodine staining technique for the identification of vaginal epithelial cells. Int J Legal Med 106(6):298–301. https://doi.org/10.1007/BF01224775

Hedman J, Dalin E, Rasmusson B, Ansell R (2011) Evaluation of amylase testing as a tool for saliva screening of crime scene trace swabs. Forensic Sci Int Genet 5(3):194–198. https://doi.org/10.1016/j.fsigen.2010.03.003

Hochmeister MN, Budowle B, Sparkes R, Rudin O, Gehrig C, Thali M, Schmidt L, Cordier A, Dirnhofer R (1999) Validation studies of an immunochromatographic 1-step test for the forensic identification of human blood. J Forensic Sci 44(3):597–601. https://doi.org/10.1520/JFS14516J

Holtkötter H, Beyer V, Schwender K, Glaub A, Johann KS, Schürenkamp M, Sibbing U, Banken S, Wiegand P, Pfeiffer H, Dennany L, Vennemann M, EUROFORGEN-NoE Consortium (2017) Independent validation of body fluid-specific CpG markers and construction of a robust multiplex assay. Forensic Sci Int Genet 29:261–268. https://doi.org/10.1016/j.fsigen.2017.05.002

Horvath S (2013) DNA methylation age of human tissues and cell types. Genome Biol 14(10):R115. https://doi.org/10.1186/gb-2013-14-10-r115

Hsu HK, Weng YI, Hsu PY, Huang TH, Huang YW (2020) Detection of DNA methylation by MeDIP and MBDCap assays: an overview of techniques. In: Methods in molecular biology, pp 225–234. https://doi.org/10.1007/978-1-0716-0223-2_12

Hussmann D, Hansen LL (2018) Methylation-sensitive high resolution melting (MS-HRM). In: Tost J (ed) DNA methylation protocols, Methods in molecular biology, vol 1708. Humana Press. https://doi.org/10.1007/978-1-4939-7481-8_28

Jones PA (2012) Functions of DNA methylation: islands, start sites, gene bodies and beyond. Nat Rev Genet 13(7):484–492. https://doi.org/10.1038/nrg3230

Kaye DH (2001) The dynamics of Daubert: methodology, conclusions, and fit in statistical and econometric studies. Va L Rev 87(6):1933–1976. https://doi.org/10.2307/1073909

Kayser M (2026) Forensic genetics in the omics era. Nat Rev Genet 27:170–186. https://doi.org/10.1038/s41576-025-00896-1

Kirschner MB, Kao SC, Edelman JJB, Armstrong NJ, Vallely MP, van Zandwijk N, Reid G (2011) Haemolysis during sample preparation alters microRNA content of plasma. PLoS One 6(9):e24145. https://doi.org/10.1371/journal.pone.0024145

Kong W, Zhao JJ, He L, Cheng JQ (2009) Strategies for profiling microRNA expression. J Cell Physiol 218(1):22–25. https://doi.org/10.1002/jcp.21577

Lee HY, Park MJ, Choi A, An JH, Yang WI, Shin KJ (2012) Potential forensic application of DNA methylation profiling to body fluid identification. Int J Legal Med 126(1):55–62. https://doi.org/10.1007/s00414-011-0569-2

Lee HY, An JH, Jung SE, Oh YN, Lee EY, Choi A, Yang WI, Shin KJ (2015) Genome-wide methylation profiling and a multiplex construction for the identification of body fluids using epigenetic markers. Forensic Sci Int Genet 17:17–24. https://doi.org/10.1016/j.fsigen.2015.03.002

Lehne B, Drong AW, Loh M, Zhang W, Scott WR, Tan S, Afzal U, Scott J, Jarvelin M, Elliott P, McCarthy MI, Kooner JS, Chambers JC (2015) A coherent approach for analysis of the Illumina HumanMethylation450 BeadChip improves data quality and performance in epigenome-wide association studies. Genome Biol 16:37. https://doi.org/10.1186/s13059-015-0600-x

Li Z, Chen D, Wang Q, Tian H, Tan M, Peng D, Tan Y, Zhu J, Liang W, Zhang L (2021) mRNA and microRNA stability validation of blood samples under different environmental conditions. Forensic Sci Int Genet 55:102567. https://doi.org/10.1016/j.fsigen.2021.102567

Li Y, Tollefsbol TO (2011) DNA methylation detection: bisulfite genomic sequencing analysis. In: Methods in molecular biology, vol 791, pp 11–21. https://doi.org/10.1007/978-1-61779-316-5_2

Li S, Tollefsbol TO (2021) DNA methylation methods: global DNA methylation and methylomic analyses. Methods 187:28–43. https://doi.org/10.1016/j.ymeth.2020.10.002

Li S, Liu J, Xu W, Zhang S, Zhao M, Miao L, Hui M, Wang Y, Hou Y, Cong B, Wang Z (2025) A multi-class support vector machine classification model based on 14 microRNAs for forensic body fluid identification. Forensic Sci Int Genet 75:103180. https://doi.org/10.1016/j.fsigen.2024.103180

Lin W, Xiao X, Zhang H, Li Y, Liu S, Sun W, Zhang X, Wu Q (2019) Whole-genome bisulfite sequencing reveals a role for DNA methylation in variants from callus culture of pineapple (*Ananas comosus* L.). Gene 10(11):877. https://doi.org/10.3390/genes10110877

Madi T, Balamurugan K, Bombardi R, Duncan G, McCord B (2012) The determination of tissue-specific DNA methylation patterns in forensic biofluids using bisulfite modification and pyrosequencing. Electrophoresis 33:1736–1745. https://doi.org/10.1002/elps.201100711

Mayne BT, Leemaqz SY, Buckberry S, Rodriguez Lopez CM, Roberts CT, Bianco-Miotto T, Breen J (2018) msgbsR: an R package for analysing methylation-sensitive restriction enzyme sequencing data. Sci Rep 8:2190. https://doi.org/10.1038/s41598-018-19655-w

Mitchell PS, Parkin RK, Kroh EM, Fritz BR, Wyman SK, Pogosova-Agadjanyan EL, Peterson A, Noteboom J, O'Briant KC, Allen A, Lin DW, Urban N, Drescher CW, Knudsen BS, Stirewalt DL, Gentleman R, Vessella RL, Nelson PS, Martin DB, Tewari M (2008) Circulating microRNAs as stable blood-based markers for cancer detection. Proc Natl Acad Sci USA 105(30):10513–10518. https://doi.org/10.1073/pnas.0804549105

Myers JR, Adkins WK (2008) Comparison of modern techniques for saliva screening. J Forensic Sci 53(4):862–867. https://doi.org/10.1111/j.1556-4029.2008.00755.x

Nakabayashi K, Yamamura M, Hasegawa K, Hata K (2023) Reduced representation bisulfite sequencing (RRBS). In: Hatada I, Horii T (eds) Epigenomics, Methods in molecular biology, vol 2577. Humana. https://doi.org/10.1007/978-1-0716-2724-2_3

Olkhov-Mitsel E, Zdravic D, Kron K, van der Kwast T, Fleshner N, Bapat B (2014) Novel multiplex MethyLight protocol for detection of DNA methylation in patient tissues and bodily fluids. Sci Rep 4:4432. https://doi.org/10.1038/srep04432

Pajares MJ, Palanca-Ballester C, Urtasun R, Alemany-Cosme E, Lahoz A, Sandoval J (2021) Methods for analysis of specific DNA methylation status. Methods 187:3–12. https://doi.org/10.1016/j.ymeth.2020.06.021

Park NJ, Zhou H, Elashoff D, Henson BS, Kastratovic DA, Abemayor E, Wong DT (2009) Salivary microRNA: discovery, characterization, and clinical utility for oral cancer detection. Clin Cancer Res 15(17):5473–5477. https://doi.org/10.1158/1078-0432.CCR-09-0736

Park JL, Kwon OH, Kim JH, Yoo HS, Lee HC, Woo KM, Kim SY, Lee SH, Kim YS (2014) Identification of body fluid-specific DNA methylation markers for use in forensic science. Forensic Sci Int Genet 13:147–153. https://doi.org/10.1016/j.fsigen.2014.07.011

Ponting CP, Oliver PL, Reik W (2009) Evolution and functions of long noncoding RNAs. Cell 136(4):629–641. https://doi.org/10.1016/j.cell.2009.02.006

Pritchard CC, Cheng HH, Tewari M (2012) MicroRNA profiling: approaches and considerations. Nat Rev Genet 13(5):358–369. https://doi.org/10.1038/nrg3198

Quinn JJ, Chang HY (2016) Unique features of long non-coding RNA biogenesis and function. Nat Rev Genet 17(1):47–62. https://doi.org/10.1038/nrg.2015.10

Raju PS (1964) Acid phosphatase reaction as a specific test for the identification of seminal fluid. J Crim L Criminol 55(2):184–188

Randall B, Riis RE (1985) Penile glycogenated epithelial cells as an indicator of recent vaginal intercourse. Am J Clin Pathol 84(4):524–526. https://doi.org/10.1093/ajcp/84.4.524

Reed K, Poulin ML, Yan L, Parissenti AM (2010) Comparison of bisulfite sequencing PCR with pyrosequencing for measuring differences in DNA methylation. Anal Biochem 397(1):96–106. https://doi.org/10.1016/j.ab.2009.10.021

Sabeeha S (2019) Forensic epigenetic analysis: the path ahead. Med Princ Pract 28(4):301–308. https://doi.org/10.1159/000499496

Saferstein R (2011) Criminalistics: an introduction to forensic science, 11th edn. Pearson

Sauer E, Reinke AK, Courts C (2016) Differentiation of five body fluids from forensic samples by expression analysis of four microRNAs using quantitative PCR. Forensic Sci Int Genet 22:89–99. https://doi.org/10.1016/j.fsigen.2016.01.018

Sensabaugh GF (1978) Isolation and characterization of a semen-specific protein from human seminal plasma: a potential new marker for semen identification. J Forensic Sci 23(1):106–115

Sensabaugh GF, Blake ET (1990) Seminal plasma protein p30: simplified purification and evidence for identity with prostate specific antigen. J Urol 144(6):1523–1526. https://doi.org/10.1016/S0022-5347(17)39790-2

Šestáková Š, Šálek C, Remešová H (2019) DNA methylation validation methods: a coherent review with practical comparison. Biol Proced Online 21:19. https://doi.org/10.1186/s12575-019-0107-z

Shearstone JR, Pop R, Bock C, Boyle P, Meissner A, Socolovsky M (2011) Global DNA demethylation during erythropoiesis in vivo. Science 334(6057):799–802. https://doi.org/10.1126/science.1207306

Silva DSBS, Antunes J, Balamurugan K, Duncan G, Alho CS, McCord B (2016) Developmental validation studies of epigenetic DNA methylation markers for the detection of blood, semen and saliva samples. Forensic Sci Int Genet 23:55–63. https://doi.org/10.1016/j.fsigen.2016.01.017

Siomi MC, Sato K, Pezic D, Aravin AA (2011) PIWI-interacting small RNAs: the vanguard of genome defence. Nat Rev Mol Cell Biol 12(4):246–258. https://doi.org/10.1038/nrm3089

Sirker M, Fimmers R, Schneider PM, Gomes I (2017) Evaluating the forensic application of 19 target microRNAs as biomarkers in body fluid and tissue identification. Forensic Sci Int Genet 27:41–49. https://doi.org/10.1016/j.fsigen.2016.11.012

Streit S, Michalski C, Erkan M, Kleeff J, Friess H (2009) Northern blot analysis for detection and quantification of RNA in pancreatic cancer cells and tissues. Nat Protoc 4:37–43. https://doi.org/10.1038/nprot.2008.216

Sun YM, Chen YQ (2020) Principles and innovative technologies for decrypting noncoding RNAs: from discovery and functional prediction to clinical application. J Hematol Oncol 13(1):109. https://doi.org/10.1186/s13045-020-00945-8

Tahiliani M, Koh KP, Shen Y, Pastor WA, Bandukwala H, Brudno Y, Agarwal S, Iyer LM, Liu DR, Aravind L, Rao A (2009) Conversion of 5-methylcytosine to 5-hydroxymethylcytosine in mammalian DNA by MLL partner TET1. Science 324(5929):930–935. https://doi.org/10.1126/science.1170116

Taiwo O, Wilson GA, Morris T, Seisenberger S, Reik W, Pearce D, Beck S, Butcher LM (2012) Methylome analysis using MeDIP-seq with low DNA concentrations. Nat Protoc 7:617–636. https://doi.org/10.1038/nprot.2012.012

Tong D, Jin Y, Xue T, Ma X, Zhang J, Ou X, Cheng J, Sun H (2015) Investigation of the application of miR-10b and miR-135b in the identification of semen stains. PLoS One 10(9):e0137067. https://doi.org/10.1371/journal.pone.0137067

Tost J, Gut IG (2007) DNA methylation analysis by pyrosequencing. Nat Protoc 2(9):2265–2275. https://doi.org/10.1038/nprot.2007.314

Van Wesenbeeck L, Janssens L, Meeuws H, Lagatie O, Stuyver L (2018) Droplet digital PCR is an accurate method to assess methylation status on FFPE samples. Epigenetics 13(3):207–213. https://doi.org/10.1080/15592294.2018.1448679

Vidaki A, Kayser M (2018) Recent progress, methods and perspectives in forensic epigenetics. Forensic Sci Int Genet 37:180–195. https://doi.org/10.1016/j.fsigen.2018.08.008

Wang Z, Luo H, Pan X, Liao M, Hou Y (2012) A model for data analysis of microRNA expression in forensic body fluid identification. Forensic Sci Int Genet 6(3):419–423. https://doi.org/10.1016/j.fsigen.2011.08.008

Wang Z, Zhang J, Luo H, Ye Y, Yan J, Hou Y (2013) Screening and confirmation of microRNA markers for forensic body fluid identification. Forensic Sci Int Genet 7(1):116–123. https://doi.org/10.1016/j.fsigen.2012.07.006

Wasserstrom A, Frumkin D, Davidson A, Shpitzen M, Herman Y, Gafny R (2013) Demonstration of DSI-semen—a novel DNA methylation-based forensic semen identification assay. Forensic Sci Int Genet 7(1):136–142. https://doi.org/10.1016/j.fsigen.2012.08.009

Weidner CI, Lin Q, Koch CM, Eisele L, Beier F, Ziegler P, Bauerschlag DO, Jöckel K-H, Erbel R, Mühleisen TW, Zenke M, Brümmendorf TH, Wagner W (2014) Aging of blood can be tracked by DNA methylation changes at just three CpG sites. Genome Biol 15(2):R24. https://doi.org/10.1186/gb-2014-15-2-r24

Williams E, Lin MH, Harbison S, Fleming R (2014) The development of a method of suspension RNA-FISH for forensically relevant epithelial cells using LNA probes. Forensic Sci Int Genet 9:85–92. https://doi.org/10.1016/j.fsigen.2013.11.007

Wojdacz TK, Dobrovic A (2007) Methylation-sensitive high resolution melting (MS-HRM): a new approach for sensitive and high-throughput assessment of methylation. Nucleic Acids Res 35(6):e41. https://doi.org/10.1093/nar/gkm013

Yin JQ, Zhao RC, Morris KV (2008) Profiling microRNA expression with microarrays. Trends Biotechnol 26(2):70–76. https://doi.org/10.1016/j.tibtech.2007.11.007

Young AP, Jackson DJ, Wyeth RC (2020) A technical review and guide to RNA fluorescence in situ hybridization. PeerJ 8:e8806. https://doi.org/10.7717/peerj.8806

Zeilinger S, Kühnel B, Klopp N, Baurecht H, Kleinschmidt A, Gieger C, Weidinger S, Lattka E, Adamski J, Peters A, Strauch K, Waldenberger M, Illig T (2013) Tobacco smoking leads to extensive genome-wide changes in DNA methylation. PLoS One 8(5):e63812. https://doi.org/10.1371/journal.pone.0063812

Zhao M, Cai M, Lei F, Yuan X, Liu Q, Fang Y, Zhu B (2025) AI-driven feature selection and epigenetic pattern analysis: a screening strategy of CpGs validated by pyrosequencing for body fluid identification. Forensic Sci Int 367:112339. https://doi.org/10.1016/j.forsciint.2024.112339

Zhou Y, Wang Y, Song M, Jiang L, Sun C, Wang S, Yao H, Wang Z, Wang X, Liu C, Luo H, Song F (2024) A high-throughput droplet digital PCR system aiming eight DNA methylation targets for age prediction. J Pharm Biomed Anal 240:115943. https://doi.org/10.1016/j.jpba.2023.115943

Zubakov D, Boersma AWM, Choi Y, van Kuijk PF, Wiemer EAC, Kayser M (2010) MicroRNA markers for forensic body fluid identification obtained from microarray screening and quantitative RT-PCR confirmation. Int J Legal Med 124:217–226. https://doi.org/10.1007/s00414-009-0402-3

Chapter 5
Epigenetic Age Estimation from Blood: Advances, Applications, and Forensic Implications

Cordulle Vanthillo and Bram Bekaert

Abstract Epigenetic age estimation using DNA methylation (DNAm) patterns in blood has emerged as a powerful tool in forensic science, offering accurate predictions of chronological age from biological samples. This chapter explores the evolution of age estimation methods using blood, from genome-wide discovery of age-related CpG sites to the development of targeted, cost-effective assays suitable for forensic casework. It highlights the transition from microarray-based approaches to advanced technologies such as pyrosequencing, SNaPshot, and massive parallel sequencing (MPS), emphasizing their respective advantages and limitations. We address the influence of biological, environmental, and technical factors, including sex, ancestry, disease, and sample degradation, on prediction accuracy. Special attention is given to pediatric models, postmortem samples as well as population-specific DNA-methylation patterns.

Keywords Age estimation · Blood · DNA methylation · Forensic epigenetics

Abbreviations

ANN	Artificial Neural Network
AR-CpG	Age-Related Cytosine-Guanine Dinucleotide
BC	Bisulfite Conversion

C. Vanthillo
Forensic Biomedical Sciences, Department of Imaging & Pathology, KU Leuven, Leuven, Belgium

B. Bekaert (✉)
Forensic Biomedical Sciences, Department of Imaging & Pathology, KU Leuven, Leuven, Belgium

Department of Forensic Medicine, Laboratory of Forensic Genetics, UZ Leuven, Leuven, Belgium
e-mail: bram.bekaert@uzleuven.be

S. C. Zapico (ed.), *Forensic Science and Epigenetics*, Epigenetics and Human Health 18, https://doi.org/10.1007/978-3-032-23085-0_5

C/T ratio	Cytosine-to-Thymine Ratio
CE	Central European
CpG	Cytosine-Guanine Dinucleotide
ddPCR	Digital Droplet Polymerase Chain Reaction
DNAm	DNA Methylation
EWAS	Epigenome-Wide Association Studies
EPIC	Illumina Infinium MethylationEPIC
EOAD	Early-onset Alzheimer's disease
HM27	Illumina Infinium HumanMethylation27
HM450	Illumina Infinium HumanMethylation 450 K
HSCT	Hematopoietic Stem Cell Transplantation
MAD	Mean Absolute Deviation
MD	Mean Difference
MPS	Massive Parallel Sequencing
MS-HRM	Methylation-Sensitive High-Resolution Melting
NGS	Next-Generation Sequencing
ME	Middle Eastern
MLR	Multivariate Linear Regression
OLS	Ordinary Least Squares
PAYA	Predictor for Adolescents and Young Adults
PCR	Polymerase Chain Reaction
QR	Quantile Regression
QRNN	Quantile Regression Neural Networks
QRSVM	Quantile Regression Support Vector Machine
RFR	Random Forest Regression
SVMs	Support Vector Machine
SVR	Support Vector Regression
WGBS	Whole-Genome Bisulfite Sequencing
WLS	Weighted Least Squares
5mC	5-Methylcytosine

5.1 Introduction

The estimation of chronological age from blood samples has become an increasingly valuable tool in forensic science, aiding criminal investigations, disaster victim identification, and other scenarios involving unidentified individuals (Refn et al. 2023). Blood is particularly well-suited for such analyses due to its frequent presence at crime scenes, ease of collection from both living and deceased individuals, and the availability of extensive methylation data from blood-derived DNA (Pokupcic 2017; Parson 2018).

Central to this approach is the study of DNA methylation (DNAm), an epigenetic modification involving the addition of methyl groups to cytosine residues, primarily at cytosine-guanine dinucleotides (CpG sites) (Jiang and Guo 2020). These methylation patterns are known to change predictably with age, making them powerful biomarkers for age estimation. While DNAm patterns are generally stable to maintain cellular identity, they also exhibit dynamic changes over time due to intrinsic biological processes and environmental influences, a phenomenon known as epigenetic drift (Teschendorff et al. 2013).

Despite the variability introduced by epigenetic drift, certain CpG sites consistently show age-associated methylation changes across individuals. These sites form the foundation of "epigenetic clocks", predictive models that estimate chronological age based on DNAm levels (Horvath and Raj 2018). Early models relied on genome-wide association studies to identify informative CpG sites, but the field has since shifted toward more targeted, cost-effective approaches that are better suited for forensic applications, where DNA samples are often limited or degraded.

This chapter explores the discovery and validation of age-informative DNAm markers in blood, the development of statistical and machine learning models for age prediction, and the practical considerations for applying these models in forensic contexts. It also examines the influence of biological sex, environmental exposures, lifestyle factors, and disease on DNAm patterns, as well as the challenges of applying age prediction models across diverse populations and sample types.

5.2 Epigenome-Wide Discovery of DNAm Markers

5.2.1 Pioneering Studies on Age-Related DNAm Changes in Human Blood

Understanding how DNAm changes with age and environmental influences is crucial for advancing medical and forensic fields. In a pioneering study, Fraga et al. examined 80 monozygotic twin pairs aged 3–74 years and identified significant epigenetic differences between young and older twins. The study analyzed lymphocyte cells and found that monozygotic twins are epigenetically indistinguishable during childhood. However, as they age, their DNAm patterns diverge considerably. These differences in DNAm patterns were attributed primarily to external factors, such as smoking habits, diet, and physical activity. However, internal factors, including the accumulation of minor defects during successive cell divisions were suggested to also contribute to these epigenetic changes (Fraga et al. 2005). Building on these findings, Bjornsson et al. conducted a prospective study to directly examine global changes in individuals over time using the luminometric methylation assay (LUMA) as well as the Illumina GoldenGate Methylation Solution method, which probes 1.505 CpG loci selected from 807 genes. DNAm patterns were analyzed in peripheral blood cells from Icelandic and Utah participants, collecting samples at two time points separated by an average of 11 and 16 years, respectively. The results showed that 8–10% of individuals in both populations experienced greater than 20% change in global methylation over an 11- to 16-year timespan.

Given the diverse applications of DNAm in both medical and forensic fields, researchers have aimed to develop robust age-prediction models. In the first step of this process, age-related DNAm markers need to be discovered. This is achieved through comprehensive scanning of DNAm patterns across the genome using technologies such as array-based DNAm analysis and whole genome bisulfite sequencing (WGBS). Second, AR-CpG (age-related cytosine-guanine nucleotide) sites are selected from these markers for validation and are used to develop age-prediction models (Bock 2009; Vidaki et al. 2013; Freire-Aradas et al. 2017). Although WGBS provides an unbiased and accurate representation of the methylome landscape of a genome as the entire sequence is mapped, its application is often limited by high sequencing costs, complex data processing, and time-intensive analysis (Bock 2009; Freire-Aradas et al. 2017). As a result, microarray-based approaches have been more widely adopted for the discovery of AR-CpG sites.

5.2.2 *Identifying AR-CpGs Sites Using Microarray Technology*

One of the first genome-scale studies of epigenomic dynamics during normal human aging in blood was reported by Rakyan et al. using the Illumina Infinium HumanMethylation27 (HM27) BeadChips array, to determine DNAm levels at 27,578 CpG sites in whole blood. A total of 360 CpG sites associated with age were identified, of which 213 were hypermethylated and 147 CpG hypomethylated ($n = 93$, aged 49–75). More than 95% of these CpGs were located within 500 bp of the transcriptional start site of the associated gene (Rakyan et al. 2010). Although aging-related DNAm dynamics were investigated previously (Fraga et al. 2005; Bjornsson et al. 2008; Christensen et al. 2009; Boks et al. 2009), it remained unclear whether these changes reflected intrinsic alterations within specific cell types or were the result of age-related shifts in the cellular composition of tissues. To address this question, Rakyan et al. conducted a genome-scale replication study using sorted CD14+ monocytes and CD4+ T-cells from an independent cohort of 25 individuals. CD14+ monocytes, part of the myeloid lineage, are short-lived cells with a lifespan of a few weeks, while CD4+ T-cells, from the lymphoid lineage, include a variety of cell types with life spans ranging from months to years. The study found that age-associated epigenetic changes in both CD4+ monocytes and CD4+ T-cells are similar to those observed in whole blood (Rakyan et al. 2010). This similarity indicates that the observed epigenetic changes are intrinsic to individual cell types rather than being driven by increased cellular heterogeneity. Consequently, DNAm age reflects epigenomic alterations rather than shifts in cell type composition. These findings were later corroborated by the epigenetic clock models developed by Hannum et al. (2013), Horvath (2013) and Aanes et al. (2023).

Around the same time, researchers discovered that AR-CpG sites overlapped across different tissues. This paved the way for the development of multi-tissue epigenetic clock models (Teschendorff et al. 2010; Bocklandt et al. 2011). For example, Bocklandt et al. identified ten out of 88 CpG sites in saliva that were previously shown to be correlated with age in whole blood (Bocklandt et al. 2011). Building upon these findings, Koch and Wagner (2011), and Horvath (2013) were the first to develop multi-tissue epigenetic clock models. Koch and Wagner used DNAm-profiles from the HM27 BeadChip array from public data repositories, selecting five CpG sites (Koch and Wagner 2011), while Horvath et al. used elastic net to select 353 CpG sites from both HM27 BeadChip and Illumina Infinium HumanMethylation 450 K (HM450) BeadChip array. Of these 193 CpGs became hypermethylated and 160 became hypomethylated with age, corresponding to positive and negative correlations, respectively (Horvath 2013). Both prediction models show considerable variability in prediction accuracy, with a mean absolute deviation (MAD) of 11.4 and 3.6 years, respectively (Koch and Wagner 2011; Horvath 2013). Although Horvath's epigenetic clock is considered universal, there are some differences in its prediction accuracy across different tissues. For instance, the MAD for whole blood is 3.7 years and 18 years for skeletal muscle in the test data. These results highlight that age-dependent DNAm changes are tissue-specific and must be considered in epigenetic clocks (Horvath 2013).

Although multi-tissue epigenetic clocks have been developed, most age prediction models are tissue-specific. In the context of blood, Hannum et al. are considered as pioneers in the development of blood-specific epigenetic age prediction models. Utilizing the HM450 BeadChip array, they identified 71 AR-CpG sites via elastic net regression to construct a model. The study was based on methylome-wide profiles of a mixed population of Caucasian and Hispanic individuals ($n = 656$), aged between 19 and 101 years. The resulting model demonstrated a strong correlation of 91% between chronological and predicted age. In a validation cohort ($n = 174$), the model achieved a MAD of 4.9 years (Hannum et al. 2013).

5.3 Development and Application of Targeted Assays

5.3.1 Development of Targeted Assays

To develop more cost-effective, faster, and user-friendly age prediction models without compromising accuracy, bisulfite-pyrosequencing was among the first techniques to be employed. The first age prediction model using this method was developed for blood and was based on three CpG sites within the genes *ASPA, ITGA2B,* and *PDE4C.* These CpG sites were selected by analyzing DNAm profiles from publicly available HM24 BeadChip datasets in the NCBI GEO database. The resulting MLR (multivariate linear regression) model achieved a MAD of 4.5 years when validated in an independent validation set ($n = 82$) (Weidner et al. 2014). This study marked a transition from genome-wide to targeted assays by employing pyrosequencing. Since then, many researchers have focused on targeted assays to analyze known age-dependent CpG sites and develop accurate age prediction models.

Following the work of Weidner et al. (2014), researchers continued to validate additional AR-CpG sites. Zbieć-Piekarska et al. developed a model for blood based on five CpG sites in *ELOVL2, TRIM59, KLF14, MIR29B2CHG* (also known as *C1orf132*), and *FHL2*. This selection was based on the correlation with age observed in DNAm analysis, in a discovery cohort ($n = 300$, aged 2–75), across 41 CpG sites within eight loci previously reported by (Hannum et al. 2013; Zbieć-Piekarska et al. 2015a). The final model, which included the five most significant AR-CpG sites, achieved MADs of 3.4 years in the discovery cohort and 3.9 years in a validation cohort ($n = 120$). Furthermore, based on the assumption that age prediction was correct when predicted and chronological age differed by a maximum of 5 years, a total of 72% correct predictions were observed in the validation set. However, the model's predictive accuracy declined in older individuals, with a MAD increasing to 5.6 years and the number of correct predictions decreasing to 50% in the 60–75 age group (Zbieć-Piekarska et al. 2015a). This reduction in accuracy among older samples is likely attributable to increased inter-individual epigenetic drift, a phenomenon observed in multiple studies (Vidaki et al. 2017; Naue et al. 2017). Due to their strong correlation with chronological age, these five genes have since become the most frequently used in age prediction models.

Other groups have expanded the range of markers and optimized statistical models for age prediction. For instance, Huang et al. developed a pyrosequencing-based assay using five CpG sites in *ASPA, ITGA2B,* and *NPXT2* (with two located in *ITGA2B* and *NPXT2*) in a cohort of 89 individuals aged 9–75. Based on previous studies, 38 candidate CpG sites in six genes were initially selected for DNAm analysis by pyrosequencing in blood samples. Following initial screening, stepwise regression analysis was performed separately in male, female, and combined cohorts, leading to the final selection of the five AR-CpG sites included in this model. The developed MLR model explained 82% of the variation in age and achieved a MAD of 7.99 years in a validation set of 40 blood samples (Huang et al. 2015).

In contrast, Bekaert et al. employed a multivariate quadratic regression model and achieved a MAD of 3.99 years in their test set ($n = 103$). This model is based on four CpGs in *ASPA, PDE4C, ELOVL2,* and *EDARADD*. These sites were selected from shortlists published by other research groups. The improved prediction accuracy is likely due to the non-linear relationship between *ELOLV2* methylation and age. Methylation data from this marker increases exponentially during childhood and plateaus later in life. Furthermore, a correct prediction rate of 79.6% within ±5 years was achieved across 206 samples. Notably, the model achieved correct prediction rates of 96.4% and 93.4% in the 0–19 and 20–39 age groups, respectively (Bekaert et al. 2015). These findings underscore the advantage of using a quadratic regression approach, which provided a better fit than MLR models for capturing changes in DNAm patterns (Bekaert et al. 2015). This observation is consistent with the results of Alisch et al. (2012) in pediatric populations, where a logarithmic model provides a better fit to the data, further underscoring the non-linear relationship between methylation and age.

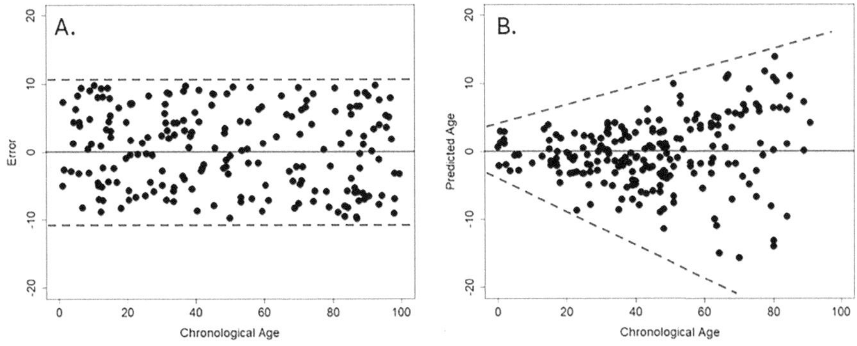

Fig. 5.1 Difference in spread of the age prediction errors with increasing chronological age. (**a**) Example of homoscedastic variation of the age prediction errors. (**b**) Heteroscedasticity with non-constant (in this case increasing) variance

A recurring challenge in DNAm-based age prediction is increased prediction error with age. This phenomenon is known as heteroscedasticity (see Fig. 5.1). To address this issue, Smeers et al. compared ordinary least squares (OLS), weighted least squares (WLS), and quantile regression (QR) models. Both WLS and QR provided more accurate, age-dependent prediction intervals than OLS, with WLS performing best in their dataset. This work underscores the importance of selecting statistical models that account for heteroscedasticity, particularly for forensic applications where accurate prediction intervals are essential (Smeers et al. 2018).

5.3.2 *Alternative Technologies for Targeted DNAm Analysis*

While pyrosequencing is widely regarded as the gold standard for DNAm analysis due to its accuracy, it presents several limitations for routine forensic use. For instance, multiplex pyrosequencing can be technically challenging, which typically restricts its application to single-reaction assays (Vidaki et al. 2017). This limitation reduces its applicability in forensic casework, where DNA samples are often limited in quantity and quality (Freire-Aradas et al. 2020). Additionally, the relatively high costs associated with pyrosequencing further limit its adoption in standard forensic laboratories (Pan et al. 2020). As a result, there is a need for the development of rapid, cost-effective, and straightforward methods for methylation analysis that are better suited to the practical demands of forensic investigations.

5.3.2.1 Agena Bioscience EpiTYPER® System

To address the challenges of pyrosequencing, researchers have investigated several alternative technologies. One such method is the Agena Bioscience EpiTYPER® system, which utilizes mass spectrometry for large-scale quantitative analysis of

DNAm (Suchiman et al. 2015; Castagnola et al. 2024). Using the EpiTYPER system, Freire-Aradas et al. developed a multivariate QR model using seven age-correlated loci (*ELOVL2, ASPA, PDE4C, FHL2, CCDC102B, MIR29B2CHG*, and chr16:85395429). These CpG sites were selected through a combination of literature review and statistical analysis of three publicly available HM450 BeadChip methylation datasets from the NCBI GEO database. The model achieved an overall MAD of 3.07 years in a training set (n = 725, aged 18–104) and 4.23 years in 104 monozygotic twins (Freire-Aradas et al. 2016).

Feng et al. used the EpiTYPER® platform to analyze DNAm in a cohort of 390 Han Chinese males, aged 15–75 years. From the initial set of 61 CpG sites identified through a bibliographic review, nine were selected as optimal predictors of chronological age using stepwise regression. The resulting MLR model achieved a MAD of 2.89 years in the training set, and 2.49 years and 3.36 years in two independent validation cohorts (n = 65 and n = 62, respectively). Although machine learning models such as support vector machines (SVMs) and artificial neural networks (ANNs) were also evaluated, they did not outperform the linear model in this study (Feng et al. 2018). This is in contrast with findings from other studies (Vidaki et al. 2017; Pan et al. 2020; Thong et al. 2021), which show the advantages of machine learning models in offering greater flexibility in modeling complex, non-linear relationships and their usefulness in narrowing down large pools of predictors (Smeers et al. 2018). Feng et al. proposed this discrepancy might stem from their study's age range, as non-linear methylation-age relationships are more prominent in older populations (Feng et al. 2018).

5.3.2.2 SNaPshot Assay

Ye et al. developed a multiplex methylation SNaPshot assay, based on 180 individuals aged 22–51. Their age prediction model included three AR-CpG sites associated with *MIR29B2CHG, ELOVL2,* and *DNAH9*, selected by a statistical analysis of a dataset from the NCBI GEO database, using the EPIC BeadChip. The model achieved a MAD of 4.04 years and explained 86.8% of the variation of age in the training set (n = 115) (Ye et al. 2022).

5.3.2.3 Massive Parallel Sequencing

One of the most promising technologies as an alternative for pyrosequencing is massive parallel sequencing (MPS). MPS enables multiplexed, high-sensitivity detection of multiple CpG sites in a single assay, making it particularly attractive for forensic applications. Additionally, it possesses the ability to process low-quality and quantity DNA (Castagnola et al. 2024). Vidaki et al. applied MPS to develop an age prediction model using a generalized regression neural network on 16 AR-CpG sites located in 16 genomic regions. These sites were selected through stepwise regression from an initial pool of 45 AR-CpG sites, which had been identified based on their reported age coefficients in a previous study and investigated using publicly available HM27 and HM450 methylation data obtained from 1156 whole blood samples (aged 2–9 years). The final model achieved a MAD 7.45 years in an independent validation set of 46 blood samples.

Similarly, Naue et al. developed a random forest regression (RFR) model using MPS data from 208 training and 104 test blood samples (ages 18–69). The model included 13 CpG markers of different loci (*DDO, ELOVL2, F5, GRM2, HOXC4, KLF14, LDB2, MEIS1-AS3, NKIRAS2, RPA2, SAMD10, TRIM59,* and *ZYG11A*), selected using eight public datasets. The DNAm data of seven markers showed an increase with age while the other six showed a decrease. The final model achieved a MAD of 3.16 years in the test set. Notably, a reduced model using only four markers (*ELOVL2, F5, KLF14,* and *TRIM59*) still performed robustly (MAD 3.24 years) (Naue et al. 2017).

A recent study on whole blood samples by Zhu et al. incorporates 25 CpG sites across *MIR29B2CHG, FHL2, TRIM59, ELOVL2,* and *KLF14.* Using elastic net regression, the model demonstrated high accuracy, achieving a MAD of 2.28 years in the training set ($n = 191$) and 2.20 years in an independent validation set ($n = 127$). Notably, 82.7% of the predictions across the entire cohort differed by a maximum of 5 years from the actual age, and a success rate of 89.6% was observed for individuals in the 50–60-year age category. While the model shows strong predictive performance, it is important to note that the training cohort was limited to individuals aged 30–70 years. This relatively narrow age range restricts the model's applicability in forensic contexts (Zhu et al. 2025).

Since numerous forensic models have been developed to estimate age from blood using DNAm markers, differing in the number of CpG sites, statistical approaches, and sequencing methods, there is need for more standardization in the development of epigenetic clocks (Verstraeten et al. 2025). To partially address this need, the VISAGE (VISible Attributes through Genomics) Consortium developed the VISAGE enhanced tool assay. First, a basic prototype tool was developed to analyze 32 CpG sites from five genes (*ELOVL2, MIR29B2CHG, FHL2, TRIM59,* and *KLF14*), initially identified by (Zbieć-Piekarska et al. 2015a; Heidegger et al. 2020). The performance of the prototype tool was evaluated using artificially methylated DNA standards to assess reproducibility and sensitivity. The tool demonstrated robust methylation quantification, with a mean standard deviation of 1.4% across ratios (Heidegger et al. 2020). In a subsequent study, the VISAGE Consortium developed a multi-tissue age prediction assay by combining the five markers of the VISAGE basic assay with *ASPA, EDARADD,* and *PDE4C.* Three tissue-specific models were built using this assay for age prediction in blood, buccal cells, and bones. The models were developed using Illumina MiSeq sequencing data in combination with the v3 reagent kit (2 × 300 bp). The blood model included six CpG sites from *ELOLV2, MIR29B2CHG, KLF14, FHL2, TRIM59,* and PDE4C and predicts age with a MAD of 3.2 years in the testing set ($n = 48$). Notably, *ELOVL2* alone explains 95% of variation in this age prediction model (Woźniak et al. 2021).

To enhance accessibility and reduce costs, Verstraeten et al. independently validated and optimized the VISAGE enhanced tool using the Illumina NovaSeq 6000 platform with the v1.5 reagent kit (2 × 150 bp). By adjusting primer concentrations, the study achieved sufficient read depths for accurate methylation calling, even for longer amplicons. Using the optimized assay, a new blood-based age prediction model was developed using 98 samples in the training set (aged 0–94). The model, which employed a QR approach, achieved a MAD of 3.22 years in a testing set ($n = 30$). Furthermore, the NovaSeq 6000 S4 flow cell offers a shorter maximum run time of about 44 h, compared to 56 h for the MiSeq v3 kit. It also enables higher samples multiplexing and a greater read output. Although the absolute cost of a NovaSeq flow cell is higher, the cost per million reads and per million bases is significantly lower, leading to a tenfold decrease in sequencing costs per sample. These results demonstrated that compared prediction accuracy to the VISAGE MiSeq-based model could be achieved with significantly reduced sequencing time and cost (Verstraeten et al. 2025).

5.3.3 Single-Locus Models: Focus on ELOVL2

Single-locus models, particularly those focused on DNAm at the *ELOVL2* gene promoter, have gained interest. *ELOVL2* has shown a strong and consistent correlation with chronological age (Garagnani et al. 2012) and is implemented in most age prediction models. Multi-locus models generally offer higher predictive accuracy than single-locus models. However, depending on the sequencing technique used, they may require more DNA and are more prone to inter-laboratory variation, which can affect the reliability of predictions (Garali et al. 2020). Consequently, it is recommended that biomarkers included in prediction models be limited to those that significantly reduce age prediction error and exhibit substantial methylation changes throughout the human lifespan (Zbieć-Piekarska et al. 2015b).

Zbieć-Piekarska et al. were the first to introduce a single-locus age prediction model based on DNAm *ELOVL2* for blood. The model is based on two CpG sites, explaining 85.9% of age variation in blood samples from individuals aged 2 to 75 years, with a MAD of 5.75 years in the testing set (n = 124). This early work demonstrated the potential of single-locus models for forensic applications (Zbieć-Piekarska et al. 2015b). Building on this, Garali et al. developed 17,018 single-locus models using DNAm data from the *ELOVL2* promoter. Data from four different studies were pooled to develop a training set of 1028 individuals aged 0–91 years. Using six statistical and machine learning methods, the best age prediction performance was obtained with SVMs. This model achieved a MAD of 4.17 years, indicating a higher accuracy than the model developed by Zbieć-Piekarska et al. (Garali et al. 2020). Manco and Correia Dias used ddPCR (digital droplet polymerase chain reaction) for *ELOVL2* methylation analysis, finding a strong correlation between DNAm and chronological age (R = 0.892). However, their single-locus model yielded a higher MAD of 10.07 years in the training set ($n = 56$). This reduced accuracy likely reflects technical factors and limited sample size rather than a fundamental limitation of the locus itself (Manco and Correia Dias 2022). Most recently,

Yamagishi et al. aimed to address the challenge of inappropriate age estimations due to deterioration in prediction accuracy caused by an increased prediction error in older groups (>70 years). To address this problem, an ANN and QR model were developed using *ELOVL2* methylation data from 1280 individuals (aged 0–91). Both models show enlarged prediction intervals in accordance with aging and overall MADs of the ANN and QR models were 5.3 and 5.6 years, respectively (Yamagishi et al. 2024).

Together, these studies show that *ELOVL2* is a robust and applicable biomarker for age estimation based on DNAm. While multi-locus models are generally more accurate, single-locus *ELOVL2* models offer a practicable alternative, especially when only limited amounts of DNA are available or when standardization between laboratories is required.

5.3.4 Application: Age Prediction in Bloodstains

As it is not uncommon to find bloodstains at crime scenes, it is important to investigate whether DNAm stays stable over time and whether accurate age prediction can still be performed. Several studies have explored the stability and accuracy of DNAm-based age prediction in bloodstains stored at room temperature conditions to represent indoor crime scene samples. For example, Zbieć-Piekarska et al. reported unchanged prediction accuracy even after 15 years of storage, even though the DNA significantly degraded (Zbieć-Piekarska et al. 2015b). Similarly, other studies did not find significant differences between the predicted ages using fresh blood and stored bloodstains (Huang et al. 2015; Thong et al. 2017; Peng et al. 2019; Ye et al. 2022; Yamagishi et al. 2024).

Other studies developed a prediction model based on fresh blood and validated their model on bloodstains. Yang et al. developed a prediction model with bloodstains stored on filter paper at room temperature for 1 year. Their study involved 241 volunteers aged 10–79 and utilized RFR, which proved superior to SVR. Based on 7 CpG sites (in *TRIM59, KLF14, MIR29B2CHG*, *ELOVL2,* and *FHL2*), the final models yielded a MAD of 2.80 and 2.93 years for males and females, respectively (Yang et al. 2023).

Han et al. further demonstrated the forensic applicability of DNAm-based age prediction models initially developed using fresh blood samples. They analyzed a bloodstain found at a crime scene of a rape-murder case to predict the suspect's age. Using a multiplex methylation SNaPshot assay targeting eight CpG sites, the suspect's age was estimated as 38.27 years with a MAD of 3.05 years via MLR and 38.59 years with a MAD of 2.34 years via SVR. Later, when the case was solved, the suspect's age was confirmed to be 38 years old (Han et al. 2022). This accurate age prediction demonstrates the potential of DNAm-based age estimation techniques in real-world forensic investigations.

These studies suggest that DNAm patterns in bloodstains remain relatively stable over time, allowing accurate age prediction. However, more studies should be done on the influence of storage conditions, such as ultraviolet light, the presence of other body fluids, and external contamination (Refn et al. 2023; Naue 2023).

5.4 Factors Influencing Age Prediction Accuracy

While the preceding sections have highlighted the potential of DNAm for age estimation in blood samples, it is important to acknowledge factors that introduce variability and may compromise prediction accuracy. Epigenetic modifications, including DNAm, are known to be sensitive to genetic factors and environmental exposures. Therefore, investigations into these influencing factors must be done to ensure the reliability and robustness of age prediction models (Fleckhaus et al. 2017; Naue 2023; Correia Dias and Manco 2024). Due to the wide range of contributing factors, only a limited number of studies can be discussed. While these factors are organized into sections, they can be interconnected and may have other smaller effects.

5.4.1 Biological Sex

As epigenetic research advances, the potential influence of biological sex on DNAm patterns is becoming increasingly relevant to the accuracy of age-prediction models. While it has been known that DNAm levels are related to sex, the extent to which these differences affect the accuracy of age-prediction models remains debated.

DNAm plays a role in X-chromosome inactivation, a process by which one of the two X-chromosomes in females is silenced to achieve dosage compensation (Boks et al. 2009; Zaghlool et al. 2015). Methylation differences observed at loci on the X-chromosome between males and females are largely attributed to this mechanism (Boks et al. 2009). However, sex-related differences in DNAm are not limited to the X-chromosome. Multiple studies have shown that men tend to have higher global DNAm levels than women. For instance, Fuke et al. reported a subtle but statistically significant higher mean methylation content in males (Fuke et al. 2004), and Zaghlool et al. identified 9630 CpG sites with genome-wide significant associations with sex in blood samples (Zaghlool et al. 2015). Similarly, Boks found replicable associations between DNAm, age, and sex in autosomal CpG loci using whole blood samples (Boks et al. 2009). However, research on specific loci has yielded inconsistent results, with some studies detecting significant differences while others have not.

Regarding age prediction, the impact of sex appears to be minimal. Zbieć-Piekarska et al. observed a slight but noticeable effect of sex on age prediction accuracy (R^2 improved by 0.001 with age as a covariate) (Zbieć-Piekarska et al. 2015a). Hannum et al. suggested that men's methylome ages approximately 4% faster than women's, though this difference did not significantly affect overall age distributions (Hannum et al. 2013). Conversely, studies like Bekaert et al., Pan et al., Hamano et al., and Filoglu et al. found no significant differences in methylation levels or age prediction accuracies between sexes (Bekaert et al. 2015; Hamano et al. 2016; Pan et al. 2020; Filoglu et al. 2024). Similarly, Freire-Aradas and Zubakov failed to detect any meaningful effect of gender on age estimation levels (Freire-Aradas et al. 2016, 2018).

Some studies suggest that while sex differences exist in DNAm patterns, they do not substantially improve the predictive power of age models. For example, Naue reported that incorporating sex into models did not enhance their performance (Naue et al. 2017). Daunay and Aliferi also concluded that gender had little to no effect on prediction accuracy, although Aliferi noted slightly higher accuracy for females in certain datasets (Daunay et al. 2019; Aliferi et al. 2022).

Overall, current evidence suggests that even if sex does exert an influence, its impact on the accuracy of age prediction appears to be minimal. However, future age-prediction models should still be evaluated for potential sex-based differences, as minor variations may exist depending on the DNAm markers included and models selected (Naue 2023).

5.4.2 Population-Specific Differences

DNAm-based age estimation is a promising tool for forensic studies, but population-specific differences can significantly impact the accuracy of prediction models. Many studies have focused on identifying novel DNAm markers for age estimation in specific populations or by combining publicly available datasets from different global populations (Cho et al. 2017; Correia Dias et al. 2020b). While general DNAm differences are documented, ancestry-related variations in age-associated DNAm patterns remain largely unexplored (Tajuddin et al. 2019; Paparazzo et al. 2023; Refn et al. 2023). Differences in DNAm across global populations are likely influenced by genetic ancestry and environmental exposures, including diet, UV radiation, air pollution, socioeconomic conditions, healthcare access, and living standards (Freire-Aradas et al. 2017; Yusipov et al. 2024). Such factors can lead to distinct DNAm patterns at the same AR-CpG site across different ancestries, underscoring the importance of considering population-specific variations in forensic applications.

5.4.2.1 Cross-Population Model Evaluations

One of the most widely studied models is the Polish five-CpG model developed by Zbieć-Piekarska et al., which includes ELOVL2, MIR29B2CHG, TRIM59, KLF14, and FHL2. This model has been tested in several populations:

- Korean Population: Cho et al. applied the Polish model to 100 Korean samples and achieved a MAD of 4.18 years (Cho et al. 2017). While the model performed reasonably well, differences in CpG site behavior were observed. For example, FHL2 showed a stronger correlation with age in Koreans, while MIR29B2CHG was more predictive in the Polish cohort. Retraining the model using Korean-specific data improved accuracy (MAD = 3.29 years).
- Singaporean Population: Thong et al. evaluated the same model in a multi-ethnic Singaporean cohort (Chinese, Malays, Indians) and reported a MAD of 4.8 years (Thong et al. 2017). Again, CpG-specific differences were noted, with ELOVL2, TRIM59, and KLF14 showing stronger correlations in Singaporeans. A retrained model using three CpGs achieved a MAD of 5.0 years, comparable to the original.
- Portuguese Population: Correia Dias et al. tested a Korean model in a Portuguese cohort and found a significant drop in accuracy (MAD = 15.26 years). CpG sites in KLF14 and TRIM59 showed weak or no correlation with age, highlighting the importance of population-specific validation (Correia Dias et al. 2020a).

These studies demonstrate that while some AR-CpG sites are broadly applicable, others may behave differently across populations, necessitating local calibration or retraining of models.

5.4.2.2 Controlled Comparative Study: Middle Eastern vs. Central European

To isolate the effects of ancestry from technical variability, Fleckhaus et al. conducted a controlled study comparing Middle Eastern (ME) and Central European (CE) populations using the same laboratory protocols. They analyzed five commonly used AR-CpG sites and found significant differences in methylation patterns, particularly in MIR29B2CHG, FHL2, and KLF14. A combined model trained on both populations yielded MADs of 3.81 years (ME) and 3.31 years (CE), but it systematically overestimated ME ages and underestimated CE ages. Population-specific models improved accuracy (MAD = 3.34 years for ME, 2.72 years for CE), emphasizing the need for ancestry-informed modeling (Fleckhaus et al. 2023).

5.4.2.3 Implications for Forensic Practice

These findings underscore the importance of considering biogeographical ancestry in DNAm-based age prediction. While models trained on diverse datasets may offer broader applicability, they may not perform as accurately as population-specific models. When ancestry is known, using tailored models can significantly improve prediction accuracy. Fortunately, forensic tools for ancestry inference are already available and can be applied to the same DNA sample used for age estimation. Integrating ancestry prediction with DNAm-based age models offers a practical solution to enhance reliability in forensic casework. Future research should include a broader range of global populations, particularly those currently underrepresented, to further improve the reliability and robustness of DNAm-based age-prediction models. Larger cohort studies are needed to validate and strengthen existing findings and explore further the need for population-specific adjustments in age prediction models.

5.4.3 Age Prediction Models for Pediatric Populations

Most epigenetic studies have focused on adults (>18 years) or have a restricted sample size of younger subjects, resulting in a knowledge gap regarding DNAm patterns in children and adolescents (Freire-Aradas et al. 2018). However, it is known that DNAm patterns undergo dynamic changes during childhood, characterized by a phase of rapid accumulation followed by stabilization in adulthood (Horvath 2013; Horvath and Raj 2018; Castagnola et al. 2024). These age-related changes in DNAm highlight the necessity of developing models tailored to children and adolescents. Forensic age estimation in children and adolescents holds potential for application such as identifying unknown remains, age assessments in asylum procedures, and determining the applicability of juvenile justice measures in legal contexts (Freire-Aradas et al. 2018). Therefore, recent developments in epigenetic research have led to the development of age prediction models optimized for children and adolescents.

Li et al. developed an age prediction model targeting individuals aged 6–17 ($n = 179$), utilizing a set of 83 AR-CpG sites from data generated using the EPIC BeadChip selected by elastic net regression. The linear model demonstrated a MAD of 1.22 years in the testing set ($n = 89$) and a strong correlation of 93% between predicted and chronological age using blood samples (Li et al. 2018).

Similarly, Freire-Aradas et al. analyzed 209 blood samples from individuals aged 2–18 years using the EpiTYPER® system to develop a QR model for age prediction. This study identified ten genomic regions covering 58 AR-CpG sites as strongly correlated with chronological age in young individuals (Freire-Aradas et al. 2018). Seven of these markers (ELOVL2, ASPA, PDE4C, FHL2, CCDC102B, MIR29B2CHG, and chr16:85,395,429) were previously included in a forensic age prediction model initially developed for adults using Illumina HM450 data (Freire-Aradas et al. 2016). The remaining three genes (KCNAB3, PRKG2, and EDARADD) were identified as highly specific for young individuals and were discovered using a dataset based on HM27 array data from Alisch et al. (2012; Freire-Aradas et al. 2018). Although broader CpG site coverage, such as that provided by HM450 or EPIC arrays, would be better for selecting suitable markers for pediatric age estimation, the availability of DNAm data for children and adolescents remains limited. The final model was developed using six AR-CpG sites located in KCNAB3, EDARADD, ELOVL2, MIR29B2CHG, CCD102B, and chr16:85,395,429 (GRCh38). These sites showed the highest correlation with age when validated using the EpiTYPER® system. The QR model achieved a MAD of 0.94 years in the training set and 1.25 years in an independent validation set ($n = 29$).

Building on this work, Freire-Aradas et al. developed a comprehensive epigenetic age prediction model covering the entire human lifespan (ages 2–104 years). This model was built using DNAm data from 895 blood samples and seven CpG sites located in ELOVL2, ASPA, PDE4C, FHL2, CCDC102B, MIR29B2CHG, and chr16:85,395,429 (GRCh38) (Freire-Aradas et al. 2022). These DNAm markers were adopted from a previously developed age prediction model for adults (Freire-Aradas et al. 2016) and were analyzed using the EpiTYPER® system in this study (Freire-Aradas et al. 2022). The DNAm data used for this analysis were derived from two previously published studies conducted by the same research group (Freire-Aradas et al. 2016, 2018). Among these markers, ELOVL2, PDE4C, and FHL2 exhibited age-associated hypermethylation, while ASPA, CCD102B, MIR29B2CHG, and chr16:85,395,429 showed hypomethylation with increasing age. Additionally, the study compared QR with two machine learning-based approaches: quantile regression neural networks (QRNN) and quantile regression support vector machines (QRSVM). QRNN and QRSVM outperformed standard QR, achieving MADs of 3.32 and 3.45 in an independent testing set ($n = 152$). These models offer age-dependent prediction intervals, with narrower errors in children and broader intervals in older adults, reflecting increased inter-individual variability with age. This unified approach enables accurate age estimation across the lifespan, making it suitable for forensic applications (Freire-Aradas et al. 2022).

More recently, Aanes et al. introduced the Predictor for Adolescents and Young Adults (PAYA), an epigenetic clock for individuals aged 12–25. The model was developed using DNAm data from publicly available datasets, all generated with the HM450 BeadChip platform. PAYA incorporates 267 AR-CpG sites and achieves a MAD of less than 0.7 years in a validation cohort of 18-year-olds. ($n = 920$) (Aanes et al. 2023). Notably, the PAYA model shares three AR-CpG sites with the earlier model by Li et al. (2018) and 14 AR-CpG sites with the Hannum et al. (2013) clock, indicating some degree of overlap in age-associated methylation markers (Aanes et al. 2023).

The enhanced accuracy of pediatric models can be attributed to the rapid changes in DNAm patterns during childhood and adolescence, allowing for more precise capture of age-related epigenetic drifts in pediatric populations compared to models developed for adults or across broader age ranges (Alisch et al. 2012; Li et al. 2018). Furthermore, children and adolescents are generally less exposed to confounding factors such as alcohol consumption and tobacco smoke, which further supports the robustness of these models (Li et al. 2018).

5.4.4 Disease-Related Variability in Age Prediction Accuracy

DNAm-based age prediction models rely on predictable changes in methylation at specific CpG sites as chronological age advances. These models have demonstrated high accuracy in healthy populations, with MAD often below 4 years. However, systemic diseases can disrupt typical DNAm patterns, leading to discrepancies between predicted and chronological age, which poses challenges for forensic and clinical applications.

Cancer is a notable example, as it induces widespread genomic hypomethylation and hypermethylation at promoter-associated CpG islands (Fernandez et al. 2012). These alterations can mimic natural aging signatures, compromising the accuracy of DNAm-based age predictions (Freire-Aradas et al. 2017; Kim et al. 2025). For instance, in blood samples of patients with chronic lymphocytic leukemia, age prediction was performed using pyrosequencing based on a previously described set of 32 CpG sites in the genes *ELOVL2, MIR29B2CHG, TRIM59*, and *KLF14.* On average, the predicted age was 10.7 years higher than the chronological age of the patients, indicating significant ($P < 1.8 \times 10^{-12}$) age acceleration (Spólnicka et al. 2018c). Similarly, cardiovascular and other aging-related diseases have been shown to affect the correlation between DNAm and age (Refn et al. 2023), further complicating age estimation, especially when an individual's medical history is unknown, or disease remains undiagnosed (Koop et al. 2020).

Neurodegenerative diseases also impact DNAm-based age prediction models, but the effects are disease- and age-group-specific. In early-onset Alzheimer's disease (EOAD), the accuracy of a five-CpG age prediction model (targeting *ELOVL2, MIR29B2CHG, TRIM59, KLF14,* and *FHL2*) was reduced. Using pyrosequencing in combination with ANN, the model achieved a MAD of 7.1 years in blood samples from EOAD patients compared to 3.8 years in healthy controls ($P = 0.002$). The inaccuracy was most pronounced in younger EOAD patients (<45 years), where predicted ages exceeded chronological ages by an average of 6.4 years, suggesting accelerated epigenetic aging in this subgroup ($P = 0.011$). Conversely, no significant reduction in prediction accuracy was observed for late-onset Alzheimer's ($P = 0.382$) or Graves' disease ($P = 0.224$), highlighting the specificity of disease effects on DNAm-based age prediction. The varying influence of these diseases can be attributed to differences in the stability of DNAm at individual CpG sites: *ELOVL2* and *MIR29B2CHG* maintained stable methylation profiles across disease groups, underscoring their robustness for forensic age estimation. At the same time, markers such as *TRIM59, KLF14,* and *FHL2* were more susceptible to disease-related changes (Spólnicka et al. 2018b).

Infectious diseases also modulate DNAm-age correlations. HIV infection accelerates epigenetic aging by approximately 5 years (Horvath and Levine 2015; Gross et al. 2016). Furthermore, infections of *Helicobacter pylori* (Gao et al. 2017), cytomegalovirus (Kananen et al. 2015), and COVID-19 (Cao et al. 2022) have been associated with increasing age prediction. These diseases may therefore affect the reliability of blood-based age prediction in real casework (Attia 2022).

Allogenic hematopoietic stem cell transplantation (HSCT) presents another example where the correlation between chronological age and DNAm signatures is disrupted. Within one-year post-transplantation, the recipient's blood DNAm patterns align with those of the donor-derived hematopoietic stem cells, reflecting the donor's epigenetic age rather than the recipient's. Both Weidner et al. (2015) and Spólnicka et al. (2016) confirm that the donor determines the recipient's post-HSCT epigenetic age, but they report different directions and causes for the observed age difference. Spólnicka et al. found that the recipient's predicted age was, on average, 3.7 years lower than the donor's, potentially due to hypermethylation of *MIR29BCHG* (Spólnicka et al. 2016). In contrast, Weidner et al. observed that the recipient's predicted age was, on average, 7 years higher than the donor's chronological age, attributing this to stimulation and accelerated aging of donor hematopoietic stem cells (Weidner et al. 2015). These discrepancies may arise from using different DNAm markers in the respective models (Spólnicka et al. 2016). Interestingly, in a separate study, Spólnicka et al. (2018b) identified *MIR29B2CHG* as a stable marker across three distinct patient groups, suggesting that in some contexts, *MIR29B2CHG* may not be a good predictor of chronological age.

Importantly, not all diseases compromise the accuracy of DNAm-based age prediction. For example, schizophrenia, rheumatoid arthritis, progressive supranuclear palsy, inflammatory bowel disease, and depression did not affect the accuracy of DNAm-based age prediction models. Both Aliferi et al. (2022) and Mathew et al. (2025) demonstrated that DNAm patterns at key CpG sites and the resulting age predictions remained consistent across control and disease groups, underscoring the robustness and reliability of these models even in the presence of various health conditions (Aliferi et al. 2022; Mathew et al. 2025).

In summary, while DNAm-based age prediction models are accurate in healthy populations, systemic diseases can significantly affect their performance, particularly those associated with aging or profound epigenetic accelerations. The extent of this impact is determined by the disease type, the specific CpGs used in the model, and the methylation stability at those sites. Careful marker selection and awareness of potential disease effects are essential for the reliable application of DNAm-based age prediction in forensic and clinical contexts.

5.4.5 Environmental Exposures and Lifestyle

Lifestyle and environmental factors can influence DNAm patterns, which is important for epigenetic age prediction in blood samples (Ambroa-Conde et al. 2024). While current forensic DNA analysis often treats discrepancies between predicted and actual ages as measurement errors, investigating the influence of these factors may improve the precision of chronological age estimation (Piniewska-Róg et al. 2021).

The impact of smoking on epigenetic age is complex, exhibiting inconsistencies across studies. Some research indicates a significant positive association ($P < 0.001/13$) between smoking and age acceleration (Gao et al. 2016; Vidaki et al. 2023). Notably, a large-scale study by Fiorito et al., which analyzed 16,245 blood samples, found that smoking was associated with an increase in epigenetic age estimates up to 0.24 years when using the Hannum et al. (2013) epigenetic clock model (Fiorito et al. 2019). However, other studies have failed to find such associations (Horvath and Ritz 2015; Vidal-Bralo et al. 2016; Gao et al. 2016; Quach et al. 2017;

Ambroa-Conde et al. 2024; Ochana et al. 2024), suggesting that the markers used in those models are not sensitive to smoking-related epigenetic changes (Ambroa-Conde et al. 2024).

Similarly, the effect of alcohol consumption on epigenetic age is complex and inconsistent across studies. While some studies propose that moderate alcohol consumption may decrease epigenetic age, potentially due to the anti-inflammatory effects (Quach et al. 2017), others report conflicting results. Beach et al. noted accelerated aging in light and heavy drinkers but decelerated aging in moderate drinkers (Beach et al. 2015). Further supporting the link between alcohol and age acceleration, Weidner et al. and Stephenson et al. observed that higher levels of alcohol consumption were associated with overestimation of chronological age (Weidner et al. 2014; Stephenson et al. 2021). Similarly, Piniewska-Róg et al. found that excessive alcohol abuse led to a small but statistically significant ($P = 0.007$) increase in predicted age, based on methylation changes at the *MIR29B2CHG* locus. However, this effect was not significant ($P = 0.582$) when using the VISAGE age prediction model (Piniewska-Róg et al. 2021). In contrast, Fiorito et al. reported that habitual alcohol consumption correlated with an increase in epigenetic age but found no significant ($P > 0.10$) difference between abstainers and occasional drinkers (Fiorito et al. 2019). Most recently, Ambroa-Conde et al. found no significant differences ($P = 0.10$) in predicted age between non-drinkers, moderate drinkers, and heavy drinkers, suggesting alcohol consumption had no effect on the methylation status of the markers included in the model (Ambroa-Conde et al. 2024).

Besides tobacco and alcohol, other lifestyle and environmental factors have been shown to influence epigenetic age. For example, exposure to pesticides (Lind et al. 2018) and fine particulate matter (Nwanaji-Enwerem et al. 2016) has been associated with increased epigenetic age, while healthy nutrition has been linked to decreased epigenetic age estimates (Quach et al. 2017). Furthermore, extreme sports may also modify aging, with a study showing accelerated DNA hypermethylation at two CpG sites in *TRIM59* and *KLF14* in elite athletes (Spólnicka et al. 2018a).

These findings suggest that epigenetic markers differ in their sensitivity to environmental factors. Therefore, selecting robust and stable markers for forensic age prediction is important to improve the accuracy of chronological age estimation (Piniewska-Róg et al. 2021). Further research should focus on identifying relevant confounding factors and developing robust models against biological influences through systematic investigations using diverse and defined cohorts (Koop et al. 2020).

5.4.6 Postmortem Blood Samples

The use of postmortem blood samples in DNAm-based age prediction models gained increasing attention. This is driven by the need to evaluate how postmortem changes might influence methylation patterns and their subsequent effect on age prediction accuracy. While most existing models rely on blood samples from living individuals, evaluating whether methylation patterns in deceased individuals differ significantly is necessary for ensuring reliable age estimation. Early studies, such as those of Bekaert et al., demonstrated that DNAm-based models applied to postmortem blood samples (0–91 years) provide prediction accuracies comparable to those from living individuals. Using pyrosequencing, Bekaert et al. developed a four-CpG model with a MAD of 3.75 years, finding no significant differences in accuracy between living and deceased samples, suggesting minimal postmortem methylation changes (Bekaert et al. 2015). These findings were supported by Hamano et al., who observed no statistically significant differences in methylation patterns in samples collected within 10 days postmortem (Hamano et al. 2016).

Further studies have supported these observations (Anaya et al. 2021; Guan et al. 2021; Grignani et al. 2024). For instance, a MLR model applied to postmortem blood samples from burned human remains achieved a MAD of 6.92 years in a test set ($n = 72$). The model was based on five AR-CpG sites (in *ELOVL2, FHL2, MIR29B2CHG, TRIM59,* and *KLF14*) selected from previous studies and used SNaPshot assay for methylation analysis. The study showed DNA integrity, assessed through the DNA degradation index, as the primary factor affecting reliability. In contrast, factors like body preservation and postmortem interval showed limited influence.

DNAm-based age prediction models show promising applicability to postmortem blood samples as postmortem-specific alterations in methylation patterns have not been detected, while DNA integrity has emerged as a factor influencing the reliability of age estimates. As there is limited research on how different stages of putrefaction impact methylation, future studies focusing on these decomposition processes are important to improve the accuracy and forensic relevance of age prediction models across various postmortem scenarios (Naue 2023).

5.5 Conclusion

Epigenetic age estimation from blood has emerged as a promising tool in forensic science. Blood, as the biological fluid most frequently found at crime scenes, offers several distinct advantages for age estimation: it is easily accessible from both living and deceased individuals, and many publicly available methylome datasets cover diverse age groups and populations. These comprehensive datasets have enabled the development of accurate prediction models, which tend to outperform those derived from other tissues (Refn et al. 2023).

Forensic age estimation has progressed from early genome-wide approaches, which provided broad insights into age-associated DNAm changes, to more targeted assays that are faster, more cost-effective, and user-friendly. While pyrosequencing has long been regarded as the gold standard for targeted DNAm analysis, its limited multiplexing capacity led to the adoption of alternative methods, such as EpiTYPER®, SNaPshot, and especially MPS. These multiplexed techniques enable the simultaneous analysis of multiple CpG sites using less DNA, an important advantage in forensic contexts where DNA quantity and quality are often restricted.

As a result of these technological advances, a wide variety of highly accurate blood-based age prediction models have been developed, differing in the number of CpG sites analyzed, the statistical approaches employed, and the sequencing method used.

Despite the advances of targeted assays for age prediction, challenges remain. Environmental factors, lifestyle, and disease can influence DNAm patterns, potentially confounding age predictions. Additionally, most current models have been developed and validated in relatively homogenous cohorts, underscoring the need for validation across diverse populations. Addressing these confounding factors will require large-scale, multi-population studies and continued technological innovation (Correia Dias et al. 2022).

In summary, for epigenetic age estimation from blood to become a standard tool in forensic laboratories, future research must focus on several key areas: (1) further validation of existing models across diverse populations and real-world forensic scenarios; (2) continued refinement of marker panels to minimize the impact of confounding variables; (3) development of standardized protocols; and (4) integration of new technologies to enhance sensitivity and robustness, especially for degraded or limited samples.

Compliance with Ethical Standards Cordulle Vanthillo declares that she has no conflict of interest. Bram Bekaert declares that he has no conflict of interest.
This chapter is a review of previously published accounts, as such, no animal or human studies were performed.

References

Aanes H, Bleka Ø, Dahlberg PS et al (2023) A new blood based epigenetic age predictor for adolescents and young adults. Sci Rep 13:2303. https://doi.org/10.1038/s41598-023-29381-7

Aliferi A, Sundaram S, Ballard D et al (2022) Combining current knowledge on DNA methylation-based age estimation towards the development of a superior forensic DNA intelligence tool. Forensic Sci Int Genet 57:102637. https://doi.org/10.1016/j.fsigen.2021.102637

Alisch RS, Barwick BG, Chopra P et al (2012) Age-associated DNA methylation in pediatric populations. Genome Res 22:623–632. https://doi.org/10.1101/gr.125187.111

Ambroa-Conde A, Casares de Cal MA, Gómez-Tato A et al (2024) Inference of tobacco and alcohol consumption habits from DNA methylation analysis of blood. Forensic Sci Int Genet 70:103022. https://doi.org/10.1016/j.fsigen.2024.103022

Anaya Y, Yew P, Roberts KA, Hardy WR (2021) DNA methylation of decedent blood samples to estimate the chronological age of human remains. Int J Legal Med 135:2163–2173. https://doi.org/10.1007/s00414-021-02650-8

Attia MAH (2022) A cautionary note on altered pace of aging in the COVID-19 era. Forensic Sci Int Genet 59:102724

Beach SRH, Dogan MV, Lei MK et al (2015) Methylomic aging as a window onto the influence of lifestyle: tobacco and alcohol use alter the rate of biological aging. J Am Geriatr Soc 63:2519–2525. https://doi.org/10.1111/jgs.13830

Bekaert B, Kamalandua A, Zapico SC et al (2015) Improved age determination of blood and teeth samples using a selected set of DNA methylation markers. Epigenetics 10:922–930. https://doi.org/10.1080/15592294.2015.1080413

Bjornsson HT, Sigurdsson MI, Fallin MD et al (2008) Intra-individual change over time in DNA methylation with familial clustering. JAMA 299:2877–2883. https://doi.org/10.1001/jama.299.24.2877

Bock C (2009) Epigenetic biomarker development. Epigenomics 1:99–110. https://doi.org/10.2217/epi.09.6

Bocklandt S, Lin W, Sehl ME et al (2011) Epigenetic predictor of age. PLoS One 6:e14821. https://doi.org/10.1371/journal.pone.0014821

Boks MP, Derks EM, Weisenberger DJ et al (2009) The relationship of DNA methylation with age, gender and genotype in twins and healthy controls. PLoS One 4(8):e6767. https://doi.org/10.1371/journal.pone.0006767

Cao X, Li W, Wang T et al (2022) Accelerated biological aging in COVID-19 patients. Nat Commun 13(1):2135. https://doi.org/10.1038/s41467-022-29801-8

Castagnola MJ, Medina-Paz F, Zapico SC (2024) Uncovering forensic evidence: a path to age estimation through DNA methylation. Int J Mol Sci 25:4917. https://doi.org/10.3390/ijms25094917

Cho S, Jung SE, Hong SR et al (2017) Independent validation of DNA-based approaches for age prediction in blood. Forensic Sci Int Genet 29:250–256. https://doi.org/10.1016/j.fsigen.2017.04.020

Christensen BC, Houseman EA, Marsit CJ et al (2009) Aging and environmental exposures alter tissue-specific DNA methylation dependent upon CPG island context. PLoS Genet 5:e1000602. https://doi.org/10.1371/journal.pgen.1000602

Correia Dias HC, Manco L (2024) Predicting age from blood by droplet digital PCR using a set of three DNA methylation markers. Forensic Sci Int 356:111950. https://doi.org/10.1016/j.forsciint.2024.111950

Correia Dias H, Cordeiro C, Corte Real F et al (2020a) Age estimation based on DNA methylation using blood samples from deceased individuals. J Forensic Sci 65:465–470. https://doi.org/10.1111/1556-4029.14185

Correia Dias H, Cordeiro C, Pereira J et al (2020b) DNA methylation age estimation in blood samples of living and deceased individuals using a multiplex SNaPshot assay. Forensic Sci Int 311:110267. https://doi.org/10.1016/j.forsciint.2020.110267

Correia Dias H, Cunha E, Corte Real F, Manco L (2022) Challenges and (un)certainties for DNAm age estimation in future. Forensic Sci 2:601–614

Daunay A, Baudrin LG, Deleuze JF, How-Kit A (2019) Evaluation of six blood-based age prediction models using DNA methylation analysis by pyrosequencing. Sci Rep 9:8862. https://doi.org/10.1038/s41598-019-45197-w

Feng L, Peng F, Li S et al (2018) Systematic feature selection improves accuracy of methylation-based forensic age estimation in Han Chinese males. Forensic Sci Int Genet 35:38–45. https://doi.org/10.1016/j.fsigen.2018.03.009

Fernandez AF, Assenov Y, Martin-Subero JI et al (2012) A DNA methylation fingerprint of 1628 human samples. Genome Res 22:407–419. https://doi.org/10.1101/gr.119867.110

Filoglu G, Sımsek SZ, Ersoy G et al (2024) Epigenetic-based age prediction in blood samples: model development. J Forensic Sci 69:869–879. https://doi.org/10.1111/1556-4029.15478

Fiorito G, McCrory C, Robinson O et al (2019) Socioeconomic position, lifestyle habits and biomarkers of epigenetic aging: a multi-cohort analysis. Aging 11:2045–2070. https://doi.org/10.18632/aging.101900

Fleckhaus J, Freire-Aradas A, Rothschild MA, Schneider PM (2017) Impact of genetic ancestry on chronological age prediction using DNA methylation analysis. Forensic Sci Int Genet Suppl Ser 6:e399–e400. https://doi.org/10.1016/j.fsigss.2017.09.162

Fleckhaus J, Bugert P, Al-Rashedi NAM, Rothschild MA (2023) Investigation of the impact of biogeographic ancestry on DNA methylation based age predictions comparing a Middle East and a Central European population. Forensic Sci Int Genet 67:102923. https://doi.org/10.1016/j.fsigen.2023.102923

Fraga MF, Ballestar E, Paz MF et al (2005) Epigenetic differences arise during the lifetime of monozygotic twins. PNAS 102:10604–10609. https://doi.org/10.1073/pnas.0500398102

Freire-Aradas A, Phillips C, Mosquera-Miguel A et al (2016) Development of a methylation marker set for forensic age estimation using analysis of public methylation data and the Agena Bioscience EpiTYPER system. Forensic Sci Int Genet 24:65–74. https://doi.org/10.1016/j.fsigen.2016.06.005

Freire-Aradas A, Phillips C, Lareu MV (2017) Forensic individual age estimation with DNA: from initial approaches to methylation tests. Forensic Sci Rev 29:121–144

Freire-Aradas A, Phillips C, Girón-Santamaría L et al (2018) Tracking age-correlated DNA methylation markers in the young. Forensic Sci Int Genet 36:50–59. https://doi.org/10.1016/j.fsigen.2018.06.011

Freire-Aradas A, Pośpiech E, Aliferi A et al (2020) A comparison of forensic age prediction models using data from four DNA methylation technologies. Front Genet 11:932. https://doi.org/10.3389/fgene.2020.00932

Freire-Aradas A, Girón-Santamaría L, Mosquera-Miguel A et al (2022) A common epigenetic clock from childhood to old age. Forensic Sci Int Genet 60:102743. https://doi.org/10.1016/j.fsigen.2022.102743

Fuke C, Shimabukuro M, Petronis A et al (2004) Age related changes in 5-methylcytosine content in human peripheral leukocytes and placentas: an HPLC-based study. Ann Hum Genet 68:196–204. https://doi.org/10.1046/j.1529-8817.2004.00081.x

Gao X, Zhang Y, Breitling LP, Brenner H (2016) Relationship of tobacco smoking and smoking-related DNA methylation with epigenetic age acceleration. Oncotarget 7:46878–46889. https://doi.org/10.18632/oncotarget.9795

Gao X, Zhang Y, Brenner H (2017) Associations of *Helicobacter pylori* infection and chronic atrophic gastritis with accelerated epigenetic ageing in older adults. Br J Cancer 117:1211–1214. https://doi.org/10.1038/bjc.2017.314

Garagnani P, Bacalini MG, Pirazzini C et al (2012) Methylation of ELOVL2 gene as a new epigenetic marker of age. Aging Cell 11:1132–1134. https://doi.org/10.1111/acel.12005

Garali I, Sahbatou M, Daunay A et al (2020) Improvements and inter-laboratory implementation and optimization of blood-based single-locus age prediction models using DNA methylation of the ELOVL2 promoter. Sci Rep 10:15652. https://doi.org/10.1038/s41598-020-72567-6

Grignani P, Bertoglio B, Monti MC et al (2024) Age estimation of burnt human remains through DNA methylation analysis. Int J Legal Med 139:175–185. https://doi.org/10.1007/s00414-024-03320-1

Gross AM, Jaeger PA, Kreisberg JF et al (2016) Methylome-wide analysis of chronic HIV infection reveals five-year increase in biological age and epigenetic targeting of HLA. Mol Cell 62:157–168. https://doi.org/10.1016/j.molcel.2016.03.019

Guan X, Ohuchi T, Hashiyada M, Funayama M (2021) Age-related DNA methylation analysis for forensic age estimation using post-mortem blood samples from Japanese individuals. Legal Med 53:101917. https://doi.org/10.1016/j.legalmed.2021.101917

Hamano Y, Manabe S, Morimoto C et al (2016) Forensic age prediction for dead or living samples by use of methylation-sensitive high resolution melting. Legal Med 21:5–10. https://doi.org/10.1016/j.legalmed.2016.05.001

Han X, Xiao C, Yi S et al (2022) Accurate age estimation from blood samples of Han Chinese individuals using eight high-performance age-related CpG sites. Int J Legal Med 136:1655–1665. https://doi.org/10.1007/s00414-022-02865-3

Hannum G, Guinney J, Zhao L et al (2013) Genome-wide methylation profiles reveal quantitative views of human aging rates. Mol Cell 49:359–367. https://doi.org/10.1016/j.molcel.2012.10.016

Heidegger A, Xavier C, Niederstätter H et al (2020) Development and optimization of the VISAGE basic prototype tool for forensic age estimation. Forensic Sci Int Genet 48:102322. https://doi.org/10.1016/j.fsigen.2020.102322

Horvath S (2013) DNA methylation age of human tissues and cell types. Genome Biol 14:3156. https://doi.org/10.1186/gb-2013-14-10-r115
Horvath S, Levine AJ (2015) HIV-1 infection accelerates age according to the epigenetic clock. J Infect Dis 212:1563–1573. https://doi.org/10.1093/infdis/jiv277
Horvath S, Raj K (2018) DNA methylation-based biomarkers and the epigenetic clock theory of ageing. Nat Rev Genet 19:371–384
Horvath S, Ritz BR (2015) Increased epigenetic age and granulocyte counts in the blood of Parkinson's disease patients. Aging 7:1130–1142. https://doi.org/10.18632/aging.100859
Huang Y, Yan J, Hou J et al (2015) Developing a DNA methylation assay for human age prediction in blood and bloodstain. Forensic Sci Int Genet 17:129–136. https://doi.org/10.1016/j.fsigen.2015.05.007
Jiang S, Guo Y (2020) Epigenetic clock: DNA methylation in aging. Stem Cells Int 2020:1047896. https://doi.org/10.1155/2020/1047896
Kananen L, Nevalainen T, Jylhävä J et al (2015) Cytomegalovirus infection accelerates epigenetic aging. Exp Gerontol 72:227–229. https://doi.org/10.1016/j.exger.2015.10.008
Kim JY, Vijayakumar KA, Cho GW (2025) Exploring the impact of methylation aging on acute myeloid leukemia: insights from the aging clock. Leuk Res 148:107620. https://doi.org/10.1016/j.leukres.2024.107620
Koch CM, Wagner W (2011) Epigenetic-aging-signature to determine age in different tissues. Aging 3:1018–1027. https://doi.org/10.18632/aging.100395
Koop BE, Reckert A, Becker J et al (2020) Epigenetic clocks may come out of rhythm-implications for the estimation of chronological age in forensic casework. Int J Legal Med 134:2215–2228. https://doi.org/10.1007/s00414-020-02375-0/Published
Li C, Gao W, Gao Y et al (2018) Age prediction of children and adolescents aged 6-17 years: an epigenome-wide analysis of DNA methylation. Aging 10:1015–1026. https://doi.org/10.18632/aging.101445
Lind PM, Salihovic S, Lind L (2018) High plasma organochlorine pesticide levels are related to increased biological age as calculated by DNA methylation analysis. Environ Int 113:109–113. https://doi.org/10.1016/j.envint.2018.01.019
Manco L, Correia Dias H (2022) DNA methylation analysis of ELOVL2 gene using droplet digital PCR for age estimation purposes. Forensic Sci Int 333:111206. https://doi.org/10.1016/j.forsciint.2022.111206
Mathew JA, Paul G, Jacob J et al (2025) A new robust AI/ML based model for accurate forensic age estimation using DNA methylation markers. Forensic Sci Med Pathol 21(3):1145–1162. https://doi.org/10.1007/s12024-025-00985-x
Naue J (2023) Getting the chronological age out of DNA: using insights of age-dependent DNA methylation for forensic DNA applications. Genes Genomics 45:1239–1261. https://doi.org/10.1007/s13258-023-01392-8
Naue J, Hoefsloot HCJ, Mook ORF et al (2017) Chronological age prediction based on DNA methylation: massive parallel sequencing and random forest regression. Forensic Sci Int Genet 31:19–28. https://doi.org/10.1016/j.fsigen.2017.07.015
Nwanaji-Enwerem JC, Colicino E, Trevisi L et al (2016) Long-term ambient particle exposures and blood DNA methylation age: findings from the VA normative aging study. Environ Epigenet 2:dvw006. https://doi.org/10.1093/eep/dvw006
Ochana B-L, Nudelman D, Cohen D et al (2024) Time is encoded by methylation changes at clustered CpG sites. bioRxiv:2024.12.03.626674
Pan C, Yi S, Xiao C et al (2020) The evaluation of seven age-related CpGs for forensic purpose in blood from Chinese Han population. Forensic Sci Int Genet 46:102251. https://doi.org/10.1016/j.fsigen.2020.102251
Paparazzo E, Lagani V, Geracitano S et al (2023) An ELOVL2-based epigenetic clock for forensic age prediction: a systematic review. Int J Mol Sci 24(3):2254
Parson W (2018) Age estimation with DNA: from forensic DNA fingerprinting to forensic (epi) genomics: a mini-review. Gerontology 64:326–332. https://doi.org/10.1159/000486239

Peng F, Feng L, Chen J et al (2019) Validation of methylation-based forensic age estimation in time-series bloodstains on FTA cards and gauze at room temperature conditions. Forensic Sci Int Genet 40:168–174. https://doi.org/10.1016/j.fsigen.2019.03.006

Piniewska-Róg D, Heidegger A, Pośpiech E et al (2021) Impact of excessive alcohol abuse on age prediction using the VISAGE enhanced tool for epigenetic age estimation in blood. Int J Legal Med 135:2209–2219. https://doi.org/10.1007/s00414-021-02665-1

Pokupcic K (2017) Blood as an important tool in criminal investigation. J Forensic Sci Crim Investig 3:555608. https://doi.org/10.19080/JFSCI.2017.03.555608

Quach A, Levine ME, Tanaka T et al (2017) Epigenetic clock analysis of diet, exercise, education, and lifestyle factors. Aging (Albany NY) 9(2):419–446

Rakyan VK, Down TA, Maslau S et al (2010) Human aging-associated DNA hypermethylation occurs preferentially at bivalent chromatin domains. Genome Res 20:434–439. https://doi.org/10.1101/gr.103101.109

Refn MR, Kampmann ML, Morling N et al (2023) Prediction of chronological age and its applications in forensic casework: methods, current practices, and future perspectives. Forensic Sci Res 8:85–97. https://doi.org/10.1093/fsr/owad021

Smeers I, Decorte R, Van de Voorde W, Bekaert B (2018) Evaluation of three statistical prediction models for forensic age prediction based on DNA methylation. Forensic Sci Int Genet 34:128–133. https://doi.org/10.1016/j.fsigen.2018.02.008

Spólnicka M, Piekarska RZ, Jaskuła E et al (2016) Donor age and C1orf132/MIR29B2C determine age-related methylation signature of blood after allogeneic hematopoietic stem cell transplantation. Clin Epigenetics 8(1):93

Spólnicka M, Pośpiech E, Adamczyk JG et al (2018a) Modified aging of elite athletes revealed by analysis of epigenetic age markers. Aging 10:241–252. https://doi.org/10.18632/aging.101385

Spólnicka M, Pośpiech E, Pepłońska B et al (2018b) DNA methylation in ELOVL2 and C1orf132 correctly predicted chronological age of individuals from three disease groups. Int J Legal Med 132:1–11. https://doi.org/10.1007/s00414-017-1636-0

Spólnicka M, Zbieć-Piekarska R, Karp M et al (2018c) DNA methylation signature in blood does not predict calendar age in patients with chronic lymphocytic leukemia but may alert to the presence of disease. Forensic Sci Int Genet 34:e15–e17. https://doi.org/10.1016/j.fsigen.2018.02.004

Stephenson M, Bollepalli S, Cazaly E et al (2021) Associations of alcohol consumption with epigenome-wide DNA methylation and epigenetic age acceleration: individual-level and co-twin comparison analyses. Alcohol Clin Exp Res 45:318–328. https://doi.org/10.1111/acer.14528

Suchiman HED, Slieker RC, Kremer D et al (2015) Design, measurement and processing of region-specific DNA methylation assays: the mass spectrometry-based method EpiTYPER. Front Genet 6:287. https://doi.org/10.3389/fgene.2015.00287

Tajuddin SM, Hernandez DG, Chen BH et al (2019) Novel age-associated DNA methylation changes and epigenetic age acceleration in middle-aged African Americans and whites. Clin Epigenetics 11:119. https://doi.org/10.1186/s13148-019-0722-1

Teschendorff AE, Menon U, Gentry-Maharaj A et al (2010) Age-dependent DNA methylation of genes that are suppressed in stem cells is a hallmark of cancer. Genome Res 20:440–446. https://doi.org/10.1101/gr.103606.109

Teschendorff AE, West J, Beck S (2013) Age-associated epigenetic drift: implications, and a case of epigenetic thrift? Hum Mol Genet 22:R7. https://doi.org/10.1093/HMG/DDT375

Thong Z, Chan XLS, Tan JYY et al (2017) Evaluation of DNA methylation-based age prediction on blood. Forensic Sci Int Genet Suppl Ser 6:e249–e251. https://doi.org/10.1016/j.fsigss.2017.09.095

Thong Z, Tan JYY, Loo ES et al (2021) Artificial neural network, predictor variables and sensitivity threshold for DNA methylation-based age prediction using blood samples. Sci Rep 11:1744. https://doi.org/10.1038/s41598-021-81556-2

Verstraeten L, Fokias K, Saerens G, Bekaert B (2025) Cost-effective optimisation and validation of the VISAGE enhanced tool assay on the Illumina NovaSeq 6000 platform. Forensic Sci Int Genet 78:103299. https://doi.org/10.1016/j.fsigen.2025.103299

Vidaki A, Daniel B, Court DS (2013) Forensic DNA methylation profiling—potential opportunities and challenges. Forensic Sci Int Genet 7:499–507. https://doi.org/10.1016/j.fsigen.2013.05.004

Vidaki A, Ballard D, Aliferi A et al (2017) DNA methylation-based forensic age prediction using artificial neural networks and next generation sequencing. Forensic Sci Int Genet 28:225–236. https://doi.org/10.1016/j.fsigen.2017.02.009

Vidaki A, Planterose Jiménez B, Poggiali B et al (2023) Targeted DNA methylation analysis and prediction of smoking habits in blood based on massively parallel sequencing. Forensic Sci Int Genet 65:102878. https://doi.org/10.1016/j.fsigen.2023.102878

Vidal-Bralo L, Lopez-Golan Y, Gonzalez A (2016) Simplified assay for epigenetic age estimation in whole blood of adults. Front Genet 7:126. https://doi.org/10.3389/fgene.2016.00126

Weidner CI, Lin Q, Koch CM et al (2014) Aging of blood can be tracked by DNA methylation changes at just three CpG sites. Genome Biol 15(2):R24. https://doi.org/10.1186/gb-2014-15-2-r24

Weidner CI, Ziegler P, Hahn M et al (2015) Epigenetic aging upon allogeneic transplantation: the hematopoietic niche does not affect age-associated DNA methylation. Leukemia 29:985–988

Woźniak A, Heidegger A, Piniewska-Róg D et al (2021) Development of the VISAGE enhanced tool and statistical models for epigenetic age estimation in blood, buccal cells and bones. Aging 13:6459–6484. https://doi.org/10.18632/aging.202783

Yamagishi T, Sakurai W, Watanabe K et al (2024) Development and comparison of forensic interval age prediction models by statistical and machine learning methods based on the methylation rates of ELOVL2 in blood DNA. Forensic Sci Int Genet 69:103004. https://doi.org/10.1016/j.fsigen.2023.103004

Yang F, Qian J, Qu H et al (2023) DNA methylation-based age prediction with bloodstains using pyrosequencing and random forest regression. Electrophoresis 44:835–844. https://doi.org/10.1002/elps.202200250

Ye Z, Jiang L, Zhao M et al (2022) Epigenome-wide screening of CpG markers to develop a multiplex methylation SNaPshot assay for age prediction. Legal Med 59:102115. https://doi.org/10.1016/j.legalmed.2022.102115

Yusipov I, Kalyakulina A, Trukhanov A et al (2024) Map of epigenetic age acceleration: a worldwide analysis. Ageing Res Rev 100:102418

Zaghlool SB, Al-Shafai M, Al Muftah WA et al (2015) Association of DNA methylation with age, gender, and smoking in an Arab population. Clin Epigenetics 7:7–6. https://doi.org/10.1186/s13148-014-0040-6

Zbieć-Piekarska R, Spólnicka M, Kupiec T et al (2015a) Development of a forensically useful age prediction method based on DNA methylation analysis. Forensic Sci Int Genet 17:173–179. https://doi.org/10.1016/j.fsigen.2015.05.001

Zbieć-Piekarska R, Spólnicka M, Kupiec T et al (2015b) Examination of DNA methylation status of the ELOVL2 marker may be useful for human age prediction in forensic science. Forensic Sci Int Genet 14:161–167. https://doi.org/10.1016/j.fsigen.2014.10.002

Zhu B, Li D, Han G et al (2025) Multiplexing and massive parallel sequencing of targeted DNA methylation to predict chronological age. Front Aging 6:1467639. https://doi.org/10.3389/fragi.2025.1467639

Chapter 6
Epigenetics of Y Chromosome for Age Estimation

Denisse Stephania Becerra-Loaiza, José Alonso Aguilar-Velázquez, and Mariano Guardado-Estrada

Abstract Although the Y chromosome is one of the smallest chromosomes in the human genome, it possesses unique characteristics that have attracted attention across multiple disciplines, including evolutionary biology, medicine, and forensic science. The Y chromosome is a haploid genome inherited patrilineally, without recombination, and carries key genes related to male development. Because it is present only in males and follows a distinct inheritance pattern, it has been widely studied in forensic contexts, particularly in cases involving male individuals, such as sexual assault investigations. Moreover, the Y chromosome is dynamic and has been shown to undergo physical changes with age, including loss in elderly males. DNA methylation, which is associated with gene regulation, has recently been correlated with ageing processes. Consequently, the analysis of DNA methylation has become increasingly relevant in forensic science as a tool for estimating chronological age. Exploring Y chromosome methylation for age prediction could extend its forensic applications. In this chapter, we review recent advances in the study of Y chromosome and DNA methylation potential for age estimation.

Keywords Y chromosome · Methylation · Age prediction · Forensics · Aging

D. S. Becerra-Loaiza
Laboratorio de Ciencias Morfológico Forenses y Medicina Molecular, Departamento de Morfología, Centro Universitario de Ciencias de la Salud, Universidad de Guadalajara, Guadalajara, Jalisco, Mexico

Departamento de Aparatos y Sistemas II, Universidad Autónoma de Guadalajara, Zapopan, Jalisco, Mexico

J. A. Aguilar-Velázquez
Laboratorio de Ciencias Morfológico Forenses y Medicina Molecular, Departamento de Morfología, Centro Universitario de Ciencias de la Salud, Universidad de Guadalajara, Guadalajara, Jalisco, Mexico

M. Guardado-Estrada (✉)
Laboratorio de Genética, Escuela Nacional de Ciencias Forenses, Universidad Nacional Autónoma de México, Ciudad de México, Mexico
e-mail: mguardado@enacif.unam.mx

S. C. Zapico (ed.), *Forensic Science and Epigenetics*, Epigenetics and Human Health 18, https://doi.org/10.1007/978-3-032-23085-0_6

Abbreviations

BGA	Biogeographical ancestry
ddPCR	Droplet digital PCR
DNA	Deoxyribonucleic acid
DNAm	DNA methylation
DTT	Dithiothreitol
EVC	External visible characters
EVCs	External visible characteristics
FDP	DNA phenotyping
HID	Human identification
Kya	Kiloyears
LR	Likelihood ratio
MAD	Mean absolute deviation
MAE	Mean absolute error
miRNAs	microRNAs
mLOY	Mosaic loss of Y chromosome
MPS	Massively parallel sequencing
mRNA	Messenger RNA
MSY	Male-specific region of the Y chromosome
PARs	Pseudoautosomal regions
PCR	Polymerase chain reaction
RM Y-STRs	Rapidly mutating Y-STRs
SNPs	Single-nucleotide polymorphism
TMRCA	Time to the most recent common ancestor
WGBS	Whole-genome bisulfite sequencing
Y-SNPs	Y-linked single nucleotide polymorphisms
Y-STRs	Y-linked short tandem repeats
YHRD	Y Chromosome Haplotype Reference Database

6.1 Introduction

The Y chromosome, a hallmark of male genetics, provides a distinctive molecular framework for investigating human aging through the lens of epigenetics. As the only haploid and strictly paternally inherited, non-recombining region of the nuclear genome, the Y chromosome accumulates lineage-specific mutations and epigenetic modifications that are uniquely traceable across generations (Bonito et al. 2021). These features, together with its evolutionary history and uniparental mode of inheritance, have made the Y chromosome a powerful tool for studying human population dynamics, migrations, and genetic ancestry (Poznik et al. 2016).

Beyond evolutionary genetics, the Y chromosome holds growing significance in forensic sciences. Its male specificity, stable mutation rate, and structural isolation make it ideal for tracing paternal lineages, resolving kinship, and analyzing complex biological samples—especially when autosomal or mitochondrial DNA is limited or degraded. Y-specific markers such as Y-linked short tandem repeats (Y-STRs) and Y-linked single nucleotide polymorphisms (Y-SNPs) have become indispensable in forensic applications, including paternity testing, identification of disaster victims, and sexual assault investigations involving mixed male-female DNA samples (Gusmão et al. 2008; Kayser 2017; Syndercombe-Court 2021).

Recent developments in the field of epigenetics have expanded the functional relevance of the Y chromosome. Epigenetic markers—particularly patterns of DNA methylation—have emerged as highly accurate indicators of biological age. These age-related changes, often referred to as "epigenetic clocks", are gaining increasing attention for their applications in clinical diagnostics, population studies, and forensic casework (Kayser 2017). Importantly, the application of epigenetic clocks to the Y chromosome presents new opportunities to estimate age in male individuals, especially in scenarios where male-specific age estimation is critical, such as mixed DNA samples from physical or sexual assault cases (Vidaki et al. 2021).

The biological isolation of the Y chromosome has allowed for the long-term preservation of molecular signatures, making it particularly well suited for calibrating molecular clocks based on mutation accumulation and haplotype clustering (Balanovsky and Zaporozhchenko 2016). In this context, epigenetic models targeting the Y chromosome hold promise for refining age estimation methodologies within male populations.

This chapter examines the role of the Y chromosome in age estimation from an epigenetic perspective. It begins by reviewing the structural and evolutionary characteristics of the Y chromosome, then addresses its utility in forensic genetics. The discussion proceeds to the phenomenon of age-related mosaic Y chromosome loss and introduces key epigenetic concepts. Finally, the chapter highlights current models and emerging research supporting the use of Y-linked epigenetic data for biological age prediction in men.

6.2 The Y Chromosome

6.2.1 Evolutionary Origins and Significance

The evolution of the Y chromosome began with mutation in the *SOX3* gene, which gave rise to the *SRY* gene, a key determinant of male sex. This marked the divergence of the human Y chromosome originated from a homologous pair of autosomes that progressively diverged following the evolution of sex-determining mechanisms (Subrini and Turner 2021). The Y-autosome fusion was approximately 115 million years ago in eutherian mammals. The accumulation of male-beneficial mutations in the vicinity of the *SRY* gene triggered selective pressure to suppress X-Y recombination, leading to the formation of non-recombining regions of the Y chromosome, leaving recombination intact only in the pseudoautosomal regions (PARs) at both chromosome ends (Subrini and Turner 2021; Charlesworth 2023; Wimmer et al. 2005). Additionally, retrotransposition and translocations from autosomes contributed to the evolution of the Y chromosome in various mammalian lineages (Subrini and Turner 2021).

Due to its lack of recombination and unique inheritance pattern father-to-son, the Y chromosome serves as a repository of "frozen accidents"—molecular signatures of ancient genomic events. These evolutionary breakpoints make the Y a valuable tool in tracing lineage history and understanding genome evolution. Its structural and functional instability, coupled with species-specific divergence, sets it apart from the X chromosome and autosomes (Wimmer et al. 2005; Bachtrog 2020). The use of the Y chromosome in evolutionary studies and forensic applications underscores its importance in understanding human genetic diversity and ancestry (Calafell and Larmuseau 2017).

Genomic studies of the human Y chromosome, initiated in the 1980s and culminating in the publication of its complete sequence in 2003, overturned the earlier "genetic wasteland" perception of the Y and revealed its biological and medical significance (Hughes and Rozen 2012). Since then, Y chromosome polymorphisms—particularly SNPs and microsatellites—have been extensively used to reconstruct male lineage history. These markers have enabled estimates of the time to the most recent common ancestor (TMRCA), known as "Y-chromosomal Adam", dated to approximately 254 kya in Africa. This was followed by a major expansion of non-African lineages consistent with the Out-of-Africa dispersal around 47–52 kya. Additionally, a pronounced bottleneck in Y chromosome diversity occurred around 10 kya, likely associated with cultural shifts that increased male reproductive variance (Jobling and Tyler-Smith 2017; Karmin et al. 2015; Hughes and Rozen 2012).

6.2.2 *Structural Features*

The human Y chromosome is unique in the realm of genetics, deviating from many fundamental genetic principles. Unlike other chromosomes, the Y chromosome is not essential for life, as females can survive without it (Jobling and Tyler-Smith 2003). Structurally, the Y chromosome spans ~60 megabases and constitutes about 2–3% of the haploid human genome. It is cytogenetically divided into five structurally and evolutionary distinct regions: the PARs (PAR1 and PAR2), euchromatic sequences, and heterochromatic blocks—each shaped by different genomic forces (Hughes and Rozen 2012; Bonito et al. 2021) (Fig. 6.1). The PARs are relatively small—covering ~2.6 Mb and 320 kb, respectively—and allow for homologous recombination. The remaining ~95% comprises the non-recombining region of the Y chromosome (NRY, MSY, or NRPY), which is transmitted exclusively from father to son (Quintana-Murci and Fellous 2001).

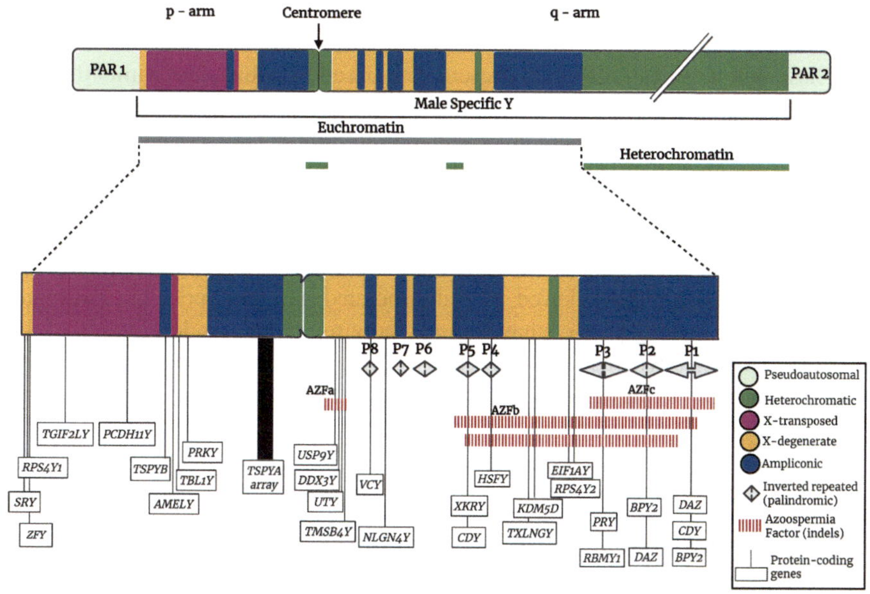

Fig. 6.1 Structural and genetic organization of the human Y chromosome. The Y chromosome is cytogenetically classified as acrocentric, with the centromere positioned near the short arm (p-arm), resulting in an extended long arm (q-arm). This schematic representation highlights key domains: the pseudoautosomal regions (PAR1 and PAR2), which recombine with the X chromosome, and the male-specific region of the Y chromosome (MSY), which is non-recombining and subdivided into: X-transposed regions, which share high sequence identity with the X chromosome, X-degenerate regions, containing relics of ancient X-Y homologs, Ampliconic regions, rich in palindromic sequences and gene families related to spermatogenesis, Heterochromatic regions, which are gene-poor and transcriptionally inactive. The euchromatic region encompasses regions with transcriptionally active genes, while the heterochromatic blocks are located predominantly on the q-arm. Below, a magnified view of the MSY illustrates the location of protein-coding genes, many of which are involved in testis function and fertility (e.g., *SRY, DAZ, CDY, TSPY, RBMY*). Azoospermia factor (AZF) regions—AZFa, AZFb, and AZFc—where microdeletions are associated with male infertility. Gray arrows indicate palindromic inverted repeats (P1–P8) that mediate intra-chromosomal recombination events, contributing to structural instability and gene conversion within the MSY. Notably, genes embedded within the palindromic arms are present in inverted orientation, with nearly identical copies located on each arm of the palindrome in sense and antisense (forward and reverse complement) orientation, allowing for gene conversion events that preserve sequence integrity despite the lack of inter-chromosomal recombination. (Created in Biorender, based on and modified from Lucotte et al. 2023; Jobling et al. 2003, 2017; Hughes and Rozen 2012; Skaletsky et al. 2003)

Y-linked genes in the NRY are typically grouped into two categories: (1) ubiquitously expressed housekeeping genes with X homologues, usually present in a single copy; and (2) testis-specific genes, often in multiple copies, which are critical for spermatogenesis and male fertility (Quintana-Murci and Fellous 2001). The male-specific region of the Y chromosome (MSY) includes three classes of euchromatic sequences: X-transposed, X-degenerated, and ampliconic. The ampliconic regions are especially notable, containing extensive intra-chromosomal duplications arranged in eight palindromes (P1–P8), which exhibit arm-to-arm nucleotide identity >99% due to frequent gene conversion events (Bonito et al. 2021).

These palindromes functionally resemble a pseudo-diploid state within the otherwise haploid MSY, suggesting an evolutionary mechanism for preserving crucial testis-specific multicopy genes from degeneration (Subrini and Turner 2021; Jobling and Tyler-Smith 2017). The evolutionary and functional structure of the Y chromosome reveals its crucial role in sex determination, with the presence of the *SRY* gene near the PAR1 boundary acting as a master switch for male development (Jobling and Tyler-Smith 2017).

Although traditionally viewed as a degenerated and gene-poor chromosome, large-scale sequencing projects have dramatically revised our understanding of the Y. The X-degenerated regions show high conservation across species, while the ampliconic regions, enriched in palindromes and multicopy genes, are subject to rapid structural evolution and lineage-specific rearrangements.

6.2.3 Y Chromosome Degeneration and Gene Retention

Multiple evolutionary models have been proposed to explain the degeneration of the male-specific region of the Y chromosome (MSY). Due to the lack of recombination with a homologous chromosome, the MSY has become a hotspot for the accumulation of mutations and structural changes—such as substitutions, duplications, and deletions—that are transmitted unchanged along the paternal lineage.

While this recombination suppression has facilitated genetic decay, the loss of ancestral genes is not entirely stochastic. Genes critical for male fertility and those that are dosage-sensitive tend to be selectively retained, reflecting functional constraints and the action of purifying selection (Bachtrog 2020; Kuroki and Fukami 2023). A major mechanism counteracting degeneration is gene conversion, particularly in palindromic regions of the MSY. These regions allow for intrachomosomal correction of deleterious mutations between homologous arms, thereby preserving the integrity of ampliconic, testis-specific gene families (Lucotte et al. 2023; Bonito et al. 2021).

Moreover, the mutation rate in palindromic gene conversion is significantly lower than in their spacer regions, supporting the view that gene conversion provides a selective advantage by reducing the mutational load and maintaining sequence fidelity. This mechanism is vital for maintaining male fertility, as it helps ensure the structural and functional preservation of key Y-linked genes over evolutionary timescales (Trombetta and Cruciani 2017).

Several theoretical models describe the processes contributing to MSY degeneration. Muller's ratchet explains the irreversible accumulation of deleterious mutations in non-recombining genomes. Genetic hitchhiking allows harmful mutations to become fixed due to linkage with advantageous alleles, while the "ruby in the rubbish" model describes the reduced efficacy of natural selection in separating beneficial mutations from linked deleterious ones. Collectively, these dynamics promote pseudogenization, expansion of repetitive DNA, and progressive reduction in Y chromosome size via large-scale deletions (Bachtrog 2013).

Despite this, the MSY has evolved compensatory mechanisms. Over 50% of its content comprises ampliconic regions embedded in palindromic structures, which facilitate gene conversion and can restore functionality to damaged gene copies. However, this same architecture predisposes the Y to intrachromosomal rearrangements and elevated mutation rates, often implicated in male infertility syndromes (Bachtrog 2013; Jobling and Tyler-Smith 2017). The high mutational activity observed in these regions underscores the balance between genomic maintenance and vulnerability (Lucotte et al. 2023).

Contrary to early hypotheses predicting gradual erosion and eventual extinction of the Y chromosome, comparative genomics across primates tells a different history. Complete MSY sequences in humans, chimpanzees, and rhesus macaques reveal that the human Y has maintained a remarkably stable gene repertoire over the past 25 million years, with gene loss largely restricted to the most recently evolved stratum (Hughes and Rozen 2012).

The "fragile Y hypothesis" adds another layer of complexity, suggesting that reduced pseudoautosomal regions (PARs) increase the risk of Y chromosome aneuploidy due to insufficient chiasma formation during male meiosis. This instability may promote gene translocation to autosomes, structural arrangements, or even alternative meiotic mechanisms such as achiasmatic segregation (Blackmon and Demuth 2015). Although complete Y loss is unlikely—given the essential role of genes like SRY—the hypothesis emphasizes the chromosome's susceptibility to structural and functional changes.

Altogether, these evolutionary forces explain why the Y chromosome has become highly specialized, reduced in gene content, and enriched in structural complexity. Far from being a passive or obsolete element, the Y chromosome has developed unique mechanisms to maintain its integrity and function under intense degenerative pressure. Its evolutionary dynamics reveal multiple bottlenecks that have shaped not only its diversity but also its enduring role in male biology.

6.3 Uses of the Y Chromosome in the Forensic Field

Analysis of the Y chromosome has been relatively neglected in forensic casework for some time, as it does not provide the same level of certainty in identification that autosomal DNA can offer. Genetic characteristics on the human Y chromosome serve as a lineage marker, in the form of a single block of alleles, named "haplotype" that is transferred directly from father to son (Jobling and Tyler-Smith 1995) (Fig. 6.2). Therefore, the uniparental nature of this marker can be useful in certain circumstances, such as sexual assault, missing persons, disaster victim identification, complex kinship analysis, and population inference (Jobling et al. 1997).

Typically, evidence for biological relationships using the Y chromosome (CY) is based on the analysis of Y-chromosomal short tandem repeats (Y-STRs). Similar to autosomal genetic systems, the genetic profiles of two or more individuals are

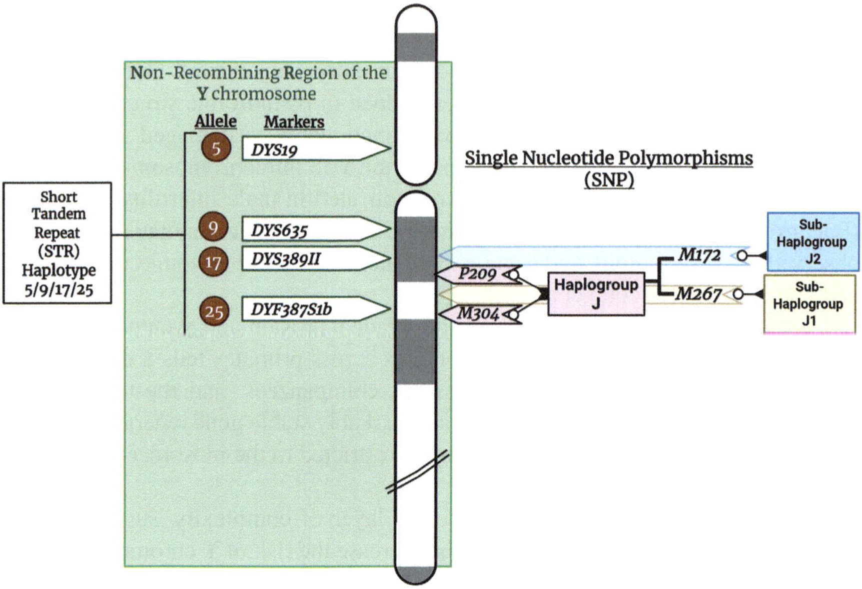

Fig. 6.2 Graphical representation of a haplotype and a haplogroup of the Y chromosome. (**a**) The image shows how the STRs located in the non-recombining region of the Y chromosome (NRY) are inherited together (as a block), constituting a haplotype. The markers DYS19 (allele 5), DYS635 (allele 9), DYS389II (allele 17), and DYF387S1b (allele 25) generate the haplotype 5/9/17/25 for the markers mentioned above, a specific allelic code of the paternal lineage of the analyzed individual; that is, all paternal relatives should have the same haplotype. (**b**) Representation of the mutations or Y-SNPs that define haplogroup J and its sub-haplogroups J1 and J2. Haplogroup J is defined by M304 and M209, and when it also possesses M172 or M267, the sub-haplogroups J1 or J2 are defined, respectively

compared to assess potential kinship (e.g., paternity, sibling relationships, grandparentage, etc.). However, since all males who share a paternal lineage possess the same Y-STR haplotype, it is not possible to distinguish specific biological relationships such as father–son, brother–brother, or uncle–nephew. Consequently, while Y-STRs can indicate a paternal connection, they cannot define the exact nature of the relationship between individuals.

Despite this limitation, Y-STRs can still be informative in specific contexts. For instance, when the putative father is deceased or unavailable, a match between the Y-STR haplotype of the child and that of another paternally related male (e.g., a paternal uncle or grandfather) can provide supporting evidence of biological relatedness. Y-STRs are also commonly used in inheritance disputes, where individuals may seek to establish a biological link to claim rights over an estate. In such cases, analyzing a large number of Y-STR markers is recommended, while avoiding rapidly mutating Y-STRs (RM Y-STRs), which can complicate interpretation. Moreover, Y-STR analysis is valuable in disaster victim identification and in the investigation of mass graves, where determining the paternal lineage of unidentified remains can aid in reestablishing family connections (Syndercombe-Court 2021).

6.3.1 Kinship Analyses

The analysis of genetic evidence for kinship using Y-STRs involves the comparison of haplotypes between male individuals. DNA is typically extracted from biological samples, and specific Y-STR loci are amplified using PCR. The resulting haplotypes are then compared to determine whether the individuals share a common paternal lineage. A match between Y-STR haplotypes supports a paternal relationship, while differences generally exclude it. Due to the lack of recombination in the Y chromosome, this method is particularly suited for tracing direct male-line ancestry across multiple generations. To evaluate the statistical weight of Y-STR results, the haplotype observed in the tested individuals is compared against population databases to determine its frequency. The rarer the haplotype in the relevant population, the greater its evidential value. This frequency is used to calculate a likelihood ratio (LR), which expresses the likelihood that the observed match is more likely under the hypothesis of relatedness than under the hypothesis of unrelatedness. However, due to the lack of recombination and inheritance as a haplotype, all male relatives in the same paternal line share the same haplotype, so the LR reflects lineage sharing rather than a specific relationship (Ge et al. 2023).

6.3.2 Criminal Investigations

Y-STRs have been used in forensic casework for nearly three decades (Prinz and Sansone 2001), with one of their key advantages being their utility in sexual assault investigations. In cases involving mixed DNA samples, Y-STRs enable the detection of a minor male DNA contribution even when it is present in a much smaller proportion than female DNA. Moreover, Y-STR analysis can sometimes reveal the presence of multiple male contributors, which may not be apparent through standard autosomal STR analysis.

Routine investigation of sexual assault typically begins with chemical tests to suggest the presence of semen, followed by microscopic confirmation of sperm cells. This is often succeeded by a differential extraction—also known as differential lysis—which separates epithelial cells from sperm cells to facilitate autosomal DNA profiling of the male-specific fraction. The differential extraction technique was first described by Peter Gill in 1985 (Gill et al. 1985), during the early development of DNA analysis and before the publication of the polymerase chain reaction (PCR) method (Mullis et al. 1986).

Gill's method, which has since been refined, involves incubating DNA extracts in a solution containing SDS, proteinase K, and dithiothreitol (DTT). This treatment preferentially lyses epithelial cells, while sperm nuclei, due to their compact chromatin structure rich in protamines, remain largely intact. This structural resilience protects sperm DNA in harsh lytic environments and plays a role in preserving genomic integrity during fertilization (Champroux et al. 2016).

Despite their utility, the use of Y-STRs in routine forensic practice remains limited. This is mainly due to their inability to identify an individual (as they represent a shared paternal lineage), as well as concerns about conserving precious forensic material, since Y-STR analysis requires separate testing. The advent of massively parallel sequencing (MPS) offers the potential to simultaneously analyze numerous markers, including Y-STRs, in a single assay. However, until MPS becomes standard in forensic laboratories, decisions regarding the use of Y-STRs will depend on institutional policies and specific case circumstances.

Additionally, without prior enrichment of male DNA, MPS-based multiplexes may still face challenges in detecting low-level male contributions due to competitive amplification by more abundant DNA. In certain scenarios—such as when semen originates from an azoospermic male or when non-sperm male DNA is present at very low concentrations due to contact or penetration—Y-STR analysis can provide critical evidence.

In criminal casework, the statistical analysis of Y-STR profiles involves estimating the rarity of the observed haplotype in a reference population. Unlike autosomal DNA, where random match probabilities can be calculated assuming independent inheritance of loci, Y-STR markers are inherited as a haplotype, limiting the statistical approach to frequency-based methods. The most common method is to determine the haplotype frequency using a relevant population database, such as the Y Chromosome Haplotype Reference Database (YHRD), applying the counting method or the upper 95% confidence interval as a conservative estimate. When a match is found, the evidential value is expressed using a likelihood ratio (LR), which compares the probability of observing the haplotype under two competing hypotheses: (1) that the DNA originated from a male in the same paternal lineage (e.g., the suspect) versus (2) an unrelated male from the population. It is essential to account for population substructure and database size, and to interpret results with caution, recognizing that Y-STRs cannot uniquely identify an individual but rather a paternal lineage (Ge et al. 2023).

6.3.3 Y Chromosome Haplotype Reference Database (YHRD)

A crucial resource in forensic practice is the creation of DNA databases, which have become one of the most effective tools in criminal investigations for identifying unknown perpetrators. Y Chromosome Haplotype Reference Database (YHRD) is a free-access (https://yhrd.org/) Y chromosome reference database established in 1999 by the Institute of Legal Medicine and Forensic Sciences at Humboldt University in Berlin. Its purpose is to collect and organize Y chromosome haplotypes from male individuals across various regions of the world. YHRD provides tools for haplotype searches, frequency estimation, and biogeographic relationship analysis through the frequency distribution of Y-STR and Y-SNP markers, made possible by the collaboration of multiple forensic laboratories worldwide (Roewer et al. 2001). Today, YHRD reports a total of 349,750 haplotypes. Among its main datasets are the following: Y12, with 309,090 haplotypes; Y17, with 289,405 haplotypes; Y23, with 103,280 haplotypes; Y27, with 106,444 haplotypes; and a category that includes all available Y-STR markers in the platform, with 46,773 registered haplotypes.

In addition to serving as a repository for Y-STR haplotype data, YHRD offers a suite of statistical tools that are essential in forensic casework, particularly for calculating haplotype frequencies in criminal investigations and kinship analyses. In cases involving unknown male perpetrators, YHRD enables forensic practitioners to determine the rarity of a given haplotype within relevant reference populations. This is accomplished through the "Estimate Haplotype Frequency" tool, which uses the counting method and incorporates the augmented counting method (also known as the count +1 method) to avoid zero frequency issues when the haplotype in question is not directly observed in the database.

YHRD also supports casework involving male lineage relationships, such as paternity testing, missing persons identification, or familial searching, where the evaluation of shared Y-STR haplotypes is central. In these contexts, the "Kinship Analysis" function allows for the comparison of haplotypes between two or more individuals and calculates the likelihood ratio (LR) based on user-defined hypotheses (e.g., whether two samples come from patrilineally related individuals). This feature provides crucial statistical support in determining whether a genetic match supports a claimed biological relationship or indicates an unrelated origin. Together, these analytical capabilities make YHRD not only a comprehensive reference database but also an indispensable tool for forensic geneticists engaged in both criminal and kinship-related casework.

6.3.4 Ancestry Inference

Biogeographical ancestry (BGA) can be defined as the genetic background shared among different human populations. BGA can be inferred at the population level and even at the individual level due to population studies that analyze the genetic diversity of different genetic biomarkers in populations worldwide. Since the origin of modern humans and their dispersal from Africa, human populations have accumulated genetic variation. This genetic variation is the result of different population phenomena, such as assortative mating, in which individuals are more likely to mate with others who have a similar genetic background rather than mating randomly. Other factors that influence genetic variation include population migration, as well as population expansion or contraction, which, together with genetic drift, can change allelic frequencies across populations and contribute to their differentiation (Bamshad et al. 2004). Regardless of whether STRs, SNPs, or other polymorphisms are analyzed, only a small proportion of the genetic variation found across populations can be used to distinguish individuals from different continents. Therefore, knowledge of these variations can be used to infer BGA for different purposes, ranging from research applications to direct-to-consumer tests for individuals who wish to know their ancestry (McNevin 2020). Among the different genetic markers that can be used for BGA analysis, the Y chromosome can also be used to infer ancestry, although this type of analysis is limited to paternal lineages and restricted to males. However, rather than using Y-STR haplotypes, it is better to estimate ancestry, population origin, or migrations through the analysis of Y-SNPs because of their stability and low mutation rates. Thanks to these analyses, phylogenetic trees have been built by analyzing Y-SNPs in different population groups to establish the core haplogroups (Chiaroni et al. 2009) (Fig. 6.2). These haplogroups are named alphabetically, and the deepest branches in this phylogeny correspond to the African haplogroups A and B. It is important to mention that results obtained using the Y chromosome for BGA may differ from those obtained when analyzing autosomal markers; thus, its usefulness may be limited depending on the population studied (Syndercombe-Court 2021).

6.4 Loss of the Y Chromosome and Age

Recent studies have shown that mosaic loss of the Y chromosome (mLOY) is a common feature in the blood of elderly men (Dumanski et al. 2015; Forsberg et al. 2014). This alteration has been associated with reduced life expectancy, various cancers, and other disorders (Dumanski et al. 2015; Forsberg 2017; Forsberg et al. 2014; Guo et al. 2020). Although rare, the presence of mLOY in younger men is likely related to infertility and developmental abnormalities (Miyado and Fukami 2019).

In recent years, multiple epidemiological studies have demonstrated that mLOY in subsets of somatic cells—occurring at various stages of postzygotic development and throughout life—is associated with increased male mortality risk and may serve as a biomarker for early disease detection (Guo et al. 2021). Genes of relevance in this context include *DDX3Y*, which modulates cell differentiation (Vakilian et al. 2015), and long non-coding RNAs that contribute to prostate cancer progression. These RNAs are also involved in processes such as fatty liver disease, inflammation associated with atherosclerosis, and cardiovascular conditions (Maan et al. 2017).

mLOY is currently recognized as the most frequent chromosomal alteration in leukocytes of elderly men (Thompson et al. 2019; Wilson et al. 2024). For a long time, it was considered a physiological process related to aging. This assumption was based on the belief that no essential genes for cell viability exist on the Y chromosome, along with the observation that women live healthily without this chromosome (Guo et al. 2020). However, growing evidence now suggests that mLOY could be involved in male infertility and is associated with neurodegenerative disorders like Alzheimer's disease, cardiovascular and autoimmune diseases, increased mortality risk, and especially with various forms of cancer (Agahozo et al. 2020; Forsberg et al. 2014; Oh et al. 2020). Despite this, there are still a few in-depth literature reviews addressing the molecular mechanisms driving mLOY (Gutiérrez-Hurtado et al. 2024).

As discussed earlier, forensic genetics typically relies on STRs for human identification (HID). Nevertheless, these markers are not always sufficient, particularly when DNA is degraded or when the genetic profile lacks enough informative loci. In such cases, forensic geneticists often turn to complementary markers, such as ancestry-informative or phenotype-inference SNPs. This brings us to forensic DNA phenotyping (FDP)—a method that uses genetic markers, mainly SNPs, to predict externally visible characteristics (EVCs) of individuals (Kayser 2015). FDP has explored a broad range of traits, including: (1) iris, hair, and skin color using SNPs in key pigmentation genes (Chaitanya et al. 2018; Walsh et al. 2014); (2) eyebrow color (Peng et al. 2019); (3) freckle presence (Hernando et al. 2018); (4) hair shape (Pośpiech et al. 2015); (5) male pattern baldness (Marcińska et al. 2015); and (6) body height (Liu et al. 2019).

In addition to these features, some studies have also attempted to infer personal habits, such as tobacco and alcohol consumption, through DNA methylation (DNAm) analysis of blood (Ambroa-Conde et al. 2024), or to detect the genital microbiome exchanged during sexual acts—termed "sexome"—which may be useful in sexual assault investigations (Dixon et al. 2023). One trait of particular interest in the forensic field is the estimation of an individual's age from biological samples. Age inference can be highly useful for narrowing down possible contributors. Furthermore, accurate age prediction enhances the interpretation of other traits (e.g., hair color, baldness, or height), since these can change significantly with age (Kayser et al. 2023).

Several approaches have been proposed for DNA-based age inference, employing different genetic targets and methodologies (Cho et al. 2017; Jung et al. 2017; Koch and Wagner 2011). Among them, DNA methylation (DNAm) stands out as the only method that has achieved sufficient accuracy to be considered viable for forensic applications (Jung et al. 2017). However, this technique requires bisulfite conversion, which is time-consuming and demands a relatively high DNA input. Other genetic mechanisms have also been proposed for age estimation, such as telomere shortening and mLOY, although these are less studied. Telomere shortening has shown potential (Márquez-Ruiz et al. 2020) and can be easily standardized in the laboratory. Still, its predictive power remains low, comparable to that of classical anthropological age estimation methods (Giuliani et al. 2016). In contrast, mLOY exhibits characteristics that may enhance its forensic value. It is the most frequent non-physiological postzygotic alteration in humans (Forsberg et al. 2017), and although it is considered part of normal aging—being observed even in healthy individuals—its prevalence increases with age (Guttenbach et al. 1995). Additionally, men with detectable mLOY in their blood cells have shown an average reduction of 5.5 years in lifespan (Forsberg et al. 2014). This supports the idea that mLOY may serve as a biomarker of biological aging and could indicate susceptibility to age-related diseases in males (Dumanski et al. 2016).

Since the Y chromosome is routinely analyzed in forensic investigations, incorporating mLOY detection could offer both advantages and disadvantages. On one hand, mLOY may compromise the detection of Y chromosomal markers in male samples, potentially affecting the analysis. On the other hand, its presence might contribute valuable information that can be presented in legal proceedings.

A recent study evaluated the potential of age prediction based on mLOY levels using droplet digital PCR (ddPCR) in 232 healthy male volunteers, with biological samples from blood, saliva, and semen. When age was analyzed as a discrete variable, the correlation with mLOY was inconsistent. However, when age groups were averaged, a stronger correlation emerged. This indicates that mLOY may not be useful for estimating individual age precisely, but it could offer insight when classifying individuals into age ranges. Interestingly, in saliva and semen samples, no significant correlations were observed between mLOY percentage and age, limiting its application to blood-derived DNA (Song et al. 2023). These findings must be interpreted with caution due to several methodological limitations: (A) a relatively small sample size (n = 232); (B) lack of population diversity—all samples were from Chinese Han individuals; (C) use of insertion/deletion markers in *AMELY* and *AMELX* via ddPCR, which are susceptible to dropouts caused by mutations in primer-binding regions (silent alleles), possibly underestimating mLOY; (D) different types of biological samples were taken from different individuals.

Future studies should consider collecting various sample types (e.g., blood, saliva, semen) from the same individuals, which would allow for a better understanding of mLOY distribution and its underlying biological mechanisms (Song et al. 2023). Before mLOY can be considered a valid forensic age-inference marker, several important limitations must be addressed. These include: (1) mLOY is exclusive to males, rendering it useless for female samples; (2) although mLOY levels increase with age, there is significant interindividual variability, which complicates precise age inference (Danielsson et al. 2020); (3) in forensic samples with mLOY, the loss of Y-chromosomal markers (Y-STRs and Y-SNPs) may hinder detection; (4) many forensic investigations—particularly sexual assault cases—rely on semen analysis, yet mLOY is absent in sperm cells (Song et al. 2023). One alternative is to detect mLOY in male epithelial cells left on the victim's body. However, these cells are usually present in very small quantities, posing a challenge for mLOY detection. To overcome these obstacles, comprehensive studies with large and diverse samples (across age groups and ethnicities) are required.

Such studies could lead to the development of more accurate algorithms and help assess the forensic viability of mLOY as an age-inference tool. Used in conjunction with other methods like DNA methylation, mLOY holds promise for enhancing human identification in forensic investigations from a broader and more integrative perspective.

6.5 Y Chromosome and Age Estimation

6.5.1 Epigenetics

Epigenetics can be defined as the chemical changes that occur in the DNA that affect the patterns of expression of the genes within (Allis and Jenuwein 2016). The result of these changes is not related to changes in the genetic code, instead they alter other mechanisms which favor or repress the expression of a gene. Epigenetics involves changes in the DNA, histones, with the participation of non-coding RNAs, as microRNAs (Heyt and Thakur 2021; Farsetti et al. 2023). The process of epigenetics is dynamic because it can be influenced by environmental factors, such as diet, stress, and toxins, among others (Webster and Phillips 2025). Unlike genetic changes, epigenetics can be modified throughout the lifespan of an individual. More astonishing, the epigenetic changes can be inherited and passed on to the following generation resulting in specific patterns of expression (Webster and Phillips 2025).

The processes implicated in gene expression changes due to epigenetic modifications are complex and are beyond this chapter, although we will summarize the main features of these processes. The first epigenetic mechanism is the presence of non-coding RNAs, which directly regulate the transcription of some genes and are important in development. The microRNAs (miRNAs) are non-coding RNA molecules about 18–20 nucleotides and regulate the expression of several genes through hybridization with target messenger RNA (mRNA) facilitating its degradation (Cai et al. 2009). The other mechanism of epigenetic regulation is the post-translational modification of the histones. At first, it was believed that histones are proteins that only function as packaging the DNA and helping to adequately segregate the genetic material. Later, it was found that in some regions of the genome, there is a higher packing of the DNA or more loose, dividing the chromatin into euchromatin (laxing) and heterochromatin (dense). Histones can have chemical modifications to facilitate or repress gene expression (Millán-Zambrano et al. 2022). This can be done by modulating the access to the DNA of the transcription factors. For activation of the transcription it can be done through the acetylation of the histones (i.e., H3K56ac, H3K64ac), or trimethylation of lysine 4 of the histone 3 (H3), while methylation of histones H3K9me3 and H4K20me3 recruits repressor proteins as HP1, helping to compact the chromatin (Millán-Zambrano et al. 2022).

The other epigenetic change happens directly in the DNA, as direct methylation of cytosines. The methylation in cytosine occurs by adding a methyl group in the 5th carbon of this nucleotide, thus named as 5mC (Jones 2012). This methylation is stable and long-term perpetuated through several cell divisions. The methylation of these cytosines does not occur randomly in the genome, rather they are specifically located in regions rich in dinucleotides of cytosine and guanine (Mao et al. 2020). The main location of methylated is the CpG island which is in the promoters of 60–70% of the human genes (Jones 2012). They can also be located in regions flanking CpG islands as shores and shelves, with an open sea (Visone et al. 2019). And finally, there is a higher proportion of methylated DNA in intergenic and repetitive regions (LINEs, SINEs, and retrotransposons) which are crucial for genomic stability (Pappalardo and Barra 2021).

The maintenance of DNA methylation is important during development, and those de novo modifications are performed by two DNA methyltransferases known as *DNMT3A* and *DNMT3B* that operate in unmethylated and hemimethylated DNA (Okano et al. 1999; Mao et al. 2020). On the other hand, the DNMT1 is a methylase that preserves the methylation patterns in replication when there is a requirement of hemimethylated DNA (Jorgensen et al. 2018). Immediately after fertilization of the oocyte, the male chromosomes lose all their epigenetic marks in a few hours. The process by which the oocyte is demethylated is slower and occurs after the first mitotic divisions after fertilization. Once the embryo is implanted, the whole genome is re-methylated (Li et al. 2018; Smith et al. 2025). This is catalyzed by the enzymes *DNMT3A* and *DNMT3B* which handle the de novo methylation after implantation. This methylation and recovery of epigenetic marks are the basis of transgenerational epigenetics, that is the patterns of methylation that come from each parent (Nagy and Turecki 2015; Boškovi´c and Rando 2025). Another example of this is the genetic imprinting which is related to the specific expression of one of the two alleles in a diploid organism depending on their parental origin. This is well described where a deletion occurs located in 15q11–13 and depending if paternal or maternal copy is lost, the resulting phenotype corresponds to Prader-Willi or Angelman, respectively (Boškovic and Rando 2025).

When there is methylation in the CpG islands of a promoter, there is a long-term repression in the gene expression which recruits remodeling proteins that compact the chromatin inhibiting the expression of that gene. For that, there are proteins such as MECP2, which have CpG-binding domain, that locate within these methylated islands and recruit other proteins (Jaenisch and Bird 2003). Another mechanism related to down-regulation of gene expression is reducing the affinity of the transcription factors as TFII. The methylation of DNA is a long-lasting process thus any aberrant methylation could be present in pathologic processes such as complex diseases or cancer (Smith et al. 2025; Li et al. 2025).

The analysis of DNA methylation in the laboratory can be laborious since there are some considerations that need to be addressed. First, the methylated DNA needs to be converted by enzymatic reactions for further analysis. This is done through the bisulfite conversion-based methods where the unmethylated cytosines are converted to uracil with the application of sodium bisulfite (Laird 2010). This reaction consists in addition of sulfonate group to unmethylated cytosine, to later hydrolyze to uracil. After this treatment, the region of interest can be further analyzed through Sanger sequencing or Massive Parallel sequencing. The islands of methylated cytosines remain unchanged in the sequencing, while the unmethylated ones change to thymine. After this, the percentage (%) of cytosines can be estimated for that region. The Sanger sequencing is a low-throughput technology since it only analyzes one region per time thus is unsuitable for analyzing extensive regions. To analyze at single-base resolution the methylation profiling of a group of cells, the best approach is the whole-genome bisulfite sequencing (WGBS), although the computational process for analysis can be exhausting. This type of analysis can be focused on CpG islands through reduced representation bisulfite sequencing (RRBS) where the DNA is previously digested, repaired, and a process which includes an adapter ligation, bisulfite conversion, amplification, and sequencing (Moser et al. 2020; Gong et al. 2022). On the other hand, there are methylation microarrays, which consist of probes that target bisulfite-converted DNA and can cover above 850,000 CpG sites distributed in the whole genome. The advantage of this microarray technology is that it can explore in large scale previously characterized methylated regions in a big cohort of individuals (Bibikova et al. 2011). Nevertheless, it is important to keep in mind that this DNA methylation analysis is restricted to the CpG islands considered in the design of the microarray, thus some regions could be underrepresented.

Recently, with the development of new sequencing technologies such as Oxford Nanopore sequencing, now it is possible to base sequencing in real time, including modified bases as 5 methyl cytosine. In principle, in this technology the DNA passes through a nanopore, a protein channel, where also an ionic current passes and it detects the changes in current depending on each base (Simpson et al. 2017). The advantage of this technology is that no previous treatment with bisulfite of the DNA is needed, since this nanopore can directly detect each base no matter if it is modified, as 5′ methyl cytosine. Moreover, in comparison with the methods mentioned above, nanopore sequencing can sequence long fragments larger than 1 Mb (Wang et al. 2021). Besides, nanopore technology has better performance in repetitive regions with lower GC bias. This allowed the full sequencing of the Y chromosome, solving about 30 Mb of unknown sequences of this chromosome (Hallast et al. 2023; Esteller-Cucala et al. 2023).

6.5.2 DNA Methylation in the Aging Process

Since methylation is important for different physiologic processes, such as cell differentiation, X chromosome inactivation and any alteration in the methylation patterns would affect the homeostasis of an individual and could be related to a pathology. One of the most commonly studied diseases where DNA methylation is important is cancer, since there is a global hypomethylation in the cells which favors the genomic stability producing chromosomal breaks (Johansson et al. 2013). This hypomethylation also has been reported with the activation of oncogenes. On the other hand, some tumor suppressor genes have been found to be methylated. Cancer is not the only disease that has been associated with defects in methylation, there are other diseases, such as neurological disease (Alzheimer's or Parkinson's), Autoimmune and inflammatory disease (lupus and rheumatoid arthritis), cardiovascular disease (hypertension), and psychiatric diseases (posttraumatic stress disorder), among others. Of course, the alterations in methylations can be modeling through the presence of environmental factors as toxins or oligoelements, and even stress (Henderson-Smith et al. 2019; Gonzalez-Jaramillo et al. 2019; Horton et al. 2024; Katrinli et al. 2024).

It has been reported that the landscape of epigenetic marks changes throughout the life span of an individual (Johansson et al. 2013). The changes in the DNA methylation patterns are explained by epigenetic drift, which can be defined as the accumulation of methylation (hypermethylation) in some genomic regions, while there are others that present hypomethylation (Tan et al. 2016). These changes have been correlated with the aging process; thus, they increase as the chronological age of an individual increases. Aging can be defined as the final destination in a human lifespan which is characterized by a functional decline and an increase of age-related degenerative diseases. The process of aging is related to the age of an individual, which has to be differentiated between biological age and chronological age (Levine and Crimmins 2018). The biological age is the accumulation of changes in the cell properly related to the aging process, which can be shortening of the telomeres, or the methylation status. On the other hand, the chronological age is just the number of years since birth.

Age prediction has an important relevance in the forensic field, since it can help to get more information about a sample found in a crime scene. A biological sample found in a crime scene is useless, unless there is a reference sample to compare with. For this reason, several DNA studies have been performed to try to obtain other information such as External Visible Characters (EVC) or the Biogeographic Ancestry. The prediction of chronological age is considered an EVC which can be included in the other characteristics validated for that purpose as hair, eyes, and skin color (Zbieć-Piekarska et al. 2015a; Verstraeten et al. 2025). The estimation of chronological age could be useful in those cases where a dead body is found for helping the anthropologic analyses of age estimation to see if the body corresponds with the missing person. Another case is where there are biological traces, thus methylation DNA studies could narrow the number of suspects according to age, even in those cases where there are multiple contributors.

With respect to the age, in one of the first reports, it was found that DNA methylation in the promoters of *EDARADD, TOM1L1,* and *NPTX2* genes is linear with age so the authors could predict the age with an accuracy of 5.2 years (Bocklandt et al. 2011). However, one limitation of this study is that the sample size was lower and they only analyzed saliva. This was circumvented by Hovarth, who analyzed 8000 samples from 82 Illumina DNA methylation array datasets, which included 51 types of tissues. He established an epigenetic clock with the analyses of 353 CpG sites with an accuracy of 3.6 years (Horvath 2013). Since then, other studies have focused not only in estimating age rather than lifespan including other biochemical markers such as Phenoage, GrimAge, and DunedinPACE (Belsky et al. 2022; Lu et al. 2019; Levine et al. 2018). However, it is important to mention that these studies, although focused on age prediction, could also be applied to aging medicine. As it has been described that the methylation patterns vary according to the analyzed tissue, other studies came into the light analyzing other tissues besides blood and saliva, such as teeth and bone. Including more tissues in the analyses, especially those that can be found in a forensic scenario, is important since it could be implemented in forensic practice.

There are two main parameters for age estimation, one is the Mean Absolute Deviation (MAD) and the other is the Mean Absolute Error (MAE), where the results are presented in years deviation from the chronological age (i.e., 3 years). The lower this estimate, the better the model for age estimation. From the whole genome DNA methylated regions, there are some genes which have been consistent in this type of analysis. For instance, methylation in the gene *ELOVL2* has been found in several reports correlated with age (Zbieć-Piekarska et al. 2015b). Other genes that have been reported to be associated with age have been *ASPA, PDE4C, FHL2, KLF14, ZNF423, CCDC102B, KCNAB3, PRKG2, MIR29B2CHG*, and *TRIM59* (Freire-Aradas et al. 2016; Zubakov et al. 2016; Hong et al. 2017; Naue et al. 2017; Goel et al. 2017; Aliferi et al. 2018; Jung et al. 2019; Han et al. 2020; Refn et al. 2023). With these genes and other CpG regions, several multivariate models have been created to predict age, although *ELOVL2* still shows the highest correlation. Interestingly, *ELOVL2* has been replicated in several populations, which suggests it is a universal marker for age estimation (Jung et al. 2019; Refn et al. 2023; Pinprasert et al. 2024; Onofri et al. 2025; Verstraeten et al. 2025). It is important to mention that although at first most of these studies were performed in blood and saliva, *FHL2*, *TRIM59,* and *KLF14* have been found associated with age in cartilage (Freire-Aradas et al. 2023; Jung et al. 2025). This indicates that the panel of CpG regions are tissue specific and must be considered in the forensic scenario. Moreover, the fewer markers included in the multivariate model would be practical in the forensic analysis.

6.5.3 Y Chromosome and Age Estimation

Although the CpG islands analyzed are represented in the 22 autosomes, including the X chromosome, the available information about the Y chromosome was limited until now. This is due to the inherent characteristics of the Y chromosome which was mentioned above. With Oxford Nanopore, it was possible to fully sequence this chromosome and fill the gaps about 30 million base pairs, which also included the complete structure of genes like *TSPY, DAZ,* and *RBMY,* as well as 41 additional protein-coding genes, mostly from the *TSPY* family. As the identification of new coding regions in this chromosome was achieved, it was also the identification of more CpG islands (Rhie et al. 2023).

The methylation study of Y chromosomes has emerged as an important tool in the forensic context and biomedical research, related to male individuals. This is so, because it can be analyzed in degraded samples or can be applied in mixtures composed of female and male DNA. The latter is important in the investigation of sexual assaults. Nevertheless, it is important to understand the difference between the methylation patterns of autosomal chromosomes and Y chromosomes. For instance, the epigenetic marks of Y chromosome are located in the ampliconic regions, which also show transcription repression (Singh et al. 2011). As in the autosomal chromosomes, the Y chromosome presents changes in its methylation status according to age. In a study comprising four cohorts of European individuals (n = 624; age range 56–94), the authors analyzed 416 CpG islands on the Y chromosome using data obtained from the HumanMethylation450 BeadChip array (Illumina, Inc., San Diego, CA, USA; Lund et al., 2020). They found between 40 and 219 CpG sites significantly methylated according to age. Of these sites, 82% of them were found to be hypermethylated, which highly contrasts with methylation patterns according to age in autosomal CpG sites. This hypermethylation was higher in the cohorts with a higher mean age (i.e., 88.51 years). Nevertheless, it is important to mention that this study was limited to middle- and older-aged individuals.

To cover more year intervals of younger people, a study included 1057 males with ages ranging from 15 to 87 years old (Vidaki et al. 2021). They analyzed the Y chromosome data of the Illumina 450 K array and selected 268 CpG islands for analysis. From these sites with variability in the methylation status, they found 22 CpG islands with hypermethylation. They found on average about 15% of increase or decrease in DNA methylation of these top hyper or hypomethylated when comparing the very young (less than 20 years) with the elderly (>70 years old). When they evaluated through several machine learning algorithms, with a supporting vector machine (SVM), they could estimate a MAD about 7.54–7.61 years. When they were reduced to 19 predictive CpGs selected by Bayesian Information Criterion, they had a MAD of 8.46 years. The age prediction error was stable through different age groups, being precisely in young and elderly, contrary to autosomal CpGs that lost precision above 60 years. Interestingly, the genes found in this set of 19 CpG islands were the genes *EIF1AY, DDX3Y, ZFY, TTTY14*, and *NLGN4Y* (Vidaki et al. 2021). These genes are related to male infertility, autoimmune processes, and prostate cancer which can also be related to age. These data suggest that the presence of methylation in the Y chromosome is correlated with those conditions that occur in elderly individuals.

In another study, focused on the methylation status of CpG islands in both sexual chromosomes, the number of CpG islands identified in the Y chromosome that change in their methylation status was 46, while there is a higher number of methylated CpGs found in the male X chromosome (1327 in the male X chromosome vs. 325 in the female X chromosome; Kananen and Marttila 2021). This suggests that the hypermethylation status of CpG islands could affect both sexual chromosomes in males. Moreover, the gene associated with the hypermethylated CpG island cg27214488 was *NLGN4Y*, whereas the hypomethylated region cg20401549 was associated with *PRKY*. Interestingly, *NLGN4Y* encodes a neuronal cell adhesion protein that belongs to the neuroligin family (Kananen and Marttila, 2021). This gene has been associated with autism and methylation found in this gene has been associated with prostate cancer. The *PRKY* is protein kinase Y linked is a pseudogene that has also been associated with being methylated in prostate cancer.

Although these works indicate that there are several CpG sites which are differentially methylated according to age, these results should be replicated in other populations (Li et al. 2022). In a study of 1,006 Chinese males, the authors identified 18 CpG islands on the Y chromosome that exhibited age-associated methylation patterns. These methylation sites were further explored in a meta-analysis leaving 14 sites. The genes associated with these sites were *TTTY15, DDX3Y, TMSB4Y, NLGN4Y, TTTY14, EIF1AY,* and *TTTY13*. Interestingly, they also found alteration in gene expression of some genes associated with these methylated sites, such as *TTTY14* and *EIF1AY* (Li et al. 2022). This indicates that Y chromosome methylation changes according to age could be replicated in other populations.

To extrapolate these studies to the forensic field, it is important to be sure that these CpG islands are male-specific, especially in cases where it can be a mixture of male-female samples. To answer this question in a study performed in 187 samples from Chinese males, they selected a set of 13 Y-CpGs age-related which were analyzed using the SNaPShot technology (Jiang et al. 2023). With this set of Y-CpGs, they obtained a MAD of 4.65 years. When they analyzed DNA proportions of male/female of 1:1 and 1:10, the deviation between real age and predicted age was below 8 years. Interestingly, in this study, they proved the sensitivity of the test where their system can detect DNA quantities above 0.5 ng and in bloodstains (Jiang et al. 2023).

Although trying to explore the methylation status in conditions that could simulate at some point forensic scenarios where there are bloodstains or can be male/female DNA mixtures, it is possible that some of these Y-CpGs could be fluid or tissue-specific. One biological sample that is commonly collected in forensic scenarios and could provide more information about the case are semen samples. However, most of the semen samples found in the forensic scenario are from unknown persons, thus the possibility of age estimation through methylation analysis could be useful. For that, in a Korean population, it was analyzed through methylation analysis of Y chromosome in semen samples (Lee et al. 2024). When they explored 9 CpGs in a multiple regression model, it showed a MAE of 5.36 years. Moreover, they performed confidence, reproducibility, and sensitivity tests to evaluate these markers. These analyses could be performed even in mixtures with female DNA or in those cases where it is not possible to perform differential extraction of semen samples, which could be useful in those cases of male forensic studies. Interestingly, in these samples, the genes *EIF1AY, PRKY,* and *NLGN4Y* were also found, which were previously reported in blood samples. In addition to these genes, new genes were found that included the genes *TSPY3, DAZ1, FAM197Y4, TSPY1,* and *USP9Y* (Lee et al. 2024). The function of these genes is related to sexual differentiation and spermatogenesis, processes related to male fertility.

It is important to mention that the study analyzing methylation of Y chromosome has been using microarray platforms, which make it difficult to implement it in the forensic scenario. Thus, a highly sensitive technique such as pyrosequencing and a selection of a few markers could be helpful to establish this approach in forensic cases. In a work done by Ji Z and colleagues., they selected a group of 5 CpG islands in the Y chromosome (*GSE41169*, *GSE87571*, *GSE52588*, *CpG4*, and *GSE42861*) to be analyzed with pyrosequencing of blood samples, dry blood samples, semen, and saliva (Ji et al. 2024). With these markers, they built a model for age prediction, obtaining a MAD between 5.50 and 6.74 years. There were no differences in the methylation between male-female DNA mixtures, and between fresh blood and dried blood. However, this was not the case for semen and saliva samples, obtaining higher prediction values of age estimation (Ji et al. 2024). Although this panel is adequate for blood samples and DNA mixture samples at low DNA concentrations and even in dried samples, it is not suitable for other fluids such as saliva and semen. The use of new sequencing technologies in the forensic field is emerging as a helpful tool in the forensic scenario, and to employ them in the DNA methylation analysis could be promising.

6.6 Concluding Remarks

The study of the methylation of Y chromosome seems a promising approach to obtain more information about a biological sample from male individuals through age estimation. The Y chromosome analysis has been useful in those cases where there is a mixture of male/female DNA, as in sexual assaults. However, this has been challenging due to the biological characteristics of its sequencing. On the other hand, besides its biological function containing the genes related to maleness, this chromosome is also implicated in aging and several diseases like cancer. For instance, there is a loss of Y chromosome according to age and some diseases. DNA methylation is a dynamic process that changes through the lifespan of an individual, being demethylated in some regions and hypermethylated in others. Due to the knowledge of these methylated regions in the genome, it is possible to predict the age of an individual. There are some studies focusing on the DNA methylation analysis of Y chromosome to predict age. However, it is important to consider that the methylation patterns in the Y chromosome are different from the autosomes, which predominates the hypermethylation as age increases. The DNA methylation of Y chromosome could predict age with consistent results. Nevertheless, it is important to mention that although this prediction is consistent, it is still lower than the prediction with autosomes. On the other hand, data suggest that the methylation patterns could be fluid-specific, finding differences between blood and semen samples. The genes that seem to be consistent in age prediction when analyzing DNA methylation in Y chromosome are *EIF1AY, PRKY,* and *NLGN4Y.* One thing to consider is that most of these CpG islands are obtained from data from microarrays to analyze DNA methylation that focus on this chromosome. This is important, since the full CpG islands in this chromosome could be underrepresented. With the newest DNA sequencing technologies that are emerging, it is now possible to sequence long reads, which facilitates sequencing the whole Y chromosome. Nanopore sequencing technologies may ease the identification of novel methylation sites on the Y chromosome, and the inclusion of these sites could enhance the accuracy of age prediction models. Finally, although there are subtle differences in the methylation patterns across the Y haplogroups, including more populations would be useful to replicate these reports and find a universal set of CpG islands that could work in populations worldwide.

Acknowledgments Mariano Guardado Estrada received funding from the National Autonomous University of Mexico (UNAM), PAPIIT grant number IA208923. José Alonso Aguilar-Velázquez received a research grant from Consejo Nacional de Humanidades, Ciencia y Tecnología (CONAHCYT) grant CF-2023-G-536.

Competing Interests The authors have no conflicts of interest to declare that are relevant to the content of this chapter.

References

Agahozo MC, Timmermans MAM, Sleddens HFBM et al (2020) Loss of Y-chromosome during male breast carcinogenesis. Cancers (Basel) 12(3):631. https://doi.org/10.3390/cancers12030631

Aliferi A, Ballard D, Gallidabino M et al (2018) DNA methylation-based age prediction using massively parallel sequencing data and multiple machine learning models. Forensic Sci Int Genet 37:215–226. https://doi.org/10.1016/j.fsigen.2018.09.003

Allis CD, Jenuwein T (2016) The molecular hallmarks of epigenetic control. Nat Rev Genet 17:487–500

Ambroa-Conde A, Casares de Cal MA, Gómez-Tato A et al (2024) Inference of tobacco and alcohol consumption habits from DNA methylation analysis of blood. Forensic Sci Int Genet 70:103022. https://doi.org/10.1016/j.fsigen.2024.103022

Bachtrog D (2013) Y-chromosome evolution: emerging insights into processes of Y-chromosome degeneration. Nat Rev Genet 14(2):113–124. https://doi.org/10.1038/nrg3366

Bachtrog D (2020) The Y chromosome as a battleground for intragenomic conflict. Trends Genet 36(7):510–522. https://doi.org/10.1016/j.tig.2020.04.008

Balanovsky OP, Zaporozhchenko VV (2016) Chromosome as a chronicler: genetic dating, historical events, and DNA-genealogic temptation. Russ J Genet 52:708–725. https://doi.org/10.1134/S1022795416070048

Bamshad M, Wooding S, Salisbury BA et al (2004) Deconstructing the relationship between genetics and race. Nat Rev Genet 5(8):598–609. https://doi.org/10.1038/nrg1401

Belsky DW, Caspi A, Corcoran DL et al (2022) DunedinPACE, a DNA methylation biomarker of the pace of aging. elife 14(11):e73420. https://doi.org/10.7554/eLife.73420

Bibikova M, Barnes B, Tsan C et al (2011) High density DNA methylation array with single CpG site resolution. Genomics 98:288–295. https://doi.org/10.1016/j.ygeno.2011.07.007

Blackmon H, Demuth JP (2015) The fragile Y hypothesis: Y chromosome aneuploidy as a selective pressure in sex chromosome and meiotic mechanism evolution. BioEssays 37(9):942–950. https://doi.org/10.1002/bies.201500040

Bocklandt S, Lin W, Sehl ME et al (2011) Epigenetic predictor of age. PLoS One 6:e14821. https://doi.org/10.1371/journal.pone.0014821

Bonito M, D'Atanasio E, Ravasini F et al (2021) New insights into the evolution of human Y chromosome palindromes through mutation and gene conversion. Hum Mol Genet 30(23):2272–2285. https://doi.org/10.1093/hmg/ddab189

Boškovic AB, Rando OJ (2025) Transgenerational epigenetic inheritance. Annu Rev Genet 23(52):21–41. https://doi.org/10.1146/annurev-genet-120417

Cai Y, Yu X, Hu S, Yu J (2009) A brief review on the mechanisms of miRNA regulation. Genomics Proteomics Bioinformatics 7:147–154

Calafell F, Larmuseau MHD (2017) The Y chromosome as the most popular marker in genetic genealogy benefits interdisciplinary research. Hum Genet 136(5):559–573. https://doi.org/10.1007/s00439-016-1740-0

Chaitanya L, Breslin K, Zuñiga S et al (2018) The HIrisPlex-S system for eye, hair and skin colour prediction from DNA: introduction and forensic developmental validation. Forensic Sci Int Genet 35:123–135. https://doi.org/10.1016/j.fsigen.2018.04.004

Champroux A, Torres-Carreira J, Gharagozloo P et al (2016) Mammalian sperm nuclear organization: resiliencies and vulnerabilities. Basic Clin Androl 26:17. https://doi.org/10.1186/s12610-016-0044-5

Charlesworth D (2023) Why and how do Y chromosome stop recombining? J Evol Biol 36(3):632–636. https://doi.org/10.1111/jeb.14137

Chiaroni J, Underhill PA, Cavalli-Sforza LL (2009) Y chromosome diversity, human expansion, drift, and cultural evolution. Proc Natl Acad Sci USA 106(48):20174–20179. https://doi.org/10.1073/pnas.0910803106

Cho S, Jung SE, Hong SR et al (2017) Independent validation of DNA-based approaches for age prediction in blood. Forensic Sci Int Genet 29:250–256. https://doi.org/10.1016/j.fsigen.2017.04.020

Danielsson M, Halvardson J, Davies H et al (2020) Longitudinal changes in the frequency of mosaic chromosome Y loss in peripheral blood cells of aging men varies profoundly between individuals. Eur J Hum Genet 28(3):349–357. https://doi.org/10.1038/s41431-019-0533-z

Dixon R, Egan S, Hughes S et al (2023) The Sexome—a proof of concept study into microbial transfer between heterosexual couples after sexual intercourse. Forensic Sci Int 348:111711. https://doi.org/10.1016/j.forsciint.2023.111711

Dumanski JP, Rasi C, Lönn M et al (2015) Mutagenesis. Smoking is associated with mosaic loss of chromosome Y. Science 347(6217):81–83. https://doi.org/10.1126/science.1262092

Dumanski JP, Lambert JC, Rasi C et al (2016) Mosaic loss of chromosome Y in blood is associated with Alzheimer disease. Am J Hum Genet 98(6):1208–1219. https://doi.org/10.1016/j.ajhg.2016.05.014

Esteller-Cucala P, Palmada-Flores M, Kuderna LFK et al (2023) Y chromosome sequence and epigenomic reconstruction across human populations. Commun Biol 6:623. https://doi.org/10.1038/s42003-023-05004-9

Farsetti A, Illi B, Gaetano C (2023) How epigenetics impacts on human diseases. Eur J Intern Med 114:15–22

Forsberg LA (2017) Loss of chromosome Y (LOY) in blood cells is associated with increased risk for disease and mortality in aging men. Hum Genet 136(5):657–663. https://doi.org/10.1007/s00439-017-1799-2

Forsberg LA, Rasi C, Malmqvist N et al (2014) Mosaic loss of chromosome Y in peripheral blood is associated with shorter survival and higher risk of cancer. Nat Genet 46(6):624–628. https://doi.org/10.1038/ng.2966

Forsberg LA, Gisselsson D, Dumanski JP (2017) Mosaicism in health and disease—clones picking up speed. Nat Rev Genet 18(2):128–142. https://doi.org/10.1038/nrg.2016.145

Freire-Aradas A, Phillips C, Mosquera-Miguel A et al (2016) Development of a methylation marker set for forensic age estimation using analysis of public methylation data and the Agena bioscience EpiTYPER system. Forensic Sci Int Genet 24:65–74. https://doi.org/10.1016/j.fsigen.2016.06.005

Freire-Aradas A, Tomsia M, Piniewska-Róg D et al (2023) Development of an epigenetic age predictor for costal cartilage with a simultaneous somatic tissue differentiation system. Forensic Sci Int Genet 67:102936. https://doi.org/10.1016/j.fsigen.2023.102936

Ge J, Crysup B, Peters D et al (2023) MPKin-YSTR: interpretation of Y chromosome STR haplotypes for missing persons cases. Electrophoresis 44:1080–1087. https://doi.org/10.1002/elps.202200225

Gill P, Jeffreys A, Werrett D (1985) Forensic application of DNA "Fingerprints". Nature 318:577–579. https://doi.org/10.1038/318577a0

Giuliani C, Cilli E, Bacalini MG et al (2016) Inferring chronological age from DNA methylation patterns of human teeth. Am J Phys Anthropol 159(4):585–595. https://doi.org/10.1002/ajpa.22921

Goel N, Karir P, Garg VK (2017) Role of DNA methylation in human age prediction. Mech Ageing Dev 166:33–41

Gong T, Borgard H, Zhang Z et al (2022) Analysis and performance assessment of the whole genome bisulfite sequencing data workflow: currently available tools and a practical guide to advance DNA methylation studies. Small Methods 6(3):e2101251. https://doi.org/10.1002/smtd.20210125

Gonzalez-Jaramillo V, Portilla-Fernandez E, Glisic M et al (2019) The role of DNA methylation and histone modifications in blood pressure: a systematic review. J Hum Hypertens 33:703–715

Guo X, Dai X, Zhou T et al (2020) Mosaic loss of human Y chromosome: what, how and why. Hum Genet 139(4):421–446. https://doi.org/10.1007/s00439-020-02114-w

Guo X, Li J, Xue J et al (2021) Loss of Y chromosome: an emerging next-generation biomarker for disease prediction and early detection? Mutat Res Rev Mutat Res 788:108389. https://doi.org/10.1016/j.mrrev.2021.108389
Gusmão L, Brión M, Gomes I (2008) The human Y chromosome male-specific polymorphisms and forensic genetics. In: Bogusz MJ (ed) Handbook of analytical separations, vol 6. Elsevier Science B.V, Amsterdam, pp 969–1000. https://doi.org/10.1016/S1567-7192(06)06034-7
Gutiérrez-Hurtado IA, Sánchez-Méndez AD, Becerra-Loaiza DS et al (2024) Loss of the Y chromosome: a review of molecular mechanisms, age inference, and implications for men's health. Int J Mol Sci 25(8):4230. https://doi.org/10.3390/ijms25084230
Guttenbach M, Koschorz B, Bernthaler U et al (1995) Sex chromosome loss and aging: in situ hybridization studies on human interphase nuclei. Am J Hum Genet 57(5):1143–1150
Hallast P, Ebert P, Loftus M et al (2023) Assembly of 43 human Y chromosomes reveals extensive complexity and variation. Nature 621:355–364. https://doi.org/10.1038/s41586-023-06425-6
Han Y, Franzen J, Stiehl T et al (2020) New targeted approaches for epigenetic age predictions. BMC Biol 18:71. https://doi.org/10.1186/s12915-020-00807-2
Henderson-Smith A, Fisch KM, Hua J et al (2019) DNA methylation changes associated with Parkinson's disease progression: outcomes from the first longitudinal genome-wide methylation analysis in blood. Epigenetics 14:365–382. https://doi.org/10.1080/15592294.2019.1588682
Hernando B, Ibañez MV, Deserio-Cuesta JA, Soria-Navarro R, Vilar-Sastre I, Martinez-Cadenas C (2018) Genetic determinants of freckle occurrence in the Spanish population: towards ephelides prediction from human DNA samples. Forensic Sci Int Genet 33:38–47. https://doi.org/10.1016/j.fsigen.2017.11.013
Heyt KM, Thakur J (2021) Regulation of epigenetic processes by non-coding RNAs. Nucleus (India) 64:285–301
Hong SR, Jung S-E, Lee EH et al (2017) DNA methylation-based age prediction from saliva: high age predictability by combination of 7 CpG markers. Forensic Sci Int Genet 29:118–125. https://doi.org/10.1016/j.fsigen.2017.04.006
Horton MK, Nititham J, Taylor KE et al (2024) Changes in DNA methylation are associated with systemic lupus erythematosus flare remission and clinical subtypes. Clin Epigenetics 16(1):181. https://doi.org/10.1186/s13148-024-01792-x
Horvath S (2013) DNA methylation age of human tissues and cell types. Genome Biol 14:R115–R115. https://doi.org/10.1186/gb-2013-14-10-r115
Hughes JF, Rozen S (2012) Genomics and genetics of human and primate y chromosomes. Annu Rev Genomics Hum Genet 13:83–108. https://doi.org/10.1146/annurev-genom-090711-163855
Jaenisch R, Bird A (2003) Epigenetic regulation of gene expression: how the genome integrates intrinsic and environmental signals. Nat Genet 33:245–254
Ji Z, Xing Y, Li J et al (2024) Male-specific age prediction based on Y-chromosome DNA methylation with blood using pyrosequencing. Forensic Sci Int Genet 71:103050. https://doi.org/10.1016/j.fsigen.2024.103050
Jiang L, Zhang K, Wei X et al (2023) Developing a male-specific age predictive model based on Y-CpGs for forensic analysis. Forensic Sci Int 343:111566. https://doi.org/10.1016/j.forsciint.2023.111566
Jobling MA, Tyler-Smith C (1995) Fathers and sons: the Y-chromosome and human evolution. TIG 11:449–456
Jobling MA, Tyler-Smith C (2003) The human Y chromosome: an evolutionary marker comes of age. Nat Rev Genet 4(8):598–612. https://doi.org/10.1038/nrg1124
Jobling MA, Tyler-Smith C (2017) Human Y-chromosome variation in the genome-sequencing era. Nat Rev Genet 18(8):485–497. https://doi.org/10.1038/nrg.2017.36
Jobling MA, Pandya A, Tyler-Smith C (1997) The Y chromosome in forensic analysis and paternity testing. Int J Legal Med 110:118–124
Johansson Å, Enroth S, Gyllensten U (2013) Continuous aging of the human DNA methylome throughout the human lifespan. PLoS One 8(6):e67378. https://doi.org/10.1371/journal.pone.0067378

Jones PA (2012) Functions of DNA methylation: islands, start sites, gene bodies and beyond. Nat Rev Genet 13:484–492
Jorgensen BG, Berent RM, Ha SE et al (2018) DNA methylation, through DNMT1, has an essential role in the development of gastrointestinal smooth muscle cells and disease. Cell Death Dis 9(5):474. https://doi.org/10.1038/s41419-018-0495-z
Jung SE, Shin KJ, Lee HY (2017) DNA methylation-based age prediction from various tissues and body fluids. BMB Rep 50(11):546–553. https://doi.org/10.5483/bmbrep.2017.50.11.175
Jung S-E, Lim S, Hong SR et al (2019) DNA methylation of the ELOVL2, FHL2, KLF14, C1orf132/MIR29B2C, and TRIM59 genes for age prediction from blood, saliva, and buccal swab samples. Forensic Sci Int Genet 38:1–8. https://doi.org/10.1016/j.fsigen.2018.09.010
Jung JY, Song YW, Jeong K et al (2025) A SNaPshot assay for epigenetic age prediction of costal cartilage. Electrophoresis 46:413–423. https://doi.org/10.1002/elps.8132
Kananen L, Marttila S (2021) Ageing-associated changes in DNA methylation in X and Y chromosomes. Epigenetics Chromatin 14(1):33. https://doi.org/10.1186/s13072-021-00407-6
Karmin M, Saag L, Vicente M et al (2015) A recent bottleneck of Y chromosome diversity coincides with a global change in culture. Genome Res 25(4):459–466. https://doi.org/10.1101/gr.186684.114
Katrinli S, Wani AH, Maihofer AX et al (2024) Epigenome-wide association studies identify novel DNA methylation sites associated with PTSD: a meta-analysis of 23 military and civilian cohorts. Genome Med 16:147. https://doi.org/10.1186/s13073-024-01417-1
Kayser M (2015) Forensic DNA phenotyping: predicting human appearance from crime scene material for investigative purposes. Forensic Sci Int Genet 18:33–48. https://doi.org/10.1016/j.fsigen.2015.02.003
Kayser M (2017) Forensic use of Y-chromosome DNA: a general overview. Hum Genet 136(5):621–635. https://doi.org/10.1007/s00439-017-1776-9
Kayser M, Branicki W, Parson W, Phillips C (2023) Recent advances in forensic DNA phenotyping of appearance, ancestry and age. Forensic Sci Int Genet 65:102870. https://doi.org/10.1016/j.fsigen.2023.102870
Koch CM, Wagner W (2011) Epigenetic-aging-signature to determine age in different tissues. Aging (Albany NY) 3(10):1018–1027. https://doi.org/10.18632/aging.100395
Kuroki Y, Fukami M (2023) Y chromosome genomic variations and biological significance in human diseases and health. Cytogenet Genome Res 163(1–2):5–13. https://doi.org/10.1159/000531933
Laird PW (2010) Principles and challenges of genome-wide DNA methylation analysis. Nat Rev Genet 11:191–203
Lee JE, Park SU, So MH, Lee HY (2024) Age prediction using DNA methylation of Y-chromosomal CpGs in semen samples. Forensic Sci Int Genet 69:103007. https://doi.org/10.1016/j.fsigen.2024.103007
Levine ME, Crimmins EM (2018) Is 60 the new 50? Examining changes in biological age over the past two decades. Demography 55:387–402. https://doi.org/10.1007/s13524-017-0644-5
Levine ME, Lu AT, Quach A et al (2018) An epigenetic biomarker of aging for lifespan and healthspan. Aging (Albany NY) 10(4):573–591. https://doi.org/10.18632/aging.101414
Li C, Fan Y, Li G et al (2018) DNA methylation reprogramming of functional elements during mammalian embryonic development. Cell Discov 4:41. https://doi.org/10.1038/s41421-018-0039-9
Li G, Wang C, Guan X et al (2022) Age-related DNA methylation on Y chromosome and their associations with total mortality among Chinese males. Aging Cell 21(3):e13563. https://doi.org/10.1111/acel.13563
Li N, Song K, Chen H, Dai M (2025) Advance and challenge of DNA methylation as cancer biomarkers for risk stratification, screening and early detection. J Natl Cancer Cent 5:108–112. https://doi.org/10.1016/j.jncc.2024.12.007
Liu F, Zhong K, Jing X et al (2019) Update on the predictability of tall stature from DNA markers in Europeans. Forensic Sci Int Genet 42:8–13. https://doi.org/10.1016/j.fsigen.2019.05.006

Lu AT, Quach A, Wilson JG et al (2019) DNA methylation GrimAge strongly predicts lifespan and healthspan. Aging (Albany NY) 11(2):303–327. https://doi.org/10.18632/aging.101684

Lucotte EA, Guðmundsdóttir VB, Jensen JM et al (2023) Characterizing the evolution and phenotypic impact of ampliconic Y chromosome regions. Nat Commun 14(1):3990. https://doi.org/10.1038/s41467-023-39644-6

Lund JB, Li S, Christensen K et al (2020) Age-dependent DNA methylation patterns on the Y chromosome in elderly males. Aging Cell 19(2):e12907. https://doi.org/10.1111/acel.12907

Maan AA, Eales J, Akbarov A, Rowland J, Xu X, Jobling MA et al (2017) The Y chromosome: a blueprint for men's health? Eur J Hum Genet 25(11):1181–1188. https://doi.org/10.1038/ejhg.2017.128

Mao SQ, Cuesta SM, Tannahill D, Balasubramanian S (2020) Genome-wide DNA methylation signatures are determined by DNMT3A/B sequence preferences. Biochemistry 59:2541–2550. https://doi.org/10.1021/acs.biochem.0c00339

Marcińska M, Pośpiech E, Abidi S, Andersen JD, van den Berge M, Carracedo Á et al (2015) Evaluation of DNA variants associated with androgenetic alopecia and their potential to predict male pattern baldness. PLoS One 10(5):e0127852. https://doi.org/10.1371/journal.pone.0127852

Márquez-Ruiz AB, González-Herrera L, Luna JD, Valenzuela A (2020) DNA methylation levels and telomere length in human teeth: usefulness for age estimation. Int J Legal Med 134(2):451–459. https://doi.org/10.1007/s00414-019-02242-7

McNevin D (2020) Forensic inference of biogeographical ancestry from genotype: the genetic ancestry lab. WIREs Forensic Sci 2:e1356. https://doi.org/10.1002/wfs2.1356

Millán-Zambrano G, Burton A, Bannister AJ, Schneider R (2022) Histone post-translational modifications—cause and consequence of genome function. Nat Rev Genet 23:563–580

Miyado M, Fukami M (2019) Losing maleness: somatic Y chromosome loss at every stage of a man's life. FASEB Bioadv 1(6):350–352. https://doi.org/10.1096/fba.2019-00006

Moser DA, Müller S, Hummel EM et al (2020) Targeted bisulfite sequencing: a novel tool for the assessment of DNA methylation with high sensitivity and increased coverage. Psychoneuroendocrinology 120:104784. https://doi.org/10.1016/j.psyneuen.2020.104784

Mullis K, Faloona F, Scharf S et al (1986) Specific enzymatic amplification of DNA in vitro: the polymerase chain reaction. Cold Spring Harb Symp Quant Biol 51(Pt 1):263–273. https://doi.org/10.1101/sqb.1986.051.01.032

Nagy C, Turecki G (2015) Transgenerational epigenetic inheritance: an open discussion. Epigenomics 7(5):781–790. https://doi.org/10.2217/epi.15.46

Naue J, Hoefsloot H, Mook O et al (2017) Chronological age prediction based on DNA methylation: massive parallel sequencing and random forest regression. Forensic Sci Int Genet 31:19–28. https://doi.org/10.1016/j.fsigen.2017.07.015

Oh S, Min K, Kim M, Lee SK (2020) Sex chromosomes are severely disrupted in gastric cancer cell lines. Int J Mol Sci 21(13):4598. https://doi.org/10.3390/ijms21134598

Okano M, Bell DW, Haber DA, Li E (1999) DNA methyltransferases Dnmt3a and Dnmt3b are essential for de novo methylation and mammalian development. Cell 99(3):247–257. https://doi.org/10.1016/s0092-8674(00)81656-6

Onofri M, Alessandrini F, Aneli S et al (2025) A Ge.F.I. Collaborative Study: evaluating reproducibility and accuracy of a DNA-methylation-based age-predictive assay for routine implementation in forensic casework. Electrophoresis 46:76–91. https://doi.org/10.1002/elps.202400190

Pappalardo XG, Barra V (2021) Losing DNA methylation at repetitive elements and breaking bad. Epigenetics Chromatin 14(1):25. https://doi.org/10.1186/s13072-021-00400-z

Peng F, Zhu G, Hysi PG et al (2019) Genome-wide association studies identify multiple genetic loci influencing eyebrow color variation in Europeans. J Invest Dermatol 139(7):1601–1605. https://doi.org/10.1016/j.jid.2018.12.029

Pinprasert P, Khacha-ananda S, Narakornsak S, Wunnapuk K (2024) Age-associated DNA methylation changes with hypermethylation ELOVL2 and ZYG11A in dental samples: a pilot study. ScienceAsia 50(1):2024.028: 1–9. https://doi.org/10.2306/scienceasia1513-1874.2024.028

Pośpiech E, Karłowska-Pik J, Marcińska et al (2015) Evaluation of the predictive capacity of DNA variants associated with straight hair in Europeans. Forensic Sci Int Genet 19:280–288. https://doi.org/10.1016/j.fsigen.2015.09.004

Poznik GD, Xue Y, Mendez FL et al (2016) Punctuated bursts in human male demography inferred from 1,244 worldwide Y-chromosome sequences. Nat Genet 48(6):593–599. https://doi.org/10.1038/ng.3559

Prinz M, Sansone M (2001) Y-chromosome specific short tandem repeats in forensic casework. Croat Med J 42:288–291

Quintana-Murci L, Fellous M (2001) The human Y chromosome: the biological role of a "functional wasteland". J Biomed Biotechnol 1(1):18–24. https://doi.org/10.1155/S1110724301000080

Refn M, Andersen M, Kampmann M et al (2023) Longitudinal changes and variation in human DNA methylation analysed with the Illumina Methylation EPIC BeadChip assay and their implications on forensic age prediction. Sci Rep 13:21658. https://doi.org/10.1038/s41598-023-49064-7

Rhie A, Nurk S, Cechova M et al (2023) The complete sequence of a human Y chromosome. Nature 621:344–354. https://doi.org/10.1038/s41586-023-06457-y

Roewer L, Krawczak M, Willuweit S et al (2001) Online reference database of European Y-chromosomal short tandem repeat (STR) haplotypes. Forensic Sci Int 118(2):106–113. https://doi.org/10.1016/S0379-0738(00)00478-3

Simpson JT, Workman RE, Zuzarte PC et al (2017) Detecting DNA cytosine methylation using nanopore sequencing. Nat Methods 14:407–410. https://doi.org/10.1038/nmeth.4184

Singh NP, Madabhushi SR, Srivastava S et al (2011) Epigenetic profile of the euchromatic region of human y chromosome. Nucleic Acids Res 39:3594–3606. https://doi.org/10.1093/nar/gkq1342

Skaletsky H, Kuroda-Kawaguchi T, Minx PJ et al (2003) The male-specific region of the human Y chromosome is a mosaic of discrete sequence classes. Nature 423(6942):825–837. https://doi.org/10.1038/nature01722

Smith ZD, Hetzel S, Meissner A (2025) DNA methylation in mammalian development and disease. Nat Rev Genet 26:7–30

Song M, Jiang L, Wang X, Zhou W, Wang N, Hou Y, Song F (2023) Loss of Y chromosome in leukocytes can be regarded as a male-specific age predictor for age group estimation in forensic genetics. Mol Gen Genomics 298(5):1073–1085. https://doi.org/10.1007/s00438-023-02039-9

Subrini J, Turner J (2021) Y chromosome functions in mammalian spermatogenesis. elife 10:e67345. https://doi.org/10.7554/eLife.67345

Syndercombe Court D (2021) The Y chromosome and its use in forensic DNA analysis. Emerg Top Life Sci 5(3):427–441. https://doi.org/10.1042/ETLS20200339

Tan Q, Heijmans BT, Hjelmborg JB et al (2016) Epigenetic drift in the aging genome: a ten-year follow-up in an elderly twin cohort. Int J Epidemiol 45:1146–1158. https://doi.org/10.1093/ije/dyw132

Thompson DJ, Genovese G, Halvardson J et al (2019) Genetic predisposition to mosaic Y chromosome loss in blood. Nature 575(7784):652–657. https://doi.org/10.1038/s41586-019-1765-3

Trombetta B, Cruciani F (2017) Y chromosome palindromes and gene conversion. Hum Genet 136(5):605–619. https://doi.org/10.1007/s00439-017-1777-8

Vakilian H, Mirzaei M, Tabar S et al (2015) DDX3Y, a male-specific region of Y chromosome gene, may modulate neuronal differentiation. J Proteome Res 14(9):3474–3483. https://doi.org/10.1021/acs.jproteome.5b00512

Verstraeten L, Fokias K, Saerens G, Bekaert B (2025) Cost-effective optimisation and validation of the VISAGE enhanced tool assay on the Illumina NovaSeq 6000 platform. Forensic Sci Int Genet 78:103299. https://doi.org/10.1016/j.fsigen.2025.103299

Vidaki A, Montiel González D, Planterose Jiménez B et al (2021) Male-specific age estimation based on Y-chromosomal DNA methylation. Aging 13(5):6442–6458. https://doi.org/10.18632/aging.202775

Visone R, Bacalini MG, Di Franco S et al (2019) DNA methylation of shelf, shore and open sea CpG positions distinguish high microsatellite instability from low or stable microsatellite status colon cancer stem cells. Epigenomics 11:587–604. https://doi.org/10.2217/epi-2018-0153

Walsh S, Chaitanya L, Clarisse L et al (2014) Developmental validation of the HIrisPlex system: DNA-based eye and hair colour prediction for forensic and anthropological usage. Forensic Sci Int Genet 9:150–161. https://doi.org/10.1016/j.fsigen.2013.12.006

Wang Y, Zhao Y, Bollas A et al (2021) Nanopore sequencing technology, bioinformatics and applications. Nat Biotechnol 39:1348–1365

Webster AK, Phillips PC (2025) Epigenetics and individuality: from concepts to causality across timescales. Nat Rev Genet 26:406–423

Wilson PC, Verma A, Yoshimura Y et al (2024) Mosaic loss of Y chromosome is associated with aging and epithelial injury in chronic kidney disease. Genome Biol 25(1):36. https://doi.org/10.1186/s13059-024-03173-2

Wimmer R, Kirsch S, Rappold GA et al (2005) Evolutionary breakpoint analysis on Y chromosomes of higher primates provides insight into human Y evolution. Cytogenet Genome Res 108(1–3):204–210. https://doi.org/10.1159/000080817

Zbieć-Piekarska R, Spólnicka M, Kupiec T et al (2015a) Development of a forensically useful age prediction method based on DNA methylation analysis. Forensic Sci Int Genet 17:173–179. https://doi.org/10.1016/j.fsigen.2015.05.001

Zbieć-Piekarska R, Spólnicka M, Kupiec T et al (2015b) Examination of DNA methylation status of the ELOVL2 marker may be useful for human age prediction in forensic science. Forensic Sci Int Genet 14:161–167. https://doi.org/10.1016/j.fsigen.2014.10.002

Zubakov D, Liu F, Kokmeijer I et al (2016) Human age estimation from blood using mRNA, DNA methylation, DNA rearrangement, and telomere length. Forensic Sci Int Genet 24:33–43. https://doi.org/10.1016/j.fsigen.2016.05.014

Chapter 7
Epigenetic for Age Estimation in Semen and Saliva

Ji Eun Lee, Sae Rom Hong, and Hwan Young Lee

Abstract Semen and saliva at a crime scene can provide important information about a donor's identity and activities that may have occurred. However, if an unknown suspect committed the crime, DNA samples should be further analyzed to obtain as much investigative information as possible. DNA methylation can vary with age, enabling estimation of the age of the sample donor. However, tissue-specific DNA methylation patterns suggest the need to develop body fluid/tissue-specific age estimation models. In the past decade, many age estimation models have been developed for semen and saliva samples. Although age-associated markers identified according to body fluid type differed, the age estimation accuracy of most models was as high as the mean absolute error of approximately 5 years. Given that semen is often collected as a mixture with vaginal epithelial cells, an age estimation model consisting of Y chromosomal CpG has also been developed. Saliva contains a variety of cells, including leukocytes and epithelium, and since most epithelial cells in saliva originate from the buccal epithelium, saliva and buccal swabs have often been considered interchangeable. However, cell type and microbiome composition between these swab types differ, and age estimation accuracy can be improved in saliva and buccal swab samples by analyzing additional cell type-specific CpG markers. In addition, a universal age estimation model was developed that can be applied to blood, saliva, and buccal swab samples, providing high

J. Eun Lee
Department of Forensic Medicine, Seoul National University College of Medicine, Seoul, South Korea

S. R. Hong
Biometrology Group, Korea Research Institute of Standards and Science, Daejeon, South Korea

H. Y. Lee (✉)
Department of Forensic Medicine, Seoul National University College of Medicine, Seoul, South Korea

Institute of Forensic and Anthropological Science, Seoul National University College of Medicine, Seoul, South Korea
e-mail: hylee192@snu.ac.kr

S. C. Zapico (ed.), *Forensic Science and Epigenetics*, Epigenetics and Human Health 18, https://doi.org/10.1007/978-3-032-23085-0_7

prediction accuracy within 4 years. Recent studies have improved accuracy, validated models across laboratories and explored population-specific differences. Future studies are needed to establish population-wide universal markers to improve age estimation accuracy as well as forensic applicability.

Keywords Age estimation · DNA methylation · Forensic science · Semen · Saliva

Abbreviations

CpG	Cytosine–phosphate–guanine
DMR	Differentially methylated region
DNAm	DNA methylation
dPCR	digital Polymerase chain reaction
EPIC	Illumina Infinium MethylationEPIC BeadChip
MAE	Mean absolute error
MPS	Massively parallel sequencing
MS-HRM	Methylation-sensitive high-resolution melting
PCR	Polymerase chain reaction
rRNA	ribosomal RNA
SBE	Single base extension
450 K	Infinium MethylationHuman450BeadChip

7.1 Introduction

Semen and saliva are the most frequently encountered biological samples found at crime scenes (Virkler and Lednev 2009). Semen plays an important role in determining whether or not sexual intercourse occurred, and is often detected mixed with the victim's vaginal epithelium (Li et al. 2023). Saliva can be detected from various sources such as lip prints, bite marks, cigarette butts, utensils, and edible items (Kayser et al. 2023), and especially when detected on a certain body part (e.g., breasts, mouth and genital area), the presence of saliva can be a supportive evidence for alleged sexual behavior such as licking or kissing (Sijen and Harbison 2021). However, if a victim was sexually assaulted by an assailant they had just met, there may be little information about the assailant, making investigation difficult.

Given the importance of semen and saliva in forensic investigations, recent research has focused on molecular markers such as DNA methylation (DNAm) to get information about the donor's age. DNA methylation at certain genomic loci exhibits age-associated changes, enabling age estimation from a variety of DNA samples collected from crime scene fluids (Bocklandt et al. 2011). Therefore, collecting semen and saliva specimens and then using DNA samples from these fluids to estimate age can constitute a practical approach for narrowing down donor pools and providing case-related activity contexts, thereby offering investigative leads. Another approach to specimen collection, buccal swabs are frequently considered as interchangeable with saliva (Theda et al. 2018) and are commonly used in forensics, including individual identification (Jovanovich et al. 2015), kinship testing (Cho et al. 2024), and DNA database construction (Hedman et al. 2008). Therefore, forensic age estimation (Eipel et al. 2016) via buccal swabs has also been studied due to their potential (e.g., estimating the legal age of refugees).

Beginning with the DNAm-based age prediction study by Bocklandt and colleagues (2011), which used saliva samples, age estimation using semen and saliva DNAm has been extensively studied for over a decade (e.g., Jung et al. 2019; Lee et al. 2025). This chapter aims to improve our understanding of the body fluid characteristics of semen and saliva and provide an overview of age-associated markers and age estimation models for better application of the developed assays and models to forensic caseworks.

7.2 Semen Age Prediction

7.2.1 Key Characteristics of Semen

Semen shows unique epigenetic patterns distinct from those of somatic tissues. In most somatic cells, aging is associated with a decrease in DNAm and telomere shortening (Li et al. 2017; Harley et al. 1990). In contrast, sperm cells undergo an increase in DNAm and an extension of telomere length with aging (Allsopp et al. 1992; Jenkins et al. 2013). Also, in terms of histone modification, histones are replaced by protamines in sperm DNA (Rathke et al. 2014).

Telomerase activity in sperm supports age-related telomere elongation. To prevent the inheritance of shortened telomeres, telomerase is activated in the testes, maintaining telomere length in germ cells. Furthermore, as sperm counts decline with age, the relative proportion of sperm benefiting from telomerase activity potentially increases (Eisenberg 2011).

Spermatogenesis involves unique, dynamic DNAm reprogramming. During this process, an initial phase of global demethylation occurs, followed by de novo methylation, which establishes the sperm-specific methylation pattern. Additional cycles of hypomethylation and remethylation during later stages further refine this pattern, resulting in a sperm methylome that is markedly distinct from that of somatic cells (Kotková and Drábek 2023; Siebert-Kuss et al. 2024; Stuppia et al. 2015).

Protamines replace histones in sperm to protect the genome and aid motility. While chromatin in somatic cells is organized with DNA wrapped around histone proteins, in sperm cells, most histones are replaced by protamines during a process known as spermiogenesis (Rathke et al. 2014). This transition leads to a much higher degree of DNA compaction. The compacted structure not only protects the paternal genome from potential damage but also enhances the hydrodynamic shape of the sperm head, thereby facilitating more efficient motility (Braun 2001).

Due to these unique epigenetic properties, age estimation models based on the epigenome developed for multiple somatic tissues cannot be directly applied to semen. For example, in Horvath's age clock study, application of the clock to sperm-derived DNA resulted in significantly underestimated chronological ages (Horvath 2013). Moreover, histone-based age predictors (de Lima Camillo et al. 2025) cannot be applied to semen due to its unique characteristic of histone-to-protamine replacement. In addition to these epigenetic characteristics, semen is not composed of sperm alone. Semen constitutes a heterogeneous mixture that also includes round cells, such as leukocytes and immature germ cells. These additional cell types can contribute their own DNAm signatures, introducing further variability into the semen methylome and making age estimation more challenging (Xiao et al. 2023). However, semen age estimation has become an essential forensic tool, particularly in sexual assault cases, where semen is often the most important biological evidence in establishing physical contact and supporting DNA profiling in sexual assault investigations. Given its forensic significance, accurate age estimation from semen can aid in narrowing down suspects and strengthening investigative leads.

7.2.2 Markers and Models for Age Estimation of Semen

Epigenetic age estimation of semen primarily relies on DNAm analysis. Since 2015, research in this field has steadily progressed (Table 7.1). A commonly used method for isolating sperm DNA, particularly in male-female mixture analysis, is *differential extraction*, which separates sperm cells based on their density (McKiernan and Danielson 2017). However, because forensic geneticists often deal with degraded or limited DNA, this method is not always effective. In about 10–20% of forensic cases, differential extraction fails, leading to potential sperm DNA loss (Gusmão et al. 2008). To address this limitation, researchers have explored age estimation using both sperm DNA and total semen DNA, improving sperm DNA analysis for forensic applications.

The first semen age estimation model was introduced in 2015 using three specific markers analyzed by a single-base extension (SBE) assay (Lee et al. 2015). Researchers identified age-related cytosine–phosphate–guanine (CpG) markers using an Infinium HumanMethylation450 BeadChip (450 K) array ($n = 12$) and designed an SBE assay for targeted analysis. Using this assay, researchers constructed a model based on just three CpGs (*TTC7B*, cg12837463, and *NOX4*). This model showed mean absolute error (MAE) of 5.4 years and was later validated with 19 forensic case samples (Lee et al. 2018). To further evaluate the model's reliability, a collaborative exercise involving 11 laboratories was conducted (Lee et al. 2022), in which the model's accuracy, reproducibility, and inter-laboratory consistency were assessed. The results confirmed the model's robustness across a variety of reagents (e.g., Gold ST*R 10X Buffer of Promega, AmpliTaq Gold® DNA polymerase of Applied Biosystems, Qiagen® Multiplex PCR Kit of Qiagen, etc.) and

Table 7.1 Forensic age estimation models for semen

Cell type	Platform		Number of markers	Accuracy		Modeling method	References
	Marker discovery	Modeling		Training	Test		
Semen	Array (450 K)	SBE	3 CpGs	MAD 4.2 ($n = 31$)	MAD 5.4 ($n = 37$)	Stepwise regression	Lee et al. (2015)
Sperm	Array (450 K)	Array (450 K)	51 regions	MAE 2.04 ($n = 329$)	MAE 2.37 ($n = 10 \times 6$)	Linear regression	Jenkins et al. (2018)
Semen	Adopted	Pyrosequencing	2 CpGs	MAD 2.974 ($n = 38$)	MAD 4.219 ($n = 17$)	Linear regression	Li et al. (2020)
Semen	Hybrid (EPIC+adopted)	MPS	6 CpGs	MAE 4.3 ($n = 125$)	MAE 5.1 ($n = 54$)	Stepwise regression	Pisarek et al. (2021)
Sperm	Array (EPIC)	Array (EPIC)	120 CpGs	MAE 1.6 ($n = 379$)	MAE 2.2 ($n = 173$)	Super learner	Pilsner et al. (2022)
			117 DMRs	MAE 1.7	MAE 2.3		
Sperm	Hybrid (EPIC+adopted)	SBE	11 CpGs	– ($n = 214$)	MAE of sperm 2.983 MAE of semen 5.715 ($n = 39$)	svmPoly	Xiao et al. (2023)
			10 CpGs	–	MAE of sperm 4.746 MAE of semen 5.281	svmPoly	
			21 CpGs	–	MAE of sperm 2.526 MAE of semen 3.890	Brnn	
			14 CpGs	–	MAE of sperm 2.918 MAE of semen 4.250	svmPoly	

(continued)

Table 7.1 (continued)

Cell type	Platform		Number of markers	Accuracy		Modeling method	References
	Marker discovery	Modeling		Training	Test		
Semen	Adopted	SBE	12 CpGs	MAE 3.81 ($n = 80$)	MAE 4.46 ($n = 35$)	Enter regression	Er et al. (2024)
			3 CpGs	MAE 4.1	MAE 4.08	Stepwise regression	
Semen	Hybrid (dRRBS+adopted)	MPS	9 CpGs	– ($n = 247$)	MAE of sperm 3.3 (CV)	Random forest	Li et al. (2025)
Semen	Hybrid (EPIC+adopted)	MPS	36 CpGs	MAE 2.97 ($n = 111$)	MAE 3.18 ($n = 48$)	Ridge regression	Lee et al. (2025)
			36 CpGs	MAE 2.86 ($n = 110$)	MAE 3.11 ($n = 43$)	Elastic net regression	
Semen	Array (EPIC)	MPS	5 Y-CpGs	MAE 4.89 ($n = 103$)	MAE 5.87 ($n = 44$)	Stepwise regression	Lee et al. (2024)
			7 Y-CpGs	MAE 4.91	MAE 5.87	Lasso regression	

svmPoly support vector machine with polynomial kernel, *brnn* Bayesian regularized neural network, *dRRBS* double-enzyme reduced representation bisulfite sequencing, *CV* cross-validation genetic analyzers (e.g., Applied Biosystems 3130, 3130XL, 3500 and 3500XL), reinforcing its applicability in forensic analysis.

A semen age estimation model based on sperm DNAm signatures was developed and tested against a wide variety of samples to assess germ line aging (Jenkins et al. 2018). Using 450 K array data from their previous studies (Jenkins et al. 2014, 2017; Aston et al. 2015), researchers analyzed samples from infertile males, heavy smokers, obese individuals, and the general population ($n = 329$). The final model, optimized for age estimation across 51 genomic regions, showed a MAE of 2.37 years. In addition to age estimation, the researchers investigated the effect of smoking on epigenetic aging and found that smokers exhibited an accelerated biological aging pattern. Specifically, estimated ages exceeded chronological ages by approximately 1.5% in smokers and 2.5% in long-term smokers (>10 years).

A targeted methylation-based model was developed for semen age estimation in Han Chinese males using a simplified CpG marker set (Li et al. 2020). Researchers took two markers (cg12837463 and *NOX4*) from a previous study (Lee et al. 2015) and designed a pyrosequencing assay. The model demonstrated MAEs of less than 4.4 years across various sample types, including liquid semen, fresh and aged semen stains, and mixed stains. Also, they tested the assay with blood, vaginal secretion and oral swabs and confirmed consistent methylation levels across different ages.

A robust semen age estimation model with inter-laboratory validation was developed by the VISAGE (Visible Attributes through Genomics) consortium using European male semen samples, highlighting its forensic utility (Pisarek et al. 2021). By analyzing Illumina Infinium MethylationEPIC BeadChip (EPIC) array data ($n = 40$), researchers identified 10 CpG sites strongly associated with age (*SYT7, TUBB3, SH2B2, ARHGEF17, EXOC3, GALR2, PPP2R2C, TBX4, PALM*, and *IFITM2*). Also, three CpG sites (*TTC7B*, cg12837463, and *NOX4*) previously identified in a Korean cohort (Lee et al. 2015) were incorporated into the analysis. The researchers developed two multiplex assays for massively parallel sequencing (MPS), each targeting 7 and 6 CpGs, respectively. Using a stepwise regression method, they selected six key contributors from five loci (*SH2B2, EXOC3, IFITM2, GALR2*, and *NOX4/FOLH1B*) and achieved a MAE of 5.1 years. Moreover, inter-laboratory validation confirmed the assays' performance, reproducibility, sensitivity, and species specificity, further supporting the model's forensic applicability (Heidegger et al. 2022). Notably, the sensitivity tests demonstrated that consistent and reliable results could be achieved with as little as 20 ng of polymerase chain reaction (PCR) input DNA, highlighting the assay's suitability for forensic casework where DNA quantity is often limited.

Two distinct age clocks for sperm were constructed based on individual CpGs and differentially methylated regions (DMRs), highlighting associations with fertility and smoking status (Pilsner et al. 2022). Using sperm EPIC array data ($n = 379$), researchers identified age-related CpGs and DMRs and constructed two age estimation clocks, where one consisted of 120 CpG sites and the other of 117 DMRs. These clocks showed high predictive accuracy with MAEs of 2.2 and 2.3 years, respectively, in terms of out-of-sample performance. Also, researchers examined the relationship between sperm epigenetic age, reproductive outcomes, and smoking. Their findings indicated that an increased sperm epigenetic age was associated with reduced pregnancy probability and lower fecundability. Furthermore, current smokers showed a higher sperm epigenetic age compared to non-smokers, consistent with earlier findings by Jenkins et al. (2018), which also suggested accelerated epigenetic aging in smokers.

Highly accurate sperm age estimation models were developed using 22 CpG markers combined in dual SBE assays, achieving MAEs as low as 2.2 years (Xiao et al. 2023). Researchers identified 19 new sperm-specific age-related CpG sites (cg19998819, cg21843517, cg18037145, cg0412357, cg12277678, cg25187042, cg11262154, cg03634854, cg13872326, cg20828122, cg20602007, cg27231587, cg01789162, cg27111970, cg04119405, cg24812634, cg19983027, cg03030301, and cg25715498) from publicly available EPIC array data and combined these with 3 CpG markers (*TTC7B*, cg12837463, and *NOX4*) from an earlier study (Lee et al. 2015) to design two SBE assays, each covering 11 CpGs. Both sperm DNA, extracted using differential methods, and semen DNA were analyzed separately. Using only sperm DNA data, researchers constructed 32 models and selected three that showed MAEs of 2.5–4.8 years for sperm DNA and 3.8–5.8 years for semen DNA. When researchers reduced the number of markers to 14, the model yielded MAEs of 2.9 years for sperm DNA and 4.3 years for semen DNA. Also, researchers developed three final models built from the complete sperm DNA dataset without splitting data into training and test sets, which showed MAEs between 2.2 and 2.8 years. Sensitivity analysis further indicated that at least 40 ng of sperm DNA is required for accurate age estimation.

A semen methylation model incorporating 12 CpGs was developed and tested under various environmental conditions to simulate forensic settings (Er et al. 2024). Researchers analyzed semen samples from male patients with infertility, premature ejaculation, and chronic prostatitis treated at Istanbul University-Cerrahpasa. Furthermore, researchers selected 12 CpGs (*PARP14, C5orf25*, cg23488376, *MXRA5, PFKFB3, DLL1, NOX4*, cg12837463, *TTC7B, KCNA7, NKX2–1,* and *SYNE4*) from previous studies (Lee et al. 2015; Pisarek et al. 2021; Li et al. 2020) and designed a multiplex panel for the SBE platform. Two models were constructed using all-points multivariate regression and stepwise regression methods, showing MAEs of 4.46 and 4.08 years for the test group. Moreover, researchers tested the robustness of semen stains across various environmental conditions including fabric type, washing conditions, hematin exposure, and UV-C light in an effort to explore the importance of evidence handling.

A random forest model for semen age estimation was developed featuring 9 CpGs and showed high accuracy (Li et al. 2025). Researchers identified fifty age-related CpGs using a double-enzyme reduced representation bisulfite sequencing (dRRBS) method on semen DNA from 21 healthy Chinese male volunteers. To enhance their analysis, researchers incorporated 27 CpG markers from previous studies (Lee et al. 2015; Pisarek et al. 2021; Xiao et al. 2023). Additionally, researchers developed a multiplex assay and applied MPS to sperm DNA isolated via differential lysis procedure. From the MPS data, researchers constructed age estimation models and ultimately selected a random forest model featuring 9 CpGs (chr2:129071885, chr12:4535197, chr10:4384066, chr21:45509586, chr13:32846238, chr16:65834073, chr20:42179997, chr12:123465627, and chrUn_GL000220v1:118890) as the final model. This final model showed a MAE of 3.30 years, demonstrating strong predictive accuracy.

The relationship between semen age estimation and population differences was explored in a study which compared European and Korean samples (Lee et al. 2025). Researchers identified eight new age-related CpGs (cg01548889, *RPL13*, cg02692808, cg04095995, *HOXD10, CAMK2A, UPF3A,* and cg23857379) using EPIC array data generated from Korean male semen samples (n = 49). Also, researchers incorporated three CpGs (*TTC7B*, cg12837463, and *NOX4*) previously identified in a Korean cohort (Lee et al. 2015) and ten CpGs (*SYT7, TUBB3, SH2B2, ARHGEF17, EXOC3, GALR2, PPP2R2C, TBX4, PALM,* and *IFITM2*) in a European cohort (Pisarek et al. 2021). With these markers, researchers developed multiplex assays and performed MPS on semen DNA. Using these data, researchers constructed highly accurate semen age estimation models with MAEs of 3–5 years. Notably, researchers found that European-derived markers were less accurate in the Korean cohort, highlighting population differences in age estimation and therefore underscoring the potential need for population-specific age estimation models for forensic applications.

A unique approach to semen age estimation was introduced using only Y-chromosomal CpG markers, enabling male-specific age estimation even in complex male-female DNA mixtures common in sexual assault cases (Lee et al. 2024). Researchers identified nine CpG loci (cg14933403, cg08258654, cg15810474, cg26432476, cg13372258, cg14446584, cg12566071, cg00063477, and cg00061679) from EPIC array data for Korean male semen samples ($n = 56$) and designed a multiplex assay for MPS analysis. Researchers' regression models achieved MAEs of less than 7 years and were validated using female samples, which showed negligible methylation effects, confirming the samples' male specificity and reliability for forensic applications.

These studies collectively demonstrate the diversity of analytical platforms applied in semen age estimation, each with distinct strengths and limitations. Researchers have utilized array-based profiling (Jenkins et al. 2018; Pilsner et al. 2022), SBE (Lee et al. 2015; Xiao et al. 2023; Er et al. 2024), pyrosequencing (Li et al. 2020), and MPS (Pisarek et al. 2021; Lee et al. 2024, 2025; Li et al. 2025), depending on their resources and the sample conditions. Arrays provide broad CpG coverage but require higher DNA input (Christiansen et al. 2022), making them less suitable for trace forensic samples. SBE is widely adopted due to its compatibility with existing capillary electrophoresis systems, but may suffer from inter-instrument variability (So and Lee 2021). Pyrosequencing offers high accuracy and quantitative data but is limited by low multiplexing capability (Roessler and Lehmann 2015; Parameswaran et al. 2007). MPS stands out for its high accuracy, sensitivity, and multiplex potential (Behjati and Tarpey 2013; Sun et al. 2025). However, the complexity of data analysis, including the need for advanced bioinformatics tools, standardized pipelines, and specialized expertise in sequence interpretation, can pose challenges (Tárraga et al. 2015; Roy et al. 2018).

7.2.3 *Potential and Limitations*

Semen age estimation research holds significant implications in forensic science, reproductive medicine, and epigenetic aging research. In forensic investigations, estimating the donor's age of semen samples found at crime scenes could enhance crime-solving capabilities by providing age-related clues that may help include or exclude individuals. Also, in reproductive medicine, understanding the relationship between semen epigenetic age and male infertility could provide valuable insights into sperm quality, reproductive health, and potential age-related fertility decline. Additionally, semen age estimation could be utilized in personalized health assessments, serving as a potential biomarker for environmental and lifestyle influences on male reproductive aging.

Despite these promising applications, several challenges associated with semen age estimation remain. Given that unlike blood or saliva, semen contains not only sperm but also round cells, such as leukocytes and immature germ cells, non-sperm cell types can introduce additional DNAm signatures. This cellular heterogeneity increases the complexity of the semen methylome and potentially reduces the accuracy and reproducibility of age estimation models. Moreover, sperm quality—particularly sperm motility and morphology—is known to decline with age, but is also susceptible to environmental factors such as air pollution, chemical exposure, and smoking (Jurewicz et al. 2018; Dissanayake et al. 2019). These factors not only affect fertility but also modulate DNAm in ways that confound age estimation models. For example, multiple studies (e.g., Jenkins et al. 2018; Pilsner et al. 2022) have reported accelerated epigenetic aging in smokers, highlighting the need to consider lifestyle factors when interpreting biological age estimation from semen. In addition to these biological complexities, practical limitations in forensic casework pose further challenges. In forensic casework, the amount of available DNA is often limited, making sensitivity testing a critical component of assay validation. Although several recent studies have addressed this by testing minimal input requirements (e.g., 20–40 ng), many others lack thorough sensitivity evaluations, leaving uncertainties about their performance in scenarios with very limited DNA input, such as trace or degraded samples. Moreover, Lee et al. (2025) explored population difference in semen age estimation models, suggesting that DNAm patterns associated with aging can vary across ethnic and genetic backgrounds. As a result, models trained on one population may not accurately predict age in another, necessitating calibration or adjustment for different population groups to improve age estimation accuracy and model applicability.

7.3 Saliva Age Estimation

7.3.1 Key Characteristics of Saliva

Saliva contains various cells, including leukocytes and epithelium. Since most epithelial cells in saliva originate from the buccal epithelium (Xie et al. 2008), saliva and buccal swabs share a common characteristic, and have been frequently regarded as interchangeable (Theda et al. 2018). However, recent studies have highlighted distinct differences in cellular composition and oral microbiota between saliva and buccal swab samples (Fig. 7.1).

Buccal swabs consistently show a higher proportion of epithelial cells compared to saliva, as evidenced via microscopy (Aps et al. 2002; Theda et al. 2018; Wong et al. 2022). In addition, the EpiDISH cell-type deconvolution algorithm (Teschendorff et al. 2017) applied to the EPIC array data revealed that buccal swabs contained approximately 83–87% epithelial cells, whereas saliva contained significantly lower epithelial fractions, around 47–58% (Wong et al. 2022). In contrast, saliva showed a higher leukocyte content, reflecting frequent interaction with immune cells in the oral environment. Moreover, cellular proportions for saliva and buccal swabs can vary depending on age and oral health status (Theda et al. 2018; Wong et al. 2022), as both aging and diseases affect leukocyte levels.

Regarding the association between age and specimen content, Theda and colleagues found that the epithelial cellular portion of saliva in children tends to be higher than that of adults (Theda et al. 2018). Similarly, a strong, negative, nonlinear age-correlation with epithelial proportion in saliva and buccal swabs was

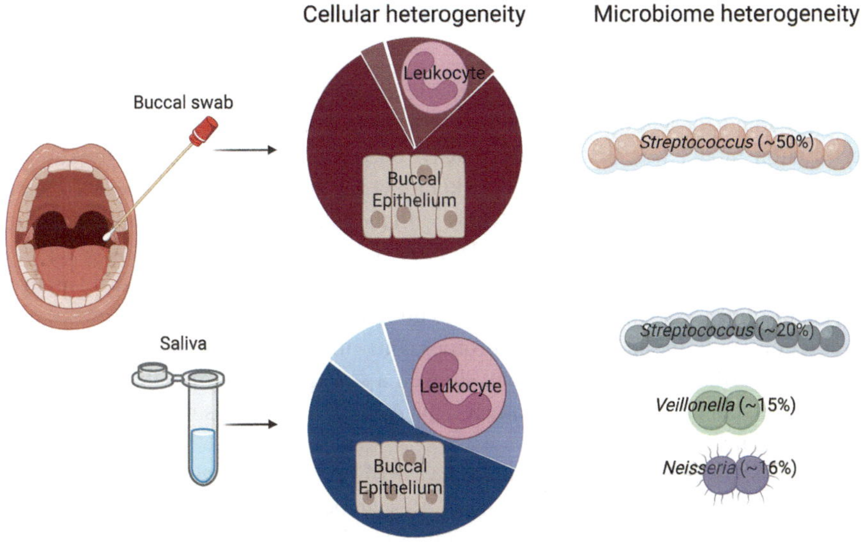

Fig. 7.1 Heterogeneity of saliva and buccal swabs

reported (Wong et al. 2022). However, the authors highlighted that high intra-population variability in epithelial proportion was observed in their study. Furthermore, the authors suggest that individual status, such as gingivitis, periodontitis, and smoking habits, could contribute to a decrease in epithelial proportion.

Considering the relation between individuals' diagnosis and leukocyte content, Aps and colleagues reported that leukocyte numbers significantly increase in individuals with gingivitis (Aps et al. 2002). Researchers used flow cytometry to measure the number of erythrocytes, leukocytes, epithelial cells, and bacteria in 1 microliter of saliva from 258 volunteers. Researchers revealed that a large number of leukocytes can indicate the presence of diseases because the origins of salivary leukocytes are the oropharyngeal area and the gingiva. Yordanova et al. (2022) showed that patients with inflammatory bowel disease exhibited higher counts of epithelial cells, leukocytes, and erythrocytes in saliva compared to healthy individuals.

Beyond cellular compositions, microbial profiles also distinguish saliva from buccal swabs. Specifically, Wong and colleagues reported that oral microbiota profiles differ significantly between saliva and buccal swabs, and that saliva exhibited higher bacterial richness and diversity than buccal mucosa (the mucous membrane lining in the inner cheeks) (Wong et al. 2022). Researchers performed 16S ribosomal RNA (rRNA) sequencing on saliva and buccal mucosa from healthy volunteers. Streptococcus (~50% of bacteria on average) was dominant in buccal mucosa, whereas saliva exhibited a more balanced flora, including Streptococcus (~20%), Veillonella (~15%), and Neisseria (~16%). These microbial differences were distinguished using a machine-learning model with an accuracy level approaching 93%.

7.3.2 Markers and Models for Age Estimation of Saliva

Saliva, the first specimen used in epigenetic age estimation studies, has been rigorously studied for age estimation using DNAm (Table 7.2). In 2011, a 3-CpG age estimation model was first reported based on the Illumina HumanMethylation27 microarray data of saliva samples (Bocklandt et al. 2011). Among 88 age-correlated loci, the authors selected the three CpGs with the highest age correlation, located in the promoter regions of the *EDARADD*, *TOM1L1*, and *NPTX* genes. A linear regression model trained using the three CpGs explained 73% of age variance, with a mean absolute error (MAE) of 5.3 years. Though this work was the first forensic age prediction study based on DNAm profiles, this milestone work exhibits a notable limitation. All three DNAm markers showed a strong correlation with age in male volunteers, but not in female volunteers; the *NPTX2* marker was not correlated with age in females.

Later, Silva et al. (2015) evaluated the *GRIA2* and *NPTX2* genes for epigenetic age estimation using saliva and blood. Researchers' pyrosequencing data from saliva samples showed that a model using 3 CpGs on the *GRIA2* genes explained 80.1% of age variance and a model using 6 CpGs on the *NPTX2* genes explained

Table 7.2 Forensic age estimation models for saliva and buccal swabs

Specimen	Cell-type marker	Gene	Accuracy	Analysis method	References
Saliva	No	EDARADD, TOM1L1, NPTX2	MAE = 5.3 years	Pyrosequencing	Bocklandt et al. (2011)
Buccal swab	No	GRIA2 (3 CpGs), NPTX2 (6 CpGs)	R^2: GRIA2 = 0.801, NPTX2 = 0.654	Pyrosequencing	Silva et al. (2015)
Buccal swab	Yes (CD6, SERPINB5)	CD6, SERPINB5	MAE = 5.0 years	Pyrosequencing	Eipel et al. (2016)
Buccal swab	No	KLF14 (single), KLF14 + SCGN (dual)	Single: MAE = 5.8 years Dual: MAE = 6.2 years	Pyrosequencing	Alghanim et al. (2017)
Saliva	Yes (PTPN7)	SST, SNGA3, KLF14, TSSK6, TBR1, SLC12A5 + PTPN7	MAE = 3.13 years	SBE	Hong et al. (2017)
Saliva	No	ELOVL2, EDARADD	MAE = 6.25 years	MS-HRM	Hamano et al. (2017)
Saliva, Buccal swab	No	ELOVL2, FHL2, KLF14, MIR29B2C, TRIM59	Saliva MAE = 3.55 years Buccal: MAE = 4.29 years	SBE	Jung et al. (2019)
Saliva	Yes (PTPN7)	SST, SNGA3, KLF14, TSSK6, TBR1, SLC12A5, PTPN7	3.19 years	SBE, MPS	Hong et al. (2019)
Buccal swab	No	PDE4C, EDARADD, KLF14	Pyrosequencing: MAE = 5.33 years SBE: MAE = 6.44 years	Pyrosequencing, SBE	Schwender et al. (2021)
Saliva, Buccal swab	Yes, but not included (HUNK, RUNX1)	MIR29B2C (cg10501210), LHFPL4, ELOVL2, PDE4C, HOXC4, OTUD7A, EDARADD	MAE = 3.66 years	SBE	Ambroa-Conde et al. (2022)
Saliva	No	SST, KLF14, TSSK6, SLC12A5	MAE = 3.3 years	dPCR	Ho Lee et al. (2022)
Saliva, Buccal swab	No	ELOVL2, FHL2, KLF14, MIR29B2C, TRIM59	Saliva: MAE = 3.5 years Buccal: MAE = 3.9 years Multi-tissue: MAE = 3.2 years	Pyrosequencing	Poussard et al. (2023)

65.4% of age variance. However, when their buccal swab models were applied to blood samples from the same donors, the accuracy of their models decreased.

Alghanim et al. (2017) also adopted a similar approach to Silva et al. (2015) by developing a single-locus model (2 CpGs on the *KLF14* gene) and a dual-locus model (2 CpGs on the *KLF14* and *SCGN* genes) for buccal swabs. These two models performed similarly, though the single-locus model included only the *KLF14* gene. The MAE of the single-locus model was 5.8 years, and the R-squared value was 0.85, while the dual-locus model showed 6.2 years of MAE and an R-squared value of 0.84. However, a blood-specific model trained based on pyrosequencing data for *KLF14* and *SCGN* CpG loci from the same donors showed lower performance with a lower age estimation accuracy than buccal-swab models. This discrepancy is associated with DNAm patterns for the *KLF14* and *SCGN* genes which differ between buccal swabs and blood, and therefore the authors suggested developing body fluid/tissue-specific models and testing universal age-correlated CpGs across a diverse range of cell types.

The importance of cell-type-specific markers in saliva and buccal swab age estimation was introduced by Eipel et al. (2016). Their research led to the proposal of a "buccal-cell-signature" consisting of 2 CpGs on the *CD6* and *SERPINB5* genes to improve age estimation accuracy in buccal swab samples. Initially, the authors retrained their former 3-CpG blood age prediction model (Weidner et al. 2014), which includes CpGs in the *ITGA2B, ASPA*, and *PDE4C* genes, on pyrosequencing data from 55 buccal swabs. The "3-CpG-swab-model" showed 7.03 years of MAE, and the selected CpGs on the *ITGA2B* and *ASPA* genes did not exhibit linearity with regard to age in public array data. However, counting leucocytes and buccal epithelium led to researchers' finding that the percentage of epithelial cells was highly correlated with DNAm markers on the *CD6* and *SERPINB5* genes. Consequently, researchers added this aforementioned "buccal-cell-signature" to the 3 CpGs, which increased the performance of the buccal swab age estimation model to a MAE of 5.0 years.

A similar approach to Eipel et al. (2016) was adopted for saliva age estimation by Hong et al. (2017). They added a cell-type-specific marker of the *PTPN7* gene to construct an age estimation model using 6 age-correlated DNAm markers (*SST, SNGA3, KLF14, TSSK6, TBR1*, and *SLC12A5*) identified from 450 K array data. The PTPN7 gene has been reported to show a substantial correlation between its DNAm level and buccal epithelium percentage (Souren et al. 2013). Furthermore, by adding the *PTPN7* gene marker to an SBE assay for detection of DNAm via 6 age-associated markers, Hong et al. (2017) proposed an age prediction model which increased age estimation accuracy from an MAE of 4.1 years to an MAE of 3.2 years.

However, Ambroa-Conde et al. (2022) reported that the inclusion of tissue-specific markers did not improve age estimation model performance. Specifically, an epigenetic age estimation model applicable to both saliva and buccal cell samples was developed by Ambroa-Conde et al. (2022) using SBE data from the same donors. Due to the cell type heterogeneity between these sample types, the authors chose two genes by cell-type specific DNAm patterns of *HUNK* and *RUNX1*, where

HUNK was hypermethylated in saliva, and *RUNX1* was hypermethylated in buccal epithelium. Researchers established a logistic regression model using two tissue-specific CpG sites to distinguish between the sample origins, achieving a classification accuracy of 88.6%. Additionally, they constructed a multivariate quantile regression-based age estimation model using seven age-correlated CpG sites (*MIR29B2C* (cg10501210), *LHFPL4, ELOVL2, PDE4C, HOXC4, OTUD7A*, and *EDARADD*), with 3.66 years of MAE.

In terms of several age prediction models using various DNAm detection methods, a saliva age estimation model using the *ELOVL2* (Zbieć-Piekarska et al. 2015a) and *EDARADD* (Bocklandt et al. 2011) genes was developed (Hamano et al. 2017), which was based on methylation-sensitive high-resolution melting (MS-HRM) data from 197 healthy individuals. The MAE of the developed model was 6.25 years for 50 independent saliva samples. Moreover, researchers collected cigarette butts from 16 smokers, whose ages were the same—40 years, and accordingly pioneered development of a saliva age estimation model for cigarette butts. Though predicted ages ranged from 29 to 51 years, this approach suggests the potential for age estimation using DNAm analysis of crime scene evidence, including cigarette butts.

In an effort to predict age using digital PCR (dPCR), Lee et al. (2022) tested DNAm from saliva using four CpG sites on the *SST, KLF14, TSSK6*, and *SLC12A5* genes. Researchers proposed a saliva age estimation model using samples from 76 individuals, showing 3.3 years of MAE. Due to the small sample size, researchers limited their work to model construction, but this study was the first to apply dPCR methods to forensic age prediction. Although high cost might hinder the adoption of dPCR for routine laboratory procedures, its exceptional sensitivity could support forensic applications involving limited sample quantities or mixture deconvolution.

Furthermore, several studies compare and develop age prediction models using DNAm data from various DNAm detection methods. For example, a subsequent study by Hong et al. (2019) expanded on their earlier work (Hong et al. 2017) by applying the previously developed SBE model to MPS data using the same seven CpG markers via amplicon sequencing. However, due to method-induced differences between SBE and amplicon sequencing, applying the SBE model to MPS data resulted in age predictions that deviated from chronological age by more than 20 years. Therefore, the author suggested adopting a platform-independent parameter to develop a new saliva age prediction model for both SBE and MPS data.

Unlike Hong et al. (2019) and Schwender et al. (2021) developed and validated two different age estimation models using buccal swab samples by employing pyrosequencing and SBE methods to measure DNAm of three CpG sites on the *PDE4C, EDARADD*, and *KLF14* genes. The pyrosequencing model yielded 5.33 years of MAE, while the SBE model yielded slightly reduced accuracy, with a MAE of 6.44 years. Both models demonstrated a high correlation between predicted and chronological ages, with negligible influence from smoking, though there was a tendency toward higher prediction accuracy for female samples. However, their higher prediction accuracy for female samples might have been influenced by gender imbalance in sampling (94 female and 47 male volunteers). The authors highlighted the utility of buccal swabs as a non-invasive source for forensic age prediction, and the SBE method could represent a cost-effective alternative to pyrosequencing for routine forensic laboratory procedures.

On the other hand, several studies have evaluated blood age estimation markers and models on saliva and buccal swab samples in an effort to develop forensically robust age estimation systems. Since saliva and buccal swabs contain leukocytes, applying a blood-based age prediction model to saliva and buccal swab samples showed potential for accurate age estimation. Particularly, *ELOVL2* has been reported to show a strong age correlation across multiple tissue types by several researchers (Bekaert et al. 2015; Giuliani et al. 2016; Zbieć-Piekarska et al. 2015a). Among many forensic age estimation models for blood, a model leveraging 5 age-predictive markers on the *ELOVL2, FHL2, KLF14, MIR29B2C*, and *TRIM59* genes is well known to provide a high age prediction accuracy in blood pyrosequencing data from a Polish population with an MAE of 3.9 years (Zbieć-Piekarska et al. 2015b). Therefore, many researchers tested these five markers using different DNAm detection methods (Dias et al. 2020; Freire-Aradas et al. 2020; Heidegger et al. 2020; Jung et al. 2019; Montesanto et al. 2020; Poussard et al. 2023), different populations (Cho et al. 2017; Dias et al. 2020), and different body fluids/tissues (Lee et al. 2020; Woźniak et al. 2021), including saliva and buccal swabs (Jung et al. 2019; Poussard et al. 2023).

A multi-tissue age estimation model using five CpG sites from the *ELOVL2, FHL2, KLF14, MIR29B2C*, and *TRIM59* genes across blood, saliva, and buccal swab samples was first developed and validated by Jung et al. (2019). All CpGs showed strong age-associated methylation patterns for saliva, and the model showed 3.55 years of MAE and an R-squared value of 0.906, indicating that the blood age-associated markers can predict age with high accuracy in saliva. Although buccal swabs demonstrated weaker age-correlations at the *FHL2* and *MIR29B2C* genes, particularly due to hypomethylated patterns, the model still showed a high predictive power (MAE = 4.29 years, R^2 = 0.884). Notably, DNAm of the CpGs on the *ELOVL2, KLF14*, and *TRIM59* genes consistently correlated with age in both saliva and buccal swabs, implying that the three CpGs could function as multi-tissue epigenetic biomarkers. The developed SBE assay effectively detected age-related methylation changes and showed potential for resolving the low-methylation signal of the *KLF14* gene.

The applicability of the age prediction models developed by Jung et al. (2019) was also evaluated using pyrosequencing on saliva and buccal swab samples in a study by Poussard et al. (2023). A total of 115 French individuals aged 0–88 years provided matched saliva and buccal samples, and DNAm at five CpG sites (*ELOVL2, FHL2, KLF14, MIR29B2C*, and *TRIM59*) were analyzed. Despite discrepancies in methylation correlation patterns between sample types and CpG loci, especially for *FHL2* and *MIR29B2C*, the models yielded promising prediction accuracy: MAEs were 3.5 years for saliva, 3.9 years for buccal swabs, and 3.2 years for the multi-tissue model, respectively. However, researchers highlighted prediction errors in the youngest (<10 years) and oldest (>70 years) age groups due to the loss of linearity in the relationship between DNAm and age, as well as potential environmental influences. Furthermore, DNAm value distributions differed between saliva and buccal swabs, likely reflecting their distinct cellular compositions. This work supports the use of DNAm profiling from oral samples for forensic age estimation, while emphasizing the need for caution in interpreting results for individuals at the extremes of the age spectrum and mixed-cell-type samples.

7.3.3 Potential and Limitations

Since saliva and buccal swabs are widely used in forensic contexts, age estimation based on saliva and buccal swabs could be useful in many cases. For example, age estimation models for saliva and buccal swabs can predict the age of an unknown suspect and verify claimed ages in cases lacking documentation without invasive or difficult collection procedures. However, several open questions concerning predicting age using saliva and buccal swabs persist: universal forensic age prediction systems, different populations, age groups, and DNAm detection methods.

As discussed, saliva and buccal swab samples originate from the same collection site, but their cellular heterogeneity presents a major challenge for age prediction. Different cellular compositions lead to different DNAm patterns and/or age correlations, even when using the same age-predictive markers. Thus, studies that exploited or highlighted cell-type-specific DNAm signatures across varying cellular compositions are often accompanied by correction factors based on cell-type-specific methylation patterns (Eipel et al. 2016; Hong et al. 2017). However, Ambroa-Conde et al. (2022) suggested a universal age prediction model for saliva and buccal swabs. This universal approach had been previously addressed by Jung et al. (2019) using a model trained with blood, saliva, and buccal swabs, which showed 3.88 years of MAE for an independent test set containing three types of samples.

Since *ELOVL2* represents a particularly promising age-predictive marker for many types of body fluids and tissues (Bekaert et al. 2015; Giuliani et al. 2016; Zbieć-Piekarska et al. 2015a), the discovery and validation of new universal markers could facilitate the establishment of well-tailored forensic age prediction systems for major body fluids and tissues. Naue and colleagues (2018) briefly tested several age-predictive markers in multiple tissue types using amplicon sequencing, and similarly, Choi and colleagues (2019) also discovered tissue-specific and tissue-independent age-predictive CpGs from public array data from various tissue types. These universal age-correlated marker studies could provide the basis of forensic age prediction systems by finding robust age prediction markers across multiple tissue types. In addition, using a moderate number of well-demonstrated universal age prediction markers can support the development of forensic age estimation systems by improving cost-effectiveness, enabling easier integration into routine laboratory work, and improving high predictive accuracy.

Cell-type-specific markers are also worthy of further investigation in forensic and medical contexts. In particular, markers reflecting the ratio between buccal epithelium and leukocytes could function as potential quality assurance indicators, with applications in assessing sample quality or monitoring the impact of lifestyle interventions. Moreover, cell type-specific markers can reveal inter-individual differences in age estimation results, and longitudinal studies of cell type-specific aging processes could be applied in both demographics research and forensic science.

Different populations and age groups also complicate age prediction in saliva and buccal swabs. Oka et al. (2022) reported that Japanese and Indonesian individuals had different DNAm patterns in saliva at the *EDARADD* and *FHL2* genes using MS-HRM data. Despite the lack of middle-aged groups, the data implied that different population groups might possess different DNAm patterns. Moreover, the accuracy of age prediction varies across age groups (Hong et al. 2017; Poussard et al. 2023). Non-linear DNAm patterns are often observed in children, and prediction errors also tend to increase for elderly individuals.

Furthermore, the decision of DNAm detection methods represents another open question for age estimation researchers. Due to method-induced DNAm differences, several researchers have attempted to develop detection method-independent age prediction models (Feng et al. 2018; Freire-Aradas et al. 2020; Hong and Shin 2023; Hong et al. 2019). Notably, a method parameter was introduced into a saliva age estimation model by Hong et al. (2019). Although the adoption of a platform-independent parameter approach enabled accurate saliva age estimation across both SBE and MPS data, several researchers have suggested that developing method-specific age prediction models may offer greater accuracy than models that integrate data across detection platforms (Hong and Shin 2023; Schwender et al. 2021). Additionally, since DNAm levels obtained via SBE can vary depending on the genetic analyzer used (So and Lee 2021), DNAm detection methods should be carefully decided, considering each laboratory's capacity and the availability of appropriate method-specific age estimation models.

7.4 Conclusion

Semen, saliva, and buccal epithelial cells are frequently found as biological evidence at crime scenes, highlighting the forensic value of age estimation based on these sample types, and current technology achieves high prediction accuracy within 5 years (e.g., Hong et al. 2017; Xiao et al. 2023). Among these specimens, semen shows unique epigenetic patterns distinct from somatic tissues (Siebert-Kuss et al. 2024), requiring specialized age estimation models. Since 2015, research has advanced from early models using a small number of CpGs to more sophisticated approaches incorporating larger marker sets (e.g., Jenkins et al. 2018), machine learning (e.g., Er et al. 2024), and various methylation analysis methods (e.g., Lee et al. 2015; Li et al. 2020; Pisarek et al. 2021; Pilsner et al. 2022). Recent studies have improved accuracy (e.g., Xiao et al. 2023), validated models across laboratories (Lee et al. 2022; Heidegger et al. 2022), and explored population-specific differences (Lee et al. 2025). Despite the forensic and clinical potential of recent age estimation models, challenges like cellular heterogeneity and population difference remain. Future research is needed to refine age estimation models, improve their predictive accuracy, and expand their forensic applicability.

Acknowledgment The authors acknowledge Richard Rose for technical editing of the manuscript.

Conflict of Interest The authors have declared no conflict of interest.

Compliance with Ethical Standards This chapter is based on previously published studies and does not involve any new studies with human participants or animals performed by the authors.

References

Alghanim H, Antunes J, Silva DSBS, Alho CS, Balamurugan K, Mccord B (2017) Detection and evaluation of DNA methylation markers found at SCGN and KLF14 loci to estimate human age. Forensic Sci Int Genet 31:81–88. https://doi.org/10.1016/j.fsigen.2017.07.011

Allsopp RC, Vaziri H, Patterson C, Goldstein S, Younglai EV, Futcher AB, Greider CW, Harley CB (1992) Telomere length predicts replicative capacity of human fibroblasts. PNAS 89(21):10114–10118. https://doi.org/10.1073/pnas.89.21.10114

Ambroa-Conde A, Girón-Santamaría L, Mosquera-Miguel A, Phillips C, Casares De Cal MA, Gómez-Tato A, Ãlvarez-Dios J, De La Puente M, Ruiz-Ramírez J, Lareu MV, Freire-Aradas A (2022) Epigenetic age estimation in saliva and in buccal cells. Forensic Sci Int Genet 61:102770. https://doi.org/10.1016/j.fsigen.2022.102770

Aps JKM, Van Den Maagdenberg K, Delanghe JR, Martens LC (2002) Flow cytometry as a new method to quantify the cellular content of human saliva and its relation to gingivitis. Clin Chim Acta 321(1):35–41. https://doi.org/10.1016/S0009-8981(02)00062-1

Aston KI, Uren PJ, Jenkins TG, Horsager A, Cairns BR, Smith AD, Carrell DT (2015) Aberrant sperm DNA methylation predicts male fertility status and embryo quality. Fertil Steril 104(6):1388–1397.e5. https://doi.org/10.1016/j.fertnstert.2015.08.019

Behjati S, Tarpey PS (2013) What is next generation sequencing? Arch Dis Child Educ Pract 98(6):236. https://doi.org/10.1136/archdischild-2013-304340

Bekaert B, Aubeline K, Zapico SC, Van de Voorde W, Decorte R (2015) Improved age determination of blood and teeth samples using a selected set of DNA methylation markers. Epigenetics 10(10):922–930. https://doi.org/10.1080/15592294.2015.1080413

Bocklandt S, Lin W, Sehl ME, Sánchez FJ, Sinsheimer JS, Horvath S, Vilain E (2011) Epigenetic predictor of age. PLoS One 6(6):e14821. https://doi.org/10.1371/journal.pone.0014821

Braun RE (2001) Packaging paternal chromosomes with protamine. Nat Genet 28(1):10–12. https://doi.org/10.1038/ng0501-10

Cho S, Jung S-E, Hong SR, Lee EH, Lee JH, Lee SD, Lee HY (2017) Independent validation of DNA-based approaches for age prediction in blood. Forensic Sci Int Genet 29:250–256. https://doi.org/10.1016/j.fsigen.2017.04.020

Cho S, Shin E, Park YG, Choi SH, Choe EK, Bae JH, Lee J-E, Lee SD (2024) A novel approach of kinship determination based on the physical length of genetically shared regions of chromosomes. Genes Genom 46(5):577–587. https://doi.org/10.1007/s13258-023-01485-4

Choi H, Joe S, Nam H (2019) Development of tissue-specific age predictors using DNA methylation data. Gene 10(11):888. https://doi.org/10.3390/genes10110888

Christiansen SN, Dyrberg AJ, Marie-Louise K, Jing L, Meyer AM, Jacob T-H, Morling N (2022) Reproducibility of the Infinium methylationEPIC BeadChip assay using low DNA amounts. Epigenetics 17(12):1636–1645. https://doi.org/10.1080/15592294.2022.2051861

De Lima Camillo LP, Asif MH, Horvath S, Larschan E, Singh R (2025) Histone mark age of human tissues and cell types. Sci Adv 11(1):eadk9373. https://doi.org/10.1126/sciadv.adk9373
Dias HC, Cordeiro C, Pereira J, Pinto C, Real FC, Cunha E, Manco L (2020) DNA methylation age estimation in blood samples of living and deceased individuals using a multiplex SNaPshot assay. Forensic Sci Int 311:110267. https://doi.org/10.1016/j.forsciint.2020.110267
Dissanayake DMIH, Keerthirathna WLR, Peiris LDC (2019) Male infertility problem: a contemporary review on present status and future perspective. Gend Genome 3:2470289719868240. https://doi.org/10.1177/2470289719868240
Eipel M, Mayer F, Arent T, Ferreira MR, Birkhofer C, Gerstenmaier U, Costa IG, Ritz-Timme S, Wagner W (2016) Epigenetic age predictions based on buccal swabs are more precise in combination with cell type-specific DNA methylation signatures. Aging (Albany NY) 8(5):1034–1048. https://doi.org/10.18632/aging.100972
Eisenberg DTA (2011) An evolutionary review of human telomere biology: the thrifty telomere hypothesis and notes on potential adaptive paternal effects. Am J Hum Biol 23(2):149–167. https://doi.org/10.1002/ajhb.21127
Er S, Abik Z, Ersoy G, Filoglu G, Ozkara H, Bulbul O (2024) A semen-specific deoxyribonucleic acid methylation model for epigenetic age estimation and its robustness under environmental challenges. Electrophoresis 45(19–20):1820–1833. https://doi.org/10.1002/elps.202400054
Feng L, Peng F, Li S, Jiang L, Sun H, Ji A, Zeng C, Li C, Liu F (2018) Systematic feature selection improves accuracy of methylation-based forensic age estimation in Han Chinese males. Forensic Sci Int Genet 35:38–45. https://doi.org/10.1016/j.fsigen.2018.03.009
Freire-Aradas A, Pośpiech E, Aliferi A, Girón-Santamaría L, Mosquera-Miguel A, Pisarek A, Ambroa-Conde A, Phillips C, Casares De Cal MA, Gómez-Tato A, Spólnicka M, Woźniak A, Ãlvarez-Dios J, Ballard D, Court DS, Branicki W, Carracedo Ã, Lareu MV (2020) A comparison of forensic age prediction models using data from four DNA methylation technologies. Front Genet 11:932. https://doi.org/10.3389/fgene.2020.00932
Giuliani C, Cilli E, Bacalini MG, Pirazzini C, Sazzini M, Gruppioni G, Franceschi C, Garagnani P, Luiselli D (2016) Inferring chronological age from DNA methylation patterns of human teeth. Am J Phys Anthropol 159(4):585–595. https://doi.org/10.1002/ajpa.22921
Gusmão L, Brión M, Gomes I (2008) Chapter 30: the human Y chromosome male-specific polymorphisms and forensic genetics. In: Bogusz MJ (ed) Handbook of analytical separations. Elsevier Science B.V, pp 969–1000
Hamano Y, Manabe S, Morimoto C, Fujimoto S, Tamaki K (2017) Forensic age prediction for saliva samples using methylation-sensitive high resolution melting: exploratory application for cigarette butts. Sci Rep 7(1):10444. https://doi.org/10.1038/s41598-017-10752-w
Harley CB, Futcher AB, Greider CW (1990) Telomeres shorten during ageing of human fibroblasts. Nature 345(6274):458–460. https://doi.org/10.1038/345458a0
Hedman J, Albinsson L, Ansell C, Tapper H, Hansson O, Holgersson S, Ansell R (2008) A fast analysis system for forensic DNA reference samples. Forensic Sci Int Genet 2(3):184–189. https://doi.org/10.1016/j.fsigen.2007.12.011
Heidegger A, Xavier C, Niederstätter H, De La Puente M, Pośpiech E, Pisarek A, Kayser M, Branicki W, Parson W (2020) Development and optimization of the VISAGE basic prototype tool for forensic age estimation. Forensic Sci Int Genet 48:102322. https://doi.org/10.1016/j.fsigen.2020.102322
Heidegger A, Pisarek A, De La Puente M, Niederstätter H, Pośpiech E, Woźniak A, Schury N, Unterländer M, Sidstedt M, Junker K, Ventayol Garcia M, Laurent FX, Ulus A, Vannier J, Bastisch I, Hedman J, Sijen T, Branicki W, Xavier C, Parson W (2022) Development and interlaboratory validation of the VISAGE enhanced tool for age estimation from semen using quantitative DNA methylation analysis. Forensic Sci Int Genet 56:102596. https://doi.org/10.1016/j.fsigen.2021.102596
Hong SR, Shin K-J (2023) Can we integrate method-specific age-predictive models?: analysis method-induced differences in detected DNA methylation status. Forensic Sci Int Genet 62:102805. https://doi.org/10.1016/j.fsigen.2022.102805

Hong SR, Jung S-E, Lee EH, Shin K-J, Yang WI, Lee HY (2017) DNA methylation-based age prediction from saliva: high age predictability by combination of 7 CpG markers. Forensic Sci Int Genet 29:118–125. https://doi.org/10.1016/j.fsigen.2017.04.006

Hong SR, Shin K-J, Jung S-E, Lee EH, Lee HY (2019) Platform-independent models for age prediction using DNA methylation data. Forensic Sci Int Genet 38:39–47. https://doi.org/10.1016/j.fsigen.2018.10.005

Horvath S (2013) DNA methylation age of human tissues and cell types. Genome Biol 14(10):3156. https://doi.org/10.1186/gb-2013-14-10-r115

Jenkins TG, Aston KI, Cairns BR, Carrell DT (2013) Paternal aging and associated intraindividual alterations of global sperm 5-methylcytosine and 5-hydroxymethylcytosine levels. Fertil Steril 100(4):945–951.e2. https://doi.org/10.1016/j.fertnstert.2013.05.039

Jenkins TG, Aston KI, Pflueger C, Cairns BR, Carrell DT (2014) Age-associated sperm DNA methylation alterations: possible implications in offspring disease susceptibility. PLoS Genet 10(7):e1004458. https://doi.org/10.1371/journal.pgen.1004458

Jenkins TG, James ER, Alonso DF, Hoidal JR, Murphy PJ, Hotaling JM, Cairns BR, Carrell DT, Aston KI (2017) Cigarette smoking significantly alters sperm DNA methylation patterns. Andrology 5(6):1089–1099. https://doi.org/10.1111/andr.12416

Jenkins TG, Aston KI, Cairns B, Smith A, Carrell DT (2018) Paternal germ line aging: DNA methylation age prediction from human sperm. BMC Genomics 19(1):763. https://doi.org/10.1186/s12864-018-5153-4

Jovanovich S, Bogdan G, Belcinski R, Buscaino J, Burgi D, Butts ELR, Chear K, Ciopyk B, Eberhart D, El-Sissi O, Franklin H, Gangano S, Gass J, Harris D, Hennessy L, Kindwall A, King D, Klevenberg J, Li Y, Mehendale N, Mcintosh R, Nielsen B, Park C, Pearson F, Schueren R, Stainton N, Troup C, Vallone PM, Vangbo M, Woudenberg T, Wyrick D, Williams S (2015) Developmental validation of a fully integrated sample-to-profile rapid human identification system for processing single-source reference buccal samples. Forensic Sci Int Genet 16:181–194. https://doi.org/10.1016/j.fsigen.2014.12.004

Jung S-E, Lim SM, Hong SR, Lee EH, Shin K-J, Lee HY (2019) DNA methylation of the ELOVL2, FHL2, KLF14, C1orf132/MIR29B2C, and TRIM59 genes for age prediction from blood, saliva, and buccal swab samples. Forensic Sci Int Genet 38:1–8. https://doi.org/10.1016/j.fsigen.2018.09.010

Jurewicz J, Dziewirska E, Radwan M, Hanke W (2018) Air pollution from natural and anthropic sources and male fertility. Reprod Biol Endocrinol 16(1):109. https://doi.org/10.1186/s12958-018-0430-2

Kayser M, Branicki W, Parson W, Phillips C (2023) Recent advances in Forensic DNA Phenotyping of appearance, ancestry and age. Forensic Sci Int Genet 65:102870. https://doi.org/10.1016/j.fsigen.2023.102870

Kotková L, Drábek J (2023) Age-related changes in sperm DNA methylation and their forensic and clinical implications. Epigenomics 15(21):1157–1173. https://doi.org/10.2217/epi-2023-0307

Lee HY, Jung S-E, Oh YN, Choi A, Yang WI, Shin K-J (2015) Epigenetic age signatures in the forensically relevant body fluid of semen: a preliminary study. Forensic Sci Int Genet 19:28–34. https://doi.org/10.1016/j.fsigen.2015.05.014

Lee JW, Choung CM, Jung JY, Lee HY, Lim S-K (2018) A validation study of DNA methylation-based age prediction using semen in forensic casework samples. Legal Med 31:74–77. https://doi.org/10.1016/j.legalmed.2018.01.005

Lee HY, Hong SR, Lee JE, Hwang IK, Kim NY, Lee JM, Fleckhaus J, Jung S-E, Lee YH (2020) Epigenetic age signatures in bones. Forensic Sci Int Genet 46:102261. https://doi.org/10.1016/j.fsigen.2020.102261

Lee JE, Lee JM, Naue J, Fleckhaus J, Freire-Aradas A, Neubauer J, Pośpiech E, Mccord B, Kalamara V, Gauthier Q, Mills C, Cao Y, Wang Z, Oh YN, Feng L, Schneider PM, Phillips C, Haas C, Pisarek A, Branicki W, Podini D, Vidaki A, Tejero NF, Ambroa-Conde A, Mosquera-Miguel A, Lareu MV, Hou Y, Lee JY, Lee HY (2022) A collaborative exercise on DNA

methylation-based age prediction and body fluid typing. Forensic Sci Int Genet 57:102656. https://doi.org/10.1016/j.fsigen.2021.102656
Lee JE, Park SU, So MH, Lee HY (2024) Age prediction using DNA methylation of Y-chromosomal CpGs in semen samples. Forensic Sci Int Genet 69:103007. https://doi.org/10.1016/j.fsigen.2024.103007
Lee JE, Cho S, So MH, Lee HY (2025) DNA methylation-based semen age prediction using the markers identified in Koreans and Europeans. Forensic Sci Int Genet 77:103243. https://doi.org/10.1016/j.fsigen.2025.103243
Li S, Christiansen L, Christensen K, Kruse T, Redmond P, Marioni R, Deary I, Tan Q (2017) Identification, replication and characterization of epigenetic remodelling in the aging genome: a cross population analysis. Sci Rep 7(1):8183. https://doi.org/10.1038/s41598-017-08346-7
Li L, Song F, Lang M, Hou J, Wang Z, Prinz M, Hou Y (2020) Methylation-based age prediction using pyrosequencing platform from seminal stains in Han Chinese males. J Forensic Sci 65(2):610–619. https://doi.org/10.1111/1556-4029.14186
Li X-N, Xu F-L, Zheng J-L, Sun M-L, Zhu X-M, Lv P, Du Z, Zhang X-P, Yao J (2023) Magnetic bead-based separation of sperm cells from semen-vaginal fluid mixed stains using an anti-ACRBP antibody. Int J Legal Med 137(2):511–518. https://doi.org/10.1007/s00414-022-02917-8
Li Y, Liu X, Chen M, Yi S, He X, Xiao C, Huang D (2025) DNA methylation-based age estimation from semen: genome-wide marker identification and model development. Forensic Sci Int Genet 76:103215. https://doi.org/10.1016/j.fsigen.2024.103215
Mckiernan HE, Danielson PB (2017) Chapter 21: molecular diagnostic applications in forensic science. In: Patrinos GP (ed) Molecular diagnostics, 3rd edn. Academic, pp 371–394
Montesanto A, D'aquila P, Lagani V, Paparazzo E, Geracitano S, Formentini L, Giacconi R, Cardelli M, Provinciali M, Bellizzi D, Passarino G (2020) A new robust epigenetic model for forensic age prediction. J Forensic Sci 65(5):1424–1431. https://doi.org/10.1111/1556-4029.14460
Naue J, Sänger T, Hoefsloot HCJ, Lutz-Bonengel S, Kloosterman AD, Verschure PJ (2018) Proof of concept study of age-dependent DNA methylation markers across different tissues by massive parallel sequencing. Forensic Sci Int Genet 36:152–159. https://doi.org/10.1016/j.fsigen.2018.07.007
Oka H, Dwi Ariani M, Akazaki T, Miyauchi M, Kitagawa M (2022) Some tips on age estimation using DNA methylation in saliva samples as an index across the Japanese and Indonesian ethnicities. Legal Med 56:102042. https://doi.org/10.1016/j.legalmed.2022.102042
Parameswaran P, Jalili R, Tao L, Shokralla S, Gharizadeh B, Ronaghi M, Fire AZ (2007) A pyrosequencing-tailored nucleotide barcode design unveils opportunities for large-scale sample multiplexing. Nucleic Acids Res 35(19):e130–e130. https://doi.org/10.1093/nar/gkm760
Pilsner JR, Saddiki H, Whitcomb BW, Suvorov A, Buck Louis GM, Mumford SL, Schisterman EF, Oluwayiose OA, Balzer LB (2022) Sperm epigenetic clock associates with pregnancy outcomes in the general population. Hum Reprod 37(7):1581–1593. https://doi.org/10.1093/humrep/deac084
Pisarek A, Pośpiech E, Heidegger A, Xavier C, Papież A, Piniewska-Róg D, Kalamara V, Potabattula R, Bochenek M, Sikora-Polaczek M (2021) Epigenetic age prediction in semen–marker selection and model development. Aging (Albany NY) 13(15):19145. https://doi.org/10.18632/aging.203399
Poussard A, Curci J-Y, Siatka C, Hermitte F, Pussiau A, Singla-Sanchez H, Hubac S (2023) Evaluation of DNA methylation-based age-prediction models from saliva and buccal swab samples using pyrosequencing data. Forensic Sci 3(2):192–204. https://doi.org/10.3390/forensicsci3020015
Rathke C, Baarends WM, Awe S, Renkawitz-Pohl R (2014) Chromatin dynamics during spermiogenesis. Biochim Biophys Acta 1839(3):155–168. https://doi.org/10.1016/j.bbagrm.2013.08.004
Roessler J, Lehmann U (2015) Quantitative DNA methylation analysis by pyrosequencing®. In: Lehmann U, Tost J (eds) Pyrosequencing: methods and protocols. Springer, New York, pp 175–188

Roy S, Coldren C, Karunamurthy A, Kip NS, Klee EW, Lincoln SE, Leon A, Pullambhatla M, Temple-Smolkin RL, Voelkerding KV, Wang C, Carter AB (2018) Standards and guidelines for validating next-generation sequencing bioinformatics pipelines: a joint recommendation of the Association for Molecular Pathology and the College of American Pathologists. JMD 20(1):4–27. https://doi.org/10.1016/j.jmoldx.2017.11.003

Schwender K, Holländer O, Klopfleisch S, Eveslage M, Danzer MF, Pfeiffer H, Vennemann M (2021) Development of two age estimation models for buccal swab samples based on 3 CpG sites analyzed with pyrosequencing and minisequencing. Forensic Sci Int Genet 53:102521. https://doi.org/10.1016/j.fsigen.2021.102521

Siebert-Kuss LM, Dietrich V, Di Persio S, Bhaskaran J, Stehling M, Cremers J-F, Sandmann S, Varghese J, Kliesch S, Schlatt S, Vaquerizas JM, Neuhaus N, Laurentino S (2024) Genome-wide DNA methylation changes in human spermatogenesis. AJHG 111(6):1125–1139. https://doi.org/10.1016/j.ajhg.2024.04.017

Silva DSBS, Antunes J, Balamurugan K, Duncan G, Sampaio Alho C, Mccord B (2015) Evaluation of DNA methylation markers and their potential to predict human aging. Electrophoresis 36(15):1775–1780. https://doi.org/10.1002/elps.201500137

So MH, Lee HY (2021) Genetic analyzer-dependent DNA methylation detection and its application to existing age prediction models. Electrophoresis 42(14–15):1497–1506. https://doi.org/10.1002/elps.202000312

Souren NYP, Lutsik P, Gasparoni G, Tierling S, Gries J, Riemenschneider M, Fryns J-P, Derom C, Zeegers MP, Walter J (2013) Adult monozygotic twins discordant for intra-uterine growth have indistinguishable genome-wide DNA methylation profiles. Genome Biol 14(5):R44. https://doi.org/10.1186/gb-2013-14-5-r44

Stuppia L, Franzago M, Ballerini P, Gatta V, Antonucci I (2015) Epigenetics and male reproduction: the consequences of paternal lifestyle on fertility, embryo development, and children lifetime health. Clin Epigenetics 7(1):120. https://doi.org/10.1186/s13148-015-0155-4

Sun W, Dong B, Chu X, Xu Q, Li H, Chen M, Jiang L, Huang A, Zhu B, Liu X (2025) Massively parallel sequencing of a forensic combined panel of 107-plex STR loci and 292-plex SNP loci in the Han Chinese population. Forensic Sci Int Genet 76:103235. https://doi.org/10.1016/j.fsigen.2025.103235

Sijen T, Harbison S (2021) On the Identification of Body Fluids and Tissues: A Crucial Link in the Investigation and Solution of Crime. Genes 12(11):1728. https://doi.org/10.3390/genes12111728

Tárraga J, Pérez M, Orduña JM, Duato J, Medina I, Dopazo J (2015) A parallel and sensitive software tool for methylation analysis on multicore platforms. Bioinformatics 31(19):3130–3138. https://doi.org/10.1093/bioinformatics/btv357

Teschendorff AE, Breeze CE, Zheng SC, Beck S (2017) A comparison of reference-based algorithms for correcting cell-type heterogeneity in Epigenome-Wide Association Studies. BMC Bioinform 18(1):105. https://doi.org/10.1186/s12859-017-1511-5

Theda C, Hwang SH, Czajko A, Loke YJ, Leong P, Craig JM (2018) Quantitation of the cellular content of saliva and buccal swab samples. Sci Rep 8(1):6944. https://doi.org/10.1038/s41598-018-25311-0

Virkler K, Lednev IK (2009) Analysis of body fluids for forensic purposes: from laboratory testing to non-destructive rapid confirmatory identification at a crime scene. Forensic Sci Int 188(1):1–17. https://doi.org/10.1016/j.forsciint.2009.02.013

Weidner CI, Lin Q, Koch CM, Eisele L, Beier F, Ziegler P, Bauerschlag DO, Jöckel K-H, Erbel R, Mühleisen TW, Zenke M, Brümmendorf TH, Wagner W (2014) Aging of blood can be tracked by DNA methylation changes at just three CpG sites. Genome Biol 15(2):R24. https://doi.org/10.1186/gb-2014-15-2-r24

Wong YT, Tayeb MA, Stone TC, Lovat LB, Teschendorff AE, Iwasiow R, Craig JM (2022) A comparison of epithelial cell content of oral samples estimated using cytology and DNA methylation. Epigenetics 17(3):327–334. https://doi.org/10.1080/15592294.2021.1950977

Woźniak A, Heidegger A, Piniewska-Róg D, Pośpiech E, Xavier C, Pisarek A, Kartasińska E, Boroń M, Freire-Aradas A, Wojtas M, De La Puente M, Niederstätter H, Płoski R, Spólnicka M, Kayser M, Phillips C, Parson W, Branicki W (2021) Development of the VISAGE enhanced tool and statistical models for epigenetic age estimation in blood, buccal cells and bones. Aging (Albany NY) 13(5):6459–6484. https://doi.org/10.18632/aging.202783

Xiao C, Li Y, Chen M, Yi S, Huang D (2023) Improved age estimation from semen using sperm-specific age-related CpG markers. Forensic Sci Int Genet 67:102941. https://doi.org/10.1016/j.fsigen.2023.102941

Xie H, Onsongo G, Popko J, De Jong EP, Cao J, Carlis JV, Griffin RJ, Rhodus NL, Griffin TJ (2008) Proteomics analysis of cells in whole saliva from oral cancer patients via value-added three-dimensional peptide fractionation and tandem mass spectrometry. Mol Cell Proteomics 7(3):486–498. https://doi.org/10.1074/mcp.M700146-MCP200

Yordanova M, Gerova D, Galunska-Kalcheva B (2022) Comparative analysis of salivary cellular composition in patients with chronic inflammatory bowel disease. BJSTR 45(3):36429–36435. https://doi.org/10.26717/bjstr.2022.45.00719710.26717/BJSTR.2022.45.007197

Zbieć-Piekarska R, Spólnicka M, Kupiec T, Makowska Ż, Spas A, Parys-Proszek A, Kucharczyk K, Płoski R, Branicki W (2015a) Examination of DNA methylation status of the ELOVL2 marker may be useful for human age prediction in forensic science. Forensic Sci Int Genet 14:161–167. https://doi.org/10.1016/j.fsigen.2014.10.002

Zbieć-Piekarska R, Spólnicka M, Kupiec T, Parys-Proszek A, Makowska Ż, Pałeczka A, Kucharczyk K, Płoski R, Branicki W (2015b) Development of a forensically useful age prediction method based on DNA methylation analysis. Forensic Sci Int Genet 17:173–179. https://doi.org/10.1016/j.fsigen.2015.05.001

Chapter 8
Epigenetics for Age Estimation in Forensic Anthropology

Maria Josefina Castagnola and Sara C. Zapico

Abstract Age estimation is a key component of the biological profile for human identification in forensic anthropology scenarios, including mass disasters, war crimes, criminal investigations, and bioarchaeological settings. In these contexts, skeletal remains are usually the only tissues recovered, with age being traditionally estimated using morphological methods. However, this determination in adults is especially challenging because it relies on degenerative changes in bones and teeth, which are influenced by diverse internal and external factors that can, as a result, hamper accurate identification. Biochemical methods, such as aspartic acid racemization, protein glycosylation, telomere shortening, or mitochondrial DNA mutations, have also been attempted for age estimation; however, they present certain limitations in terms of precision or applicability. Building on previous research in criminalistics, DNA methylation has emerged as a potential tool for estimating age from bones, teeth, and even cartilage. Researchers have explored multiple techniques and candidate DNA methylation markers to develop age estimation models to achieve higher accuracy than traditional anthropological techniques. These findings demonstrate the potential of epigenetics to improve human identification, opening the door to the integration of this discipline into forensic anthropology approaches.

Keywords Bone · Teeth · Cartilage · Ancient DNA · Epigenetics · Forensic anthropology · Human identification

M. J. Castagnola
Department of Chemistry and Environmental Sciences, New Jersey Institute of Technology, Newark, NJ, USA

S. C. Zapico (✉)
Department of Chemistry and Environmental Sciences, New Jersey Institute of Technology, Newark, NJ, USA

Anthropology Department and Laboratories of Analytical Biology, National Museum of Natural History, Smithsonian Institution, Washington, DC, USA
e-mail: sc338@njit.edu; casado-zapicos@si.edu

S. C. Zapico (ed.), *Forensic Science and Epigenetics*, Epigenetics and Human Health 18, https://doi.org/10.1007/978-3-032-23085-0_8

Abbreviations

aDNA	Ancient DNA
CpG	Cytosine-phosphate-guanine dinucleotide
DNAm	DNA methylation
ERLs	Environmentally Responsive Epigenetic Loci
HPLC	High-performance liquid chromatography
LOOCV	Leave-one-out cross-validation
MAD	Mean absolute deviation
MAE	Mean absolute error
MALDI TOF	Matrix-assisted laser desorption/ionization—time of flight
MedAE	Median absolute error
MPS	Massive parallel sequencing
PMI	Postmortem interval
r	Pearson's correlation coefficient
R^2	Coefficient of determination
RMSE	Root mean squared error
RT-MSP	Real-time methylation-specific PCR
VISAGE	VISible Attributes through GEnomics
Ybp	Years before present

8.1 Introduction

The application of epigenetics in forensic anthropology contexts, such as mass disasters, war crimes, and humanitarian and human rights scenarios, could improve the prospects for human identification. Forensic anthropologists create a "biological profile", which includes the estimation of age, biological sex, population affinity, and stature. Determining phenotypic traits is valuable for narrowing the search for unidentified individuals when antemortem data are lacking. Although epigenetic studies typically focus on tissues commonly used in criminalistics (blood, saliva, and semen), skeletal remains are the primary tissues recovered in forensic anthropology because they exhibit greater resilience to time and environment due to their structural properties and composition (Villagran et al. 2017).

Age estimation has traditionally relied on the analysis of morphological traits. In children, it typically focuses on growth and development measurements. In contrast, adult age estimation is based on degenerative changes in bones and teeth. These changes are influenced by endogenous and exogenous factors such as pathological conditions, environment, and lifestyle, which contribute to a greater degree of interindividual variability in the outcomes. As a result, traditional morphological methods often present discrepancies of approximately ±10 years between chronological and estimated age (Ubelaker 1987; Ubelaker and Khosrowshahi 2019). New methodologies based on the inherent processes of aging could complement existing methods to improve the accuracy of age estimation.

Several biomolecular approaches have been explored for age estimation, including aspartic acid racemization (Zapico and Ubelaker 2022; Ubelaker and Khosrowshahi 2019), mitochondrial DNA mutations (Amorim et al. 2019; Lujan et al. 2020), telomere shortening (Marquez-Ruiz et al. 2020; Vaiserman and Krasnienkov 2020), and T-cell receptor analysis (Zubakov et al. 2016). However, several constraints exist in the techniques. For instance, aspartic acid racemization is influenced by high temperatures and cannot be applied to burned remains (Zapico et al. 2019), while telomere shortening is affected by genetic, epigenetic, and environmental factors, which increases variability and results in a high standard error (Bekaert et al. 2005; Zapico and Ubelaker 2013).

Based on previous criminalistics studies, DNA methylation (DNAm) has emerged as a promising tool for age estimation. Researchers employ diverse methodologies and DNAm markers; some studies analyze a large number of CpG sites to investigate candidate genes, while others aim to develop models with a minimal number of markers to simplify the methodology for its applicability in forensic contexts (reviewed in Castagnola et al. 2024). Although most research focuses on skeletal remains, recent studies have also examined cartilage, a tissue that can be present until full skeletonization. Attempts to identify universal CpGs across tissues have been made; however, age determination remains highly tissue-specific. Additionally, analyzing epigenetic markers in historical or ancient samples is challenging because spontaneous deamination in ancient DNA could complicate methylation analysis.

This chapter reviews the main studies on DNAm for age estimation relevant to forensic anthropology. It outlines the approaches and markers investigated, discusses current challenges, and establishes a basis for integrating epigenetics into the field.

8.2 Skeletal Remains in Forensics

In forensic genetics, higher bone density is associated with improved DNA preservation (Becker et al. 2025). Consequently, compact cortical bones are typically selected for analysis because they are less susceptible to degradation than trabecular or spongy bones (Christensen et al. 2019; Prinz et al. 2007). Cortical bones from the lower extremities, including the femur, tibia, and fibula, have shown the highest DNA yield success rates (Milos et al. 2007). In addition, the petrous part of the temporal bone, which is a highly compact region, has demonstrated considerable DNA recovery in ancient remains (Gamba et al. 2014).

Bones are composed of approximately 60% inorganic extracellular matrix, primarily hydroxyapatite crystals, and 40% organic matrix, which includes collagen and non-collagenous proteins (Lin et al. 2020). While the organic matrix provides structural support, it can hinder DNA extraction. Therefore, demineralization is frequently incorporated into protocols to facilitate the release of genetic material (Iorio and Cooley 2023; Loreille et al. 2007).

The bone matrix contains several distinct cell types. Osteocytes, the most abundant and longest-living cells in mammalian bone, are essential for bone remodeling and metabolic maintenance (Le et al. 2021; Manolagas and Parfitt 2010). Notably, bones with a high density of osteocytic lacunae, such as the petrous bone, exhibit increased resistance to diagenesis and better preservation of endogenous DNA (Ibrahim et al. 2022). In addition to osteocytes, osteoblasts are responsible for producing the bone matrix, while osteoclasts mediate bone resorption. Furthermore, the bone marrow contains mesenchymal stem cells, endothelial cells, and blood cells (Corrado et al. 2020; Park-Min 2019), all of which can be useful as additional sources of DNA.

Compared to bone, teeth have lower porosity and are protected within the jaws, offering greater resistance to external contamination, environmental changes, and postmortem degradation (Higgins and Austin 2013). The crown is covered by enamel, which is impermeable to aqueous liquids (Hillson 2005). The internal tissues of genetic interest include cementum, dentin, and pulp, each with distinct mineral and cellular compositions. Cementum, the outermost layer, is avascular, contains 50% mineral content, and includes tissue-specific cells called cementocytes (Higgins and Austin 2013).

Dentin contains the highest mineral content among the three dental tissues, at approximately 70%. It also includes structural proteins and parallel tubules that enclose cytoplasmic extensions from odontoblasts and nerve fibers. Odontoblastic cells, which may become trapped during reparative dentin formation, as well as cell bodies that can be incorporated into dentinal tubules and occasional blood vessels from the canals, can also be present in the dentin (Higgins and Austin 2013; Zaslansky et al. 2010). Dentin's highly bradytrophic nature provides greater structural stability than bone, especially trabecular bone (MacDougall and Javed 2010; Pignolo et al. 2021).

Dental pulp is highly vascularized and contains the highest cellular content among dental tissues, primarily composed of fibroblasts and odontoblasts. It also comprises stromal tissue, nerves, blood vessels, and lymphatics. These characteristics make dental pulp the preferred source for DNA extraction from teeth (Zapico et al. 2021). Some studies refer to pulp and dentin collectively as the pulp/dentin complex, which represents the most cellular region of the tooth (Higgins and Austin 2013). Multirooted molars are frequently chosen for DNA recovery due to their larger pulp volume, which allows for greater DNA yield when preservation conditions are favorable (Kondo et al. 2020). However, the reduction in pulp volume associated with aging can hinder DNA extraction. This limitation highlights the importance of evaluating alternative dental sources for genetic studies (Castagnola et al. 2025; Higgins and Austin 2013).

Obtaining high-quality and high-quantity DNA is fundamental for genetic and epigenetic studies. Common methods for bones and teeth include organic extraction with phenol-chloroform, silica-based column purification, manual extraction using magnetic beads, and fully automated magnetic DNA extraction systems (Haarkötter et al. 2023). The selection of the extraction strategy is determined by the availability of resources, the sample type, and the purpose of the analysis. Since no single method fulfills all criteria for efficiency, yield, cost, automation, and compatibility with degraded DNA, laboratories must evaluate and implement the most suitable approach for each specific application.

8.2.1 Bones in Forensic Epigenetics

One of the first attempts to estimate age in bones was the application of Horvath's multi-tissue epigenetic clock (Horvath 2013), which was not previously tested in bone tissue but in bone marrow, to 48 trabecular bone samples from osteoporosis patients aged 48 to 104 (Horvath et al. 2015). The correlation obtained between epigenetic and chronological age was $r = 0.88$. DNAm age, or epigenetic age, considers both the chronological and biological age of an individual (Horvath and Raj 2018). Building on this, Horvath et al. introduced the "Skin and Blood Clock", which predicts lifespan and is associated with various age-related conditions (Horvath et al. 2018). This model, based on 391 CpGs from different tissues, was also applied to the same trabecular bone samples from the previous study (Horvath et al. 2015), and achieved a correlation (r) of 0.83 between DNAm age and chronological age.

Later, Gopalan et al. (2019) employed the Illumina Human Methylation 450 k array to identify bone-specific CpG sites and then developed an age estimation model using a reduced set of loci. Their dataset included 165 bone samples from living and deceased individuals aged 49–112 years: 133 from previously published data (Horvath et al. 2015; Reppe et al. 2017) and 32 from new samples (trabecular femur bone and preserved petrous portions of the temporal bone). Initially, 108 CpG sites were significantly associated with age, and 37 of them were selected for the final bone-specific model, which included two CpG sites in ELOVL2, a gene previously known for age-related methylation patterns across several tissues. One study limitation mentioned is related to whether the model can accurately estimate age in younger adults, since the dataset included only individuals aged 49 years or older. Also, this work's point of view differs from previous studies based on biological age estimation (Horvath et al. 2015), focusing instead on chronological age estimation.

In addition, a high-throughput multi-CpG approach evaluated the use of DNAm markers for age estimation using massive parallel sequencing (MPS) across different tissues (Naue et al. 2018). A total of 144 samples from 29 deceased individuals aged 0 to 87 years were analyzed, including blood, bone, muscle, brain, and buccal swabs. Thirteen loci previously identified in blood were investigated: DDO, ELOVL2, F5, GRM2, HOXC4, KLF14, LDB2, NKIRAS2, RPA2, SAMD10, TRIM59, MEIS1, and ZYG11A (Naue et al. 2017). Seven of these loci showed to be age-related across all the studied tissues: DDO, ELOVL2, KLF14, NKIRAS2, RPA2, TRIM59, and ZYG11A. In bone, the CpG sites most strongly correlated with age were in the RPA2, DDO, and KLF14 genes. This research highlighted a recurring consideration across studies: the importance of increasing sample size. Larger datasets are essential for better characterizing tissue-specific differences.

In contrast, Lee et al. applied two distinct techniques to estimate age (Lee et al. 2020). One approach utilized Illumina's Infinium MethylationEPIC BeadChip array to conduct a broader analysis of DNA methylation. The other approach was a multiplex SNaPshot assay developed to estimate age employing fewer DNAm markers. This method requires lower amounts of DNA and is more suitable for routine forensic laboratories than the microarray. For the microarray, 32 femoral bone samples were analyzed, with ages at death between 31 and 96 years old. Only 12 of the 32 samples provided sufficient data quality to estimate age using the "Skin and Blood Clock" (Horvath et al. 2018), achieving a mean absolute deviation (MAD) of 6.4 years. Then, 13 candidate DNAm genes were selected from the microarray's dataset, and five additional age-associated markers (ELOVL2, FHL2, KLF14, C1orf132/MIR29B2C, and TRIM59) were included based on a previous SNaPshot assay for blood, saliva, and buccal swabs (Jung et al. 2019). Thirty bone samples (independent of those used in the microarray) with ages ranging from 29 to 82 years were used for the multiplex SNaPshot assay. The results showed correlations with age, ranging from moderate to weak, and the genes TMEM51, TRIM59, ELOVL2, and EPHA6 demonstrated the strongest associations (Lee et al. 2020).

Correia Dias et al. (2020) followed a similar strategy, targeting methods routinely used in forensic genetics to estimate age by DNAm. They analyzed between 29 and 31 samples from deceased individuals using both Sanger sequencing and the SNaPshot assay. The Sanger-based model included 6 CpG sites from the genes ELOVL2, EDARADD, and C1orf132, yielding a MAD of 2.56 in the training set and a MAD of 3.77 after cross-validation. As in Lee et al. (2020), the SNaPshot assay was also based on the primers and conditions previously described by Jung et al. (2019). In this case, the best-performing bone model incorporated CpGs in FHL2 and KLF14, showing a MAD of 7.18 for the training set and a MAD of 7.84 after cross-validation. CpG sites in ELOVL2 and C1orf132 exhibited the highest correlations with age using Sanger sequencing, while CpG sites in FHL2 showed the greatest correlation in the SNaPshot assay.

Subsequently, the VISAGE consortium (VISible Attributes through Genomics) employed MPS to develop age-prediction models applicable to multiple somatic tissues, including blood, bone, and buccal cells. Woźniak et al. (2021) examined 161 occipital and femoral bone samples from individuals aged 19 to 93 years, analyzing 44 CpG sites from 8 age-informative DNAm markers. These included five markers that had been previously identified by VISAGE (Heidegger et al. 2020) and 3 additional new markers (EDARADD, PDE4C, and ASPA). The resulting bone-specific model, which incorporated six CpG sites from four genes (ELOVL2, KLF14, PDE4C, and ASPA), achieved a mean absolute error (MAE) of 3.3 years in the training set and 3.4 years in the test set. Compared with earlier studies, MAE remained consistent across the entire age range. ELOVL2 displayed a non-linear relationship, as reported in other studies (Bekaert et al. 2015; Ogata et al. 2022). This methodology was optimized for use with highly degraded forensic samples and produced reliable results with as little as 20 ng of DNA.

More recently, Becker et al. (2025) analyzed 190 parietal skull bone samples from individuals aged 19 to 96 years old to develop age prediction models based on three molecular markers. They tested different approaches and their combinations under the hypothesis that employing multiple biomarkers could potentially reduce both intraindividual and interindividual variation. The epigenetic model employed MPS for 6 CpG sites in the genes ELOVL2, KLF14, PDE4C, RPA2, TRIM59, and ZYG11A, which had previously shown good correlation with age in bone (Becker et al. 2020; Naue et al. 2018; Woźniak et al. 2021). The other two models used high-performance liquid chromatography (HPLC) to measure the levels of pentosidine (Pen) and D-aspartate (D-Asp). The accumulation of D-aspartic acid (D-Asp) with age is attributed to the spontaneous non-enzymatic conversion of L-asparagine and L-aspartic acid into their respective D-forms (Zapico and Ubelaker 2022). Pentosidine (Pen) is an advanced glycation end-product (AGE) that accumulates as a fluorescent cross-link between arginine and lysine residues in proteins such as collagen (Li and Yu 2018). Samples were classified as decomposed or non-decomposed based on criteria from Megyesi et al. (2005). Eighty-six non-decomposed samples were used for the training set, and validation was carried out on two independent sets: one consisting of 44 non-decomposed samples and the other of 48 decomposed samples. The authors attempted to cover the entire age range across training and validation sets.

In the non-decomposed set, the DNAm model performed better (MAE = 5.5 years) compared with the Pen (MAE = 7.7) and D-Asp (MAE = 11.7) models for age estimation. Combining methods did not substantially improve accuracy over DNAm alone. In decomposed samples, overall accuracy decreased across all methods (MAE 7.4–11.8 years). A model trained exclusively on decomposed bones, validated by leave-one-out cross-validation (LOOCV), achieved MAEs of 6.6–12 years. In this context, the combination of the three molecular approaches was primarily useful for reducing deviations in outlier cases and improving predictions for samples with advanced decomposition and low DNA content. The limitations stated by the researchers included that the age distribution of the decomposed samples was predominantly from individuals over 40 years old (only nine were between 18 and 39 years old) and the observation of a decline in DNAm model accuracy with increasing age.

8.2.2 *Teeth in Forensic Epigenetics*

One of the earliest studies to employ DNAm for tooth age estimation was carried out by Bekaert et al. (2015). Pyrosequencing was used to examine four genes in dentin: ASPA, PDE4C, ELOVL2, and EDARADD. 29 tooth samples from individuals ranging in age from 19 to 70 were analyzed. A multivariate quadratic regression model for dentin was developed, including seven CpGs from PDE4C, ELOVL2, and EDARADD, achieving a MAD of 4.86. The quadratic approach was based on the observation that ELOVL2 methylation levels increased exponentially before reaching a plateau. This study was one of the first to utilize a non-linear model to estimate tooth age using DNAm.

Giuliani et al. (2016) constructed individual and combined age estimation models using the three primary dental tissues (pulp, dentin, and cementum). They employed MALDI-TOF (matrix-assisted laser desorption/ionization—time of flight) mass spectrometry to assess the DNAm levels of three genes (ELOVL2, FHL2, and PENK) in 21 modern teeth (molars and premolars) belonging to individuals ranging from 17 to 77 years old. Each tissue's methylation pattern was determined, and both single and combination models were created. The pulp model (MedAE = 2.25) was the most accurate, followed by the cementum model (MedAE = 2.45). Each model employed 13 CpGs. Dentin (5 CpGs) had the lowest correlation with age (MedAE = 7.07), possibly due to variability related to dentin composition. Furthermore, a combined model incorporating pulp and cementum data (8 CpGs identified in the genes ELOVL2 and FHL2) yielded a MedAE of 1.20. Each model had between 17 and 21 samples. Twelve whole teeth were also analyzed to identify the marker most strongly correlated with age, which was a CpG in FHL2. As one of the earliest studies in this field, it established a foundation by demonstrating that pulp offers the highest accuracy from the tooth tissues.

Later studies expanded the scope to include analysis of the whole tooth, rather than a specific tissue, and assessed novel biomarkers and their combinations. In this sense, Márquez-Ruiz et al. (2020) used pyrosequencing to examine DNAm patterns in three genes (ELOVL2, ASPA, and PDE4C) in 65 molars belonging to living donors who ranged from 15 to 85 years old. They also evaluated relative telomere length using quantitative real-time PCR. The best DNAm model consisted of nine CpG sites from two genes (ELOVL2 and PDE4C), achieving a MAE of 5.08 in the training set; ASPA was excluded since it did not show a significant correlation with age. Adding telomere length only slightly improved models based on DNAm (MAE = 5.04 for the combined model vs. 5.08 for methylation alone). It is important to note that in this study, the greatest predictions were achieved for samples of individuals aged 15–35 years, while accuracy decreased with increasing age. There were no variations between sexes or tooth types (incisors, canines, premolars, and molars).

Similarly, Correia Dias et al. (2020) used the whole tooth to build prediction models employing techniques routinely used in forensic laboratories, such as Sanger sequencing and Snapshot assay. They employed between 23 and 30 teeth from both living and deceased individuals (aged 26–94 years) for both approaches. Applying Sanger sequencing, they were unable to provide a multilocus DNAm model for teeth, and the best prediction model was achieved with a single FHL2 CpG. However, the high MAD of 11.35 indicated limited use. In comparison, the SNaPshot-based model achieved a MAD of 7.07 by combining two CpGs from ELOVL2 and KLF14. The CpG with the highest individual correlation was KLF14. The use of the whole tooth rather than specific tissues may explain the comparatively high MAD values seen in both methods. SNaPshot, a semi-quantitative approach, can be a cost-effective tool for methylation research. However, more research is needed to see whether this technique may provide higher MAD values.

Real-time methylation-specific PCR (RT-MSP), another semi-quantitative method employed in conventional forensic laboratories, was investigated for its application in DNAm analysis in two studies by the same study group. Kondo et al. (2020) employed RT-MSP to analyze the CpG island of the ELOVL2 gene to estimate age on 29 teeth (mainly molars, with two premolars) from individuals aged 20 to 79 years. Methylation levels increased with age, as did the standard error, and the linear regression model revealed a MAD of 8.94 years employing 3 CpG sites from ELOVL2. There were no significant differences by sex, as in earlier research (Marquez-Ruiz et al. 2020). Similarly, Ogata et al. (2022) used RT-MSP to evaluate 6 CpG sites, each from the ELOVL2 and the EDARADD genes, in the analysis. They employed 99 teeth (59 for training and 40 for validation) from Japanese individuals aged between 20 and 85 years old, primarily molars but also premolars and anterior teeth. The final model incorporated a quadratic function for the ELOVL2 gene and a logarithmic function for the EDARADD gene, yielding a MAE of 6.69 years for the training group and 8.28 years for the testing set, which increased with age. Both studies emphasized the need to select the appropriate regression model (linear versus nonlinear).

Returning to specific tissue research, Zapico et al. (2021) focused on estimating the age of dental pulp by pyrosequencing. The study included 20 third molars from adults aged 22 to 70 and evaluated 46 CpG sites in five genes: two previously used for age estimation and DNAm for teeth (ELOVL2 and FHL2) and three new to this type of sample (KLF14, NPTX2, and SCGN). They developed four multivariate regression models and validated them with LOOCV. CpG sites on the ELOVL2 gene showed the strongest correlation with age. The two more promising multivariate linear regression models contained four genes (ELOVL2, FHL2, NPTX2, and SCGN) with MAE values of 1.7 and 1.9 years after cross-validation. The first model employed 13 CpG sites, while the second used 11 CpGs. Although the models were more accurate than previous ones described in this chapter, they also included more CpG sites, which must be considered in the analysis following the pyrosequencing. This work's greater accuracy can be attributed to focusing on a specific tissue, the dental pulp, and also, the inclusion of genes more related to the function of the pulp on the tooth, such as NPTX2, which reflects the presence of nerve cells in the pulp.

Finally, attempts were made to construct models (Correia Dias et al. 2021) that could be used to estimate age in multiple tissues at the same time. They employed samples of blood, bone (rib), and teeth (molars) from living and deceased individuals (postmortem intervals of up to 5 days). As in prior work by this group (Correia Dias et al. 2020), they used two methods for age estimation: Sanger sequencing and SNaPshot assay, and created two independent stepwise linear regression models. The Sanger model was developed employing 185 samples (23 teeth, 29 bones, and the remaining were blood samples), and included seven CpGs from ELOVL2, EDARADD, PDE4C, FHL2, and C1orf132 genes, achieving a MAD of 6.06 years in the training set and 6.27 years in the cross-validation set. The final SNaPshot model included three CpGs from ELOVL2, KLF14, and C1orf132 genes and was tested in 168 samples (31 bone and 23 teeth), achieving a MAD of 6.49 years for the training set and 6.73 years for the cross-validation set. In both models, ELOVL2 CpGs had the strongest correlation with age. Prediction accuracies were lower at both age extremes: over 80 and under 30.

The accuracy of this study could have been affected by the fact that the model was multi-tissue. As mentioned previously, methylation patterns tend to be tissue-specific, which could have hindered the precision of predictions. It has also been noted in this section that using the whole tooth rather than a single tissue impacts the age estimates obtained using DNAm. Furthermore, other common challenges discussed across studies include age range, limited sample sizes, and missing donor information (such as medical records), all of which researchers suggest could influence the models' accuracy.

8.3 Cartilage in Forensic Epigenetics

In recent years, cartilage has also been considered as a potential target for age estimation by employing epigenetic markers (Freire-Aradas et al. 2023; Tomsia et al. 2022). Cartilage is a type of connective tissue in which the predominant cells are chondrocytes, responsible for secreting an abundant amorphous extracellular matrix that protects them even from postmortem degradation (reviewed in Tomsia et al. 2022). One advantage of this tissue is that its extraction can be faster and more cost-effective compared to bones and teeth (Freire-Aradas et al. 2023; Tomsia et al. 2021).

In addition, cartilage plays an essential role as a scaffold for bone formation during endochondral ossification. At this stage, blood vessels invade the cartilage, introducing cells that promote bone formation and maturation, followed by mineralization and calcification (Mackie et al. 2008). Based on this process, it has been hypothesized that bone and cartilage may display distinct epigenetic markers, as their composition and structure diverge during ossification (Jung et al. 2024).

The first attempt to estimate age from cartilage employed Horvath's first epigenetic clock (Horvath 2013), which was based on microarrays and analyzed 353 CpG sites, limiting its routine application in forensic laboratories (Horvath 2013). Most of the studies conducted to date have focused on costal hyaline cartilage, which connects the ribs to the sternum and provides rigidity to the thoracic cage (Hardy et al. 2020). It is also the thickest cartilage in the body, making it less susceptible to degradation (Freire-Aradas et al. 2023).

Woźniak et al. (2021) employed age prediction models developed with the VISAGE Enhanced Tool for cartilage. The age estimation accuracy for cartilage applying blood, buccal cells, and bone models was significantly higher, with MAEs of 13.1, 12.3, and 25.8, respectively. Researchers discovered that methylation patterns in cartilage seemed to be more comparable to those identified in blood and saliva than in bone.

Then, the VISAGE Enhanced Tool (Woźniak et al. 2021) was employed to develop a model for estimating the age of 181 costal cartilage samples from European individuals aged 19 to 75 years old, with a death-to-autopsy time of 1–5 days (Freire-Aradas et al. 2023). First, this approach was used to determine the methylation levels at 44 CpG sites in eight relevant biomarkers. A CpG site in the FHL2 gene had the strongest association with age, but sites in C1orf132, EDARADD, and ASPA had the weakest correlations and were excluded from the final age prediction model. A quantile regression model that included CpGs from FHL2, TRIM59, and KLF14 genes provided the best results, with a MAE of 4.41 years for the trained set ($n = 109$) after cross-validation and 4.26 years for an independent test set ($n = 72$). Additionally, 392 samples of blood, buccal cells, bone, and cartilage were analyzed to identify tissue-specific DNA patterns. C1orf132 and EDARADD were removed from the age prediction model, but were incorporated in the tissue classification model, which included one CpG site from each of the eight genes explored in this study. Cartilage was successfully differentiated from other tissues with a classification rate of 98.72%. The analysis revealed that the model slightly overestimated age in younger individuals and underestimated age in older ones, consistent with findings from other studies on skeletal tissues. This is important to consider when evaluating potential age estimate models, because covering the entire lifespan could be critical for forensic scenarios.

In a subsequent study, Jung et al. (2024) also developed an age prediction model for costal cartilage, using MPS and based on genes previously investigated by the VISAGE consortium (Woźniak et al. 2021) and by Lee et al. (2020). The analysis included 85 samples from deceased Korean individuals aged 26 to 89 years, with postmortem intervals (PMI) of less than 1 year. An initial MPS library targeted 45 CpG sites from nine genes (TMEM51, C1orf132, EDARADD, FHL2, TRIM59, ELOVL2, KLF14, ASPA, and PDE4C). Among these, CpGs in FHL2 and TRIM59 showed the strongest individual correlations with age, while C1orf132 and ASPA had weaker correlations but improved overall model performance when included. Based on these findings, the researchers developed a stepwise multiple regression model that utilized nine CpG sites from six genes (C1orf132, FHL2, TRIM59, ELOVL2, KLF14, and ASPA), yielding a MAE of 4.97 years following LOOCV validation. These samples demonstrated good DNA preservation; therefore, further evaluation is necessary to assess performance in more degraded remains for broader application in forensic contexts. Importantly, this same model was then cross-validated on the Freire-Aradas et al. dataset (Freire-Aradas et al. 2023), providing a MAE of 6.06 years, demonstrating that the model developed in this study could be used to estimate the age of individuals from other populations. Lastly, the authors found that age estimation accuracy was lower for individuals above 70 years old, which emphasizes the difficulty of developing a single model that can be used for all age intervals, which is consistent with earlier findings (Freire-Aradas et al. 2023).

More recently, Jung et al. (2025) used the SNaPshot assay on costal cartilage to create a more accessible age prediction model for conventional forensic laboratories. The study included 136 Korean samples (83 for the training set and 53 for the testing set), aged 28–84 years, and assessed 11 CpG sites in six age-related genes. CpGs in TRIM59 and FHL2 had the most significant individual correlations with age, and the best-performing model (based on four CpG sites in C1orf132, FHL2, TRIM59, and KLF14 genes) achieved a MAE of 4.08 years for the training set and a MAE of 4.60 years for the validation set.

The model was then applied to blood, buccal swabs, and femoral bone; however, the MAEs were more than 20 years old, emphasizing the importance of tissue-specific predictors. Age estimation accuracy decreased in individuals younger than 30 years old and older than 70 years. A sensitivity assessment revealed that 4 ng of bisulfite-converted DNA was adequate for predicting age with accuracy in this study. Given the limited availability of DNA in forensic cases, this finding is significant for further investigation.

The best-performing age-prediction models based on a limited number of CpG markers across bone, tooth, and cartilage are summarized in Table 8.1.

8.4 Ancient DNA

Ancient DNA (aDNA) epigenetics research is useful at both the population and individual levels. At the population scale, methylation biomarkers may help reconstruct how past communities adapted to their environments, providing insights into their diets, cultural practices, and different exposures (Gokhman et al. 2017). In addition, epigenetics enables the study of specific biological traits, such as age-at-death estimation, contributing to human identification.

In the case of aDNA research, it is necessary to consider a common type of DNA damage that occurs postmortem: hydrolytic deamination. This spontaneous chemical reaction converts unmethylated cytosines into uracil and methylated cytosines into thymine, complicating the accurate interpretation of methylation profiles (Brotherton et al. 2007; Gokhman et al. 2017). Normally, bisulfite treatment converts unmethylated cytosines into uracil while leaving methylated cytosines intact. In degraded DNA, the overlap between this laboratory-induced conversion and natural deamination can lead to divergent sequences after PCR, which in turn can generate misleading results and further complicate the DNAm interpretation (Fig. 8.1). Thus, some studies analyze methylation patterns obtained exclusively through spontaneous deamination (Gokhman et al. 2014), whereas others apply bisulfite conversion followed by sequencing before analyzing DNA methylation (Schmidt et al. 2021).

Until now, it remains unclear to what extent the postmortem interval influences methylation patterns (Smith et al. 2015). Consequently, several studies have focused on the stability of epigenetic marks over time. For instance, Llamas et al. (2012) investigated whether methylation patterns could be preserved in mammalian aDNA, analyzing bones of *Bison priscus* (Pleistocene, ~26,360 ± 220 years) and comparing them with fresh bovine tissue and mummified skin. Using Sanger sequencing, researchers showed that methylation patterns in ancient bones closely matched those in fresh tissue and mummified samples, concluding that spontaneous deamination had a limited impact. Later, Smith et al. (2015) analyzed aDNA from 30 Native American skeletal remains (200–4500 ybp), and concluded that sample preservation was critical for accurate estimation, regardless of postmortem interval or geographic location. Overall, findings so far indicate that methylation patterns can remain stable in well-preserved samples.

Table 8.1 Best-performing age-prediction models with a limited number of CpG markers (<20) reported in published studies across different tissues and methods.

Publication	No. of markers	Genes	Tissue/sample type	Sample size	Age (years)	Error metrics (MAD, MAE, MedAE) (years)	Method
Correia Dias et al. (2020)	6	ELOVL2, EDARADD, C1orf132	Bone (not specified)	29	26–81	MAD 2.56–3.77	Sanger sequencing
	2	FHL2, KLF14	Bone (not specified)	31	26–81	MAD 7.18–7.84	SNaPshot
Woźniak et al. (2021)	6	ELOVL2, KLF14, PDE4C, ASPA	Bone (occipital and femoral)	161 (112 training, 49 test)	19–93	MAE 3.3–3.4	MPS (VISAGE)
Becker et al. (2025)	6	ELOVL2, KLF14, PDE4C, RPA2, TRIM59, ZYG11A	Bone (skull, parietal)	Non-decomposed: 86 training, 44 test Decomposed: 48 test	19–96	Non-decomposed: MAE 4.95–5.48 Decomposed: MAE 7.38	MPS
Bekaert et al. (2015)	7	PDE4C, ELOVL2, EDARADD	Dentin (molars)	29	19–70	MAD 4.86	Pyrosequencing
Giuliani et al. (2016)	13 (pulp), 5 (dentin), 13 (cementum)	ELOVL2, FHL2, PENK	Pulp, cementum, dentin (molars, premolars)	17–21	17–77	MedAE 2.25 (pulp), 2.45 (cementum), 7.07 (dentin)	MALDI-TOF
Marquez-Ruiz et al. (2020)	9	ELOVL2, PDE4C	Whole tooth (incisors, canines, premolars, molars)	65	15–85	MAE 4.84–6.43	Pyrosequencing
Correia Dias et al. (2020)	2	ELOVL2, KLF14	Whole tooth (not specified)	23	26–94	MAD 7.07–7.33	SNaPshot
	1	FHL2	Whole tooth (not specified)	30	26–94	MAD 11.35	Sanger sequencing
Kondo et al. (2020)	3	ELOVL2	Whole tooth (molars and premolars)	29	20–79	MAD 8.94	RT-MSP

(continued)

Table 8.1 (continued)

Publication	No. of markers	Genes	Tissue/sample type	Sample size	Age (years)	Error metrics (MAD, MAE, MedAE) (years)	Method
Zapico et al. (2021)	11–14	ELOVL2, FHL2, NPTX2, SCGN,KLF14	Pulp (molars)	20	22–70	MAE 1.5–2.13	Pyrosequencing
Ogata et al. (2022)	12	ELOVL2, EDARADD	Whole tooth (molars, premolars, anterior teeth)	99	20–85	MAE 6.69–8.28	RT-MSP
Correia Dias et al. (2021)	7	ELOVL2, EDARADD, PDE4C, FHL2, C1orf132	Blood, bone (rib), whole tooth (molars)	185 (23 teeth, 29 bone, blood 65 + 68)	1–94	MAD 6.06–6.27	Sanger
	3	ELOVL2, KLF14, C1orf132	Blood, bone (rib), whole tooth (molars)	168 (31 bone, 23 teeth) blood 55 + 59	1–94	MAD 6.49–6.73	SNaPshot
Freire-Aradas et al. (2023)	3	FHL2, TRIM59, KLF14	Cartilage (costal)	181 (109 training, 72 test)	19–75	MAE 4.26–4.41	MPS
Jung et al. (2024)	9	C1orf132, FHL2, TRIM59, ELOVL2, KLF14, ASPA	Cartilage (costal)	85	26–89	MAE 4.17–4.97	MPS
Jung et al. (2025)	3–11	C1orf132, FHL2, TRIM59, KLF14, ELOVL2, ASPA	Cartilage (costal)	136 (83 training, 53 test)	28–84	MAE 4.03–5.06	SNaPshot

Note: MAD, MAE, and MedAE are not directly comparable; values are reported as in the original publications, indicating here the ranges of the training and test sets

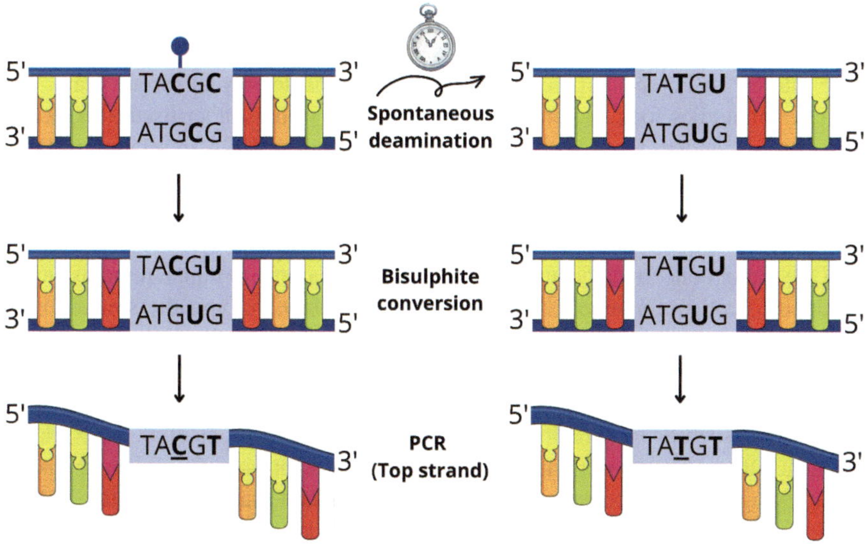

Fig. 8.1 Spontaneous deamination of cytosine and 5-methylcytosine in ancient DNA. Intact methylated DNA (left); DNA with spontaneous deamination (right). Both undergo bisulfite conversion: unmethylated cytosines become uracil (read as T in PCR), while 5-methylcytosines remain as C. Only changes in the top strand are shown

Schmidt et al. (2021) applied the Illumina EPIC BeadChip microarray to study DNAm patterns in two eighteenth-century mummies with known identities. Even though DNA extraction was challenging due to the low quantity and quality of the samples, the researchers reported that a subset of CpGs in the lung tissue produced detectable methylation signals. Using this information, they investigated whether it was possible to determine the tissue type and compared the 22,778 CpGs detected in the lung sample to 301 public datasets from nine distinct tissue types. The pulmonary sample was clustered with reference pulmonary DNAm profiles and identified candidate CpGs for tissue differentiation. Age prediction was carried out using Horvath's multi-tissue epigenetic clock (Horvath et al. 2018). However, only seven of the 353 CpGs in Horvath's model were identified and therefore analyzed in the study data. A model was rebuilt with these seven CpGs and public profiles, resulting in an R^2 of 0.71 in the training set and 0.43 in the validation set. Applied to the lung sample, the model predicted an age of 33 years, which is only 5 years older than the documented chronological age of 28. This study holds great promise for future research, as it employed current technologies to extract meaningful epigenetic information from mummified human remains. Despite the challenges associated with the low methylation signal intensity of the samples, this study demonstrated that epigenetic markers can be preserved for centuries and used to determine biological traits in the identification of ancient remains.

8.5 Conclusion

The interest in Forensic epigenetics has grown considerably in recent years. The number of studies focusing on skeletal remains, as well as other tissues like cartilage, has expanded, along with studies related to ancient DNA in relation to DNA methylation. The research described here demonstrates the potential value of epigenetics in human identification in forensic contexts; however, several challenges still need to be addressed.

One major concern is the number and quality of samples. Human remains can be difficult to collect for research investigations, and they are typically sourced from complex contexts, resulting in DNA of poor quality. Additionally, it should be noted that bisulfite conversion continues to be applied as a pretreatment for DNA methylation analysis, which is known to degrade samples even further. Nowadays, new techniques, such as enzymatic methods, are emerging (Simons et al. 2025); however, bisulfite conversion remains the gold standard for now. Recent methods for quantifying bisulfite-converted DNA may offer valuable support for future validations in degraded samples (Hong and Shin 2021; Karkala et al. 2024). Working with aDNA presents additional limitations because DNA damage can be substantially greater, making analysis and interpretation more challenging. Understanding how different postmortem intervals influence DNAm patterns will be essential to improving precision in these contexts.

Further challenges are related to methodological aspects. As described, many authors employed different approaches and techniques, which complicate comparative analysis across studies. For the future application of epigenetics in forensics, it is necessary to develop age prediction models that are practical, effective, and allow for rapid results within a routine laboratory. This involves selecting the most informative DNAm markers and their optimal combinations, while simplifying the model as much as possible, for instance by reducing the number of CpG sites required for age prediction. As previously mentioned, methylation tends to be tissue-specific; thus, there is a need to select appropriate DNAm markers based on the tissue of origin. Given this limitation, the question remains whether it will be possible in the future to develop a universal age prediction model that is sufficiently accurate to be applied across all tissues. Furthermore, age estimation models should be assessed for all age intervals, as multiple studies have shown that variability and error increase with age, thereby affecting the accuracy of the determination.

The ultimate goal is to validate epigenetic age estimation methods for their application in forensic anthropology. This includes evaluating how different techniques and models perform across laboratories, while also exploring the potential of integrating multiple approaches, such as morphological and biochemical methods, with epigenetics to enhance accuracy. In this context, interdisciplinarity becomes essential for strengthening biological profile reconstruction and advancing human identification.

Acknowledgment Maria Josefina Castagnola gratefully acknowledges funding from the Fulbright Foreign Student Program through the Ministry of Education of Argentina.

Conflict of Interest The authors declare no conflict of interest.

Ethical Approval This chapter is a review of previously published accounts, as such, no animal or human studies were performed.

References

Amorim A, Fernandes T, Taveira N (2019) Mitochondrial DNA in human identification: a review. PeerJ 7:e7314. https://doi.org/10.7717/peerj.7314

Becker J, Mahlke NS, Reckert A, Eickhoff SB, Ritz-Timme S (2020) Age estimation based on different molecular clocks in several tissues and a multivariate approach: an explorative study. Int J Legal Med 134(2):721–733. https://doi.org/10.1007/s00414-019-02054-9

Becker J, Buhren V, Schmelzer L, Reckert A, Eickhoff SB, Ritz S, Naue J (2025) Molecular age prediction using skull bone samples from individuals with and without signs of decomposition: a multivariate approach combining analysis of posttranslational protein modifications and DNA methylation. Int J Legal Med 139(1):157–174. https://doi.org/10.1007/s00414-024-03314-z

Bekaert S, De Meyer T, Van Oostveldt P (2005) Telomere attrition as ageing biomarker. Anticancer Res 25(4):3011–3021

Bekaert B, Kamalandua A, Zapico SC, Van de Voorde W, Decorte R (2015) Improved age determination of blood and teeth samples using a selected set of DNA methylation markers. Epigenetics 10(10):922–930. https://doi.org/10.1080/15592294.2015.1080413

Brotherton P, Endicott P, Sanchez JJ, Beaumont M, Barnett R, Austin J, Cooper A (2007) Novel high-resolution characterization of ancient DNA reveals C > U-type base modification events as the sole cause of post mortem miscoding lesions. Nucleic Acids Res 35(17):5717–5728. https://doi.org/10.1093/nar/gkm588

Castagnola MJ, Medina-Paz F, Zapico SC (2024) Uncovering forensic evidence: a path to age estimation through DNA methylation. Int J Mol Sci 25(9):4917. https://doi.org/10.3390/ijms25094917

Castagnola MJ, Hassan M, Pujols A, Feirstein S, Cabo L, Adserias-Garriga J, Zapico SC (2025) The truth in the lye: optimizing DNA, RNA, and protein extraction from human teeth under different incubation times and environmental conditions. EC Dent Sci 24(4):1–18. https://ecronicon.net/ecde/the-truth-in-the-lye-optimizing-dna-rna-and-protein-extraction-from-human-teeth-under-different-incubation-times-and-environmental-conditions.php

Christensen AM, Passalacqua NV, Bartelink EJ (2019) Chapter 2—human osteology and odontology. In: Christensen AM, Passalacqua NV, Bartelink EJ (eds) Forensic anthropology, 2nd edn. Academic, pp 33–76. https://doi.org/10.1016/B978-0-12-815734-3.00002-6

Corrado A, Cici D, Rotondo C, Maruotti N, Cantatore FP (2020) Molecular basis of bone aging. Int J Mol Sci 21(10):3679. https://www.mdpi.com/1422-0067/21/10/3679

Correia Dias H, Corte-Real F, Cunha E, Manco L (2020) DNA methylation age estimation from human bone and teeth. Aust J Forensic Sci 54(2):163–176. https://doi.org/10.1080/00450618.2020.1805011

Correia Dias H, Manco L, Corte Real F, Cunha E (2021) A blood-bone-tooth model for age prediction in forensic contexts. Biology (Basel) 10(12):1312. https://doi.org/10.3390/biology10121312

Freire-Aradas A, Tomsia M, Piniewska-Rog D, Ambroa-Conde A, Casares de Cal MA, Pisarek A, Gomez-Tato A, Alvarez-Dios J, Pospiech E, Parson W, Kayser M, Phillips C, Branicki W (2023) Development of an epigenetic age predictor for costal cartilage with a simultaneous somatic tissue differentiation system. Forensic Sci Int Genet 67:102936. https://doi.org/10.1016/j.fsigen.2023.102936

Gamba C, Jones ER, Teasdale MD, McLaughlin RL, Gonzalez-Fortes G, Mattiangeli V, Domboróczki L, Kővári I, Pap I, Anders A, Whittle A, Dani J, Raczky P, Higham TF, Hofreiter M, Bradley DG, Pinhasi R (2014) Genome flux and stasis in a five millennium transect of European prehistory. Nat Commun 5:5257. https://doi.org/10.1038/ncomms6257

Giuliani C, Cilli E, Bacalini MG, Pirazzini C, Sazzini M, Gruppioni G, Franceschi C, Garagnani P, Luiselli D (2016) Inferring chronological age from DNA methylation patterns of human teeth. Am J Phys Anthropol 159(4):585–595. https://doi.org/10.1002/ajpa.22921

Gokhman D, Lavi E, Prüfer K, Fraga MF, Riancho JA, Kelso J, Pääbo S, Meshorer E, Carmel L (2014) Reconstructing the DNA methylation maps of the Neandertal and the Denisovan. Science 344(6183):523–527. https://doi.org/10.1126/science.1250368

Gokhman D, Malul A, Carmel L (2017) Inferring past environments from ancient epigenomes. Mol Biol Evol 34(10):2429–2438. https://doi.org/10.1093/molbev/msx211

Gopalan S, Gaige J, Henn BM (2019) DNA methylation-based forensic age estimation in human bone. bioRxiv:801647. https://doi.org/10.1101/801647

Haarkötter C, Gálvez X, Vinueza-Espinosa DC, Medina-Lozano MI, Saiz M, Lorente JA, Álvarez JC (2023) A comparison of five DNA extraction methods from degraded human skeletal remains. Forensic Sci Int 348:111730. https://doi.org/10.1016/j.forsciint.2023.111730

Hardy J, Chrosciany S, Bernard J-P, Mabit C, Marcheix P-S (2020) The human costal cartilage: anatomical and radiological study of macro-vascularization and micro-vascularization and its clinical relevance regarding vascularized chondrocostal free flap surgery. Ann Anat Anat Anz 232:151581. https://doi.org/10.1016/j.aanat.2020.151581

Heidegger A, Xavier C, Niederstätter H, de la Puente M, Pośpiech E, Pisarek A, Kayser M, Branicki W, Parson W (2020) Development and optimization of the VISAGE basic prototype tool for forensic age estimation. Forensic Sci Int Genet 48:102322. https://doi.org/10.1016/j.fsigen.2020.102322

Higgins D, Austin JJ (2013) Teeth as a source of DNA for forensic identification of human remains: a review. Sci Justice 53(4):433–441. https://doi.org/10.1016/j.scijus.2013.06.001

Hillson S (2005) Teeth, 2nd edn. Cambridge University Press. https://doi.org/10.1017/CBO9780511614477

Hong SR, Shin KJ (2021) Bisulfite-converted DNA quantity evaluation: a multiplex quantitative real-time PCR system for evaluation of bisulfite conversion. Front Genet 12:618955. https://doi.org/10.3389/fgene.2021.618955

Horvath S (2013) DNA methylation age of human tissues and cell types. Genome Biol 14(10):R115. https://doi.org/10.1186/gb-2013-14-10-r115

Horvath S, Raj K (2018) DNA methylation-based biomarkers and the epigenetic clock theory of ageing. Nat Rev Genet 19(6):371–384. https://doi.org/10.1038/s41576-018-0004-3

Horvath S, Mah V, Lu AT, Woo JS, Choi OW, Jasinska AJ, Riancho JA, Tung S, Coles NS, Braun J, Vinters HV, Coles LS (2015) The cerebellum ages slowly according to the epigenetic clock. Aging (Albany NY) 7(5):294–306. https://doi.org/10.18632/aging.100742

Horvath S, Oshima J, Martin GM, Lu AT, Quach A, Cohen H, Felton S, Matsuyama M, Lowe D, Kabacik S, Wilson JG, Reiner AP, Maierhofer A, Flunkert J, Aviv A, Hou L, Baccarelli AA, Li Y, Stewart JD et al (2018) Epigenetic clock for skin and blood cells applied to Hutchinson Gilford Progeria Syndrome and ex vivo studies. Aging (Albany NY) 10(7):1758–1775. https://doi.org/10.18632/aging.101508

Ibrahim J, Brumfeld V, Addadi Y, Rubin S, Weiner S, Boaretto E (2022) The petrous bone contains high concentrations of osteocytes: one possible reason why ancient DNA is better preserved in this bone. PLoS One 17(10):e0269348. https://doi.org/10.1371/journal.pone.0269348

Iorio BL, Cooley AM (2023) DNA extraction of bone through demineralization. Methods Mol Biol 2685:93–102. https://doi.org/10.1007/978-1-0716-3295-6_6

Jung S-E, Lim SM, Hong SR, Lee EH, Shin K-J, Lee HY (2019) DNA methylation of the ELOVL2, FHL2, KLF14, C1orf132/MIR29B2C, and TRIM59 genes for age prediction from blood, saliva, and buccal swab samples. Forensic Sci Int Genet 38:1–8. https://doi.org/10.1016/j.fsigen.2018.09.010

Jung JY, So MH, Jeong KS, Kim SI, Kim EJ, Park JH, Kim E, Lee HY (2024) Epigenetic age prediction using costal cartilage for the investigation of disaster victims and missing persons. J Forensic Sci 69(5):1578–1586. https://doi.org/10.1111/1556-4029.15470

Jung JY, Song YW, Jeong KS, Park HC, So MH, Lee HY (2025) A SNaPshot assay for epigenetic age prediction of costal cartilage. Electrophoresis 46(7–8):413–423. https://doi.org/10.1002/elps.8132

Karkala F, Simons RB, Claessens F, Kalamara V, Kayser M, Vidaki A (2024) qBiCo: a method to assess global DNA conversion performance in epigenetics via single-copy genes and repetitive elements. bioRxiv:2024.2008.2029.610354. https://doi.org/10.1101/2024.08.29.610354

Kondo M, Aboshi H, Yoshikawa M, Ogata A, Murayama R, Takei M, Aizawa S (2020) A newly developed age estimation method based on CpG methylation of teeth-derived DNA using real-time methylation-specific PCR. J Oral Sci 63(1):54–58. https://doi.org/10.2334/josnusd.20-0138

Le TN, Dzenita M, Handt O, Henry J, Linacre A (2021) A novel approach for rapid cell assessment to estimate DNA recovery from human bone tissue. Forensic Sci Med Pathol 17(4):649–659. https://doi.org/10.1007/s12024-021-00428-3

Lee HY, Hong SR, Lee JE, Hwang IK, Kim NY, Lee JM, Fleckhaus J, Jung SE, Lee YH (2020) Epigenetic age signatures in bones. Forensic Sci Int Genet 46:102261. https://doi.org/10.1016/j.fsigen.2020.102261

Li H, Yu S-J (2018) Review of pentosidine and pyrraline in food and chemical models: formation, potential risks and determination. J Sci Food Agric 98(9):3225–3233. https://doi.org/10.1002/jsfa.8853

Lin X, Patil S, Gao YG, Qian A (2020) The bone extracellular matrix in bone formation and regeneration. Front Pharmacol 11:757. https://doi.org/10.3389/fphar.2020.00757

Llamas B, Holland ML, Chen K, Cropley JE, Cooper A, Suter CM (2012) High-resolution analysis of cytosine methylation in ancient DNA. PLoS One 7(1):e30226. https://doi.org/10.1371/journal.pone.0030226

Loreille OM, Diegoli TM, Irwin JA, Coble MD, Parsons TJ (2007) High efficiency DNA extraction from bone by total demineralization. Forensic Sci Int Genet 1(2):191–195. https://doi.org/10.1016/j.fsigen.2007.02.006

Lujan SA, Longley MJ, Humble MH, Lavender CA, Burkholder A, Blakely EL, Alston CL, Gorman GS, Turnbull DM, McFarland R, Taylor RW, Kunkel TA, Copeland WC (2020) Ultrasensitive deletion detection links mitochondrial DNA replication, disease, and aging. Genome Biol 21(1):248. https://doi.org/10.1186/s13059-020-02138-5

MacDougall MJ, Javed A (2010) Dentin and bone: similar collagenous mineralized tissues. In: Bronner F, Farach-Carson MC, Roach HI (eds) Bone and development. Springer, London, pp 183–200. https://doi.org/10.1007/978-1-84882-822-3_11

Mackie EJ, Ahmed YA, Tatarczuch L, Chen KS, Mirams M (2008) Endochondral ossification: how cartilage is converted into bone in the developing skeleton. Int J Biochem Cell Biol 40(1):46–62. https://doi.org/10.1016/j.biocel.2007.06.009

Manolagas SC, Parfitt AM (2010) What old means to bone. Trends Endocrinol Metab 21(6):369–374. https://doi.org/10.1016/j.tem.2010.01.010

Marquez-Ruiz AB, Gonzalez-Herrera L, Luna JD, Valenzuela A (2020) DNA methylation levels and telomere length in human teeth: usefulness for age estimation. Int J Legal Med 134(2):451–459. https://doi.org/10.1007/s00414-019-02242-7

Megyesi MS, Nawrocki SP, Haskell NH (2005) Using accumulated degree-days to estimate the postmortem interval from decomposed human remains. J Forensic Sci 50(3):618–626

Milos A, Selmanović A, Smajlović L, Huel RL, Katzmarzyk C, Rizvić A, Parsons TJ (2007) Success rates of nuclear short tandem repeat typing from different skeletal elements. Croat Med J 48(4):486–493

Naue J, Hoefsloot HCJ, Mook ORF, Rijlaarsdam-Hoekstra L, van der Zwalm MCH, Henneman P, Kloosterman AD, Verschure PJ (2017) Chronological age prediction based on DNA methylation: massive parallel sequencing and random forest regression. Forensic Sci Int Genet 31:19–28. https://doi.org/10.1016/j.fsigen.2017.07.015

Naue J, Sanger T, Hoefsloot HCJ, Lutz-Bonengel S, Kloosterman AD, Verschure PJ (2018) Proof of concept study of age-dependent DNA methylation markers across different tissues by massive parallel sequencing. Forensic Sci Int Genet 36:152–159. https://doi.org/10.1016/j.fsigen.2018.07.007

Ogata A, Kondo M, Yoshikawa M, Okano M, Tsutsumi T, Aboshi H (2022) Dental age estimation based on DNA methylation using real-time methylation-specific PCR. Forensic Sci Int 340:111445. https://doi.org/10.1016/j.forsciint.2022.111445

Park-Min KH (2019) Metabolic reprogramming in osteoclasts. Semin Immunopathol 41(5):565–572. https://doi.org/10.1007/s00281-019-00757-0

Pignolo RJ, Law SF, Chandra A (2021) Bone aging, cellular senescence, and osteoporosis. JBMR Plus 5(4):e10488. https://doi.org/10.1002/jbm4.10488

Prinz M, Carracedo A, Mayr WR, Morling N, Parsons TJ, Sajantila A, Scheithauer R, Schmitter H, Schneider PM (2007) DNA Commission of the International Society for Forensic Genetics (ISFG): recommendations regarding the role of forensic genetics for disaster victim identification (DVI). Forensic Sci Int Genet 1(1):3–12. https://doi.org/10.1016/j.fsigen.2006.10.003

Reppe S, Lien TG, Hsu YH, Gautvik VT, Olstad OK, Yu R, Bakke HG, Lyle R, Kringen MK, Glad IK, Gautvik KM (2017) Distinct DNA methylation profiles in bone and blood of osteoporotic and healthy postmenopausal women. Epigenetics 12(8):674–687. https://doi.org/10.1080/15592294.2017.1345832

Schmidt M, Maixner F, Hotz G, Pap I, Szikossy I, Palfi G, Zink A, Wagner W (2021) DNA methylation profiling in mummified human remains from the eighteenth-century. Sci Rep 11(1):15493. https://doi.org/10.1038/s41598-021-95021-7

Simons RB, Karkala F, Kukk MM, Adams HHH, Kayser M, Vidaki A (2025) Comparative performance evaluation of bisulfite- and enzyme-based DNA conversion methods. Clin Epigenetics 17(1):56. https://doi.org/10.1186/s13148-025-01855-7

Smith RW, Monroe C, Bolnick DA (2015) Detection of cytosine methylation in ancient DNA from five native American populations using bisulfite sequencing. PLoS One 10(5):e0125344. https://doi.org/10.1371/journal.pone.0125344

Tomsia M, Droździok K, Javan GT, Skowronek R, Szczepański M, Chełmecka E (2021) Costal cartilage ensures low degradation of DNA needed for genetic identification of human remains retrieved at different decomposition stages and different postmortem intervals. Postepy Hig Med Dosw 75(1):852–858. https://doi.org/10.2478/ahem-2021-0035

Tomsia M, Cieśla J, Pilch-Kowalczyk J, Banaszek P, Chełmecka E (2022) Cartilage tissue in forensic science—state of the art and future research directions. PRO 10(11):2456. https://www.mdpi.com/2227-9717/10/11/2456

Ubelaker DH (1987) Estimating age at death from immature human skeletons: an overview. J Forensic Sci 32(5):1254–1263

Ubelaker DH, Khosrowshahi H (2019) Estimation of age in forensic anthropology: historical perspective and recent methodological advances. Forensic Sci Res 4(1):1–9. https://doi.org/10.1080/20961790.2018.1549711

Vaiserman A, Krasnienkov D (2020) Telomere length as a marker of biological age: state-of-the-art, open issues, and future perspectives. Front Genet 11:630186. https://doi.org/10.3389/fgene.2020.630186

Villagran XS, Huisman DJ, Mentzer SM, Miller CE, Jans MM (2017) Bone and other skeletal tissues. In: Archaeological soil and sediment micromorphology, pp 9–38. https://doi.org/10.1002/9781118941065.ch1

Woźniak A, Heidegger A, Piniewska-Róg D, Pośpiech E, Xavier C, Pisarek A, Kartasińska E, Boroń M, Freire-Aradas A, Wojtas M, de la Puente M, Niederstätter H, Płoski R, Spólnicka M, Kayser M, Phillips C, Parson W, Branicki W (2021) Development of the VISAGE enhanced tool and statistical models for epigenetic age estimation in blood, buccal cells and bones. Aging (Albany NY) 13(5):6459–6484. https://doi.org/10.18632/aging.202783

Zapico SC, Ubelaker DH (2013) Applications of physiological bases of ageing to forensic sciences. Estimation of age-at-death. Ageing Res Rev 12(2):605–617. https://doi.org/10.1016/j.arr.2013.02.002

Zapico SC, Ubelaker DH (2022) Application of aspartic acid racemization for age estimation in a Spanish sample. Biology (Basel) 11(6):856. https://doi.org/10.3390/biology11060856

Zapico SC, DeGaglia CM, Adserias-Garriga J (2019) Chapter 14—age estimation based on chemical approaches. In: Adserias-Garriga J (ed) Age estimation. Academic, pp 199–211. https://doi.org/10.1016/B978-0-12-814491-6.00014-5

Zapico SC, Gauthier Q, Antevska A, McCord BR (2021) Identifying methylation patterns in dental pulp aging: application to age-at-death estimation in forensic anthropology. Int J Mol Sci 22(7):3717. https://doi.org/10.3390/ijms22073717

Zaslansky P, Zabler S, Fratzl P (2010) 3D variations in human crown dentin tubule orientation: a phase-contrast microtomography study. Dent Mater 26(1):e1–e10. https://doi.org/10.1016/j.dental.2009.09.007

Zubakov D, Liu F, Kokmeijer I, Choi Y, van Meurs JBJ, van Ijcken WFJ, Uitterlinden AG, Hofman A, Broer L, van Duijn CM, Lewin J, Kayser M (2016) Human age estimation from blood using

Chapter 9
Epigenetics Applications to Postmortem Interval Estimation

Mariano Guardado-Estrada and Sara C. Zapico

Abstract Postmortem interval (PMI) estimation is one of the most challenging tasks for a forensic scientist, particularly during the latest stages of decomposition and in skeletonized remains, as it depends on various internal and external factors. Several techniques have been proposed to address the issue from different scientific fields, including molecular biology. Among them, epigenetics could be a potential solution to this problem, and more specifically, the use of microRNAs (miRNAs) to determine PMI in a quantifiable manner. miRNAs are small, single-stranded, non-coding RNAs, which are essential players in the post-transcriptional layer of gene expression regulation. Over the years, different studies demonstrated the presence and role of miRNAs in different tissues during PMI times, leading to considering them as potential tools to estimate time since death. Besides, their stability under harsh environmental conditions and their straightforward expression analysis through reverse transcription-quantitative polymerase chain reaction (RT-QPCR) enhance their usefulness and application in forensic contexts by crime labs.

Keywords MicroRNA · Postmortem interval · Gene expression · Reverse Transcription Quantitative Polymerase Chain Reaction

---

M. Guardado-Estrada
Laboratorio de Genética, Escuela Nacional de Ciencias Forenses, Universidad Nacional Autónoma de México, Ciudad de México, Mexico

S. C. Zapico (✉)
Department of Chemistry and Environmental Sciences, New Jersey Institute of Technology, Newark, NJ, USA

Anthropology Department and Laboratories of Analytical Biology, National Museum of Natural History, Smithsonian Institution, Washington, DC, USA
e-mail: sc338@njit.edu; casado-zapicos@si.edu

S. C. Zapico (ed.), *Forensic Science and Epigenetics*, Epigenetics and Human Health 18, https://doi.org/10.1007/978-3-032-23085-0_9

Abbreviations

AATF	Apoptosis Antagonizing Transcription Factor
ACTB	Actin Beta
BMF	Bcl-2 Modifying Factor
CPR	Cardiopulmonary Resuscitation
DNA	Deoxyribonucleic Acid
EPC1	Enhancer of Polycomb Homolog 1
ePMI	early Postmortem Interval
GAPDH	Glyceraldehyde 3-phosphate dehydrogenase
miRNAs	microRNAs
miRNome	microRNA transcriptome
NGS	Next-Generation Sequencing
PMI	Postmortem Interval
pre-miRNA	precursor microRNA
pri-miRNA	primary microRNA
RISC	RNA-induced Silencing Complex
RNA	Ribonucleic Acid
RPS29	Ribosomal Protein (RP) S29 Mrna
rRNA	ribosomal RNA
RT-PCR	Reverse Transcription-Quantitative Polymerase Chain Reaction
SIRT1	Sirtuin 1
TGFBR2	Transforming Growth Factor, Beta Receptor II
TSD	Time-since-death

9.1 Introduction

Postmortem interval (PMI) or time-since-death (TSD) is the time between an individual's death and the discovery of their body (Cianci et al. 2024). The estimation of this parameter is not straightforward because it depends on various internal and external factors (Babapulle and Jayasundera 1993), making it the most challenging determination. However, it is crucial for confirming alibis, elucidating possible criminal acts, and defining appropriate civil repercussions (Adserias-Garriga and Zapico 2022).

Estimation of PMI is based on changes undergone by a corpse after death, including early changes involving physical processes (body cooling and hypostasis), metabolic processes (supravital reactions), autolysis (loss of selective membrane permeability, diffusion), and physicochemical processes (rigor mortis) (Henssge and Madea 2004). However, the assessment of these changes is only effective in the first hours of death, between 3 and 72 h, which is called "early postmortem interval (ePMI)" (Saber et al. 2021). Additionally, these methods could be affected by intrinsic and extrinsic variables, like environmental conditions, scavenger activity, body mass, the health status of the individual previous to death, among others (Wescott 2018).

Beyond the ePMI, these techniques are less effective, and other methods come into play, like forensic entomology, which is the study of insect activity (larval development or insect succession) on a corpse (Amendt et al. 2004). This could study PMIs' ranges of days or the first few months after death. However, it is not exempt from drawbacks. It depends on insects' access to the body (Amendt et al. 2011).

Starting in the 1960s, new research in this area aimed to improve PMI estimates in a more quantifiable way, initiating the field called "thanatochemistry". It refers to the study of postmortem chemical changes, like the quantification of volatile fatty acids and other compounds, and developing formulas based on these parameters also taking into account humidity and temperature (Vass 2011; Vass et al. 2002).

Current research is focused on the study of the degradation of biomolecules and/or changes in signaling pathways of cell death (DNA, RNA, and proteins) to improve PMI estimates, leading to the field called "thanatobiology" (Adserias-Garriga and Zapico 2022; Martinez et al. 2019; Zapico et al. 2014). Another field called "thanatomicrobiome" studies the changes of microbial communities in our body because of the decomposition process (Javan et al. 2016a, b).

Despite the variety of research approaches, finding a method/formula to enhance PMI estimates is elusive. For that reason, some researchers have turned to study epigenetics. Epigenetics is defined as the study of processes that produce heritable phenotypes that do not strictly depend on the DNA sequence (Lieb et al. 2006). There are three types of epigenetic modifications: DNA methylation, chromatin remodeling and histone modifications, and non-coding RNA regulation.

For PMI estimation applying epigenetics, the studies have been focused on non-coding RNAs, and particularly, microRNAs (miRNAs), as potential targets to improve TSD determination (Glynn 2020).

9.2 MicroRNAs

RNAs are divided into two large families: coding RNAs (translating into proteins) and non-coding RNAs (not translating into proteins). The microRNAs (miRNAs) belong to the second group (Cianci et al. 2024).

The miRNAs are small, single-stranded, non-coding RNAs of 18–24 nucleotides in length. They are essential players in the post-transcription layer of gene expression regulation (Guo et al. 2015). They play a critical role in a variety of biological (development, differentiation, proliferation, and apoptosis) and pathological processes, being tissue-specific (Glynn 2020).

The first miRNA was identified in *Caenorhabditis elegans* by Lee et al. in 1993. They discovered a non-coding RNA, lin-4, which regulated larval development by repressing the expression of the lin-14 gene (Lee et al. 1993). Beginning in the early 2000s, additional miRNAs such as lin-7 were characterized, leading to the recognition of miRNAs as a distinct class of post-transcriptional regulators with broad biological significance (Rocchi et al. 2020).

miRNAs originate from RNA transcript regions that form short hairpin structures. These are initially transcribed in the nucleus as primary miRNAs (pri-miRNAs), which are approximately 70 nucleotides long and may encompass up to six individual miRNA precursors. The nuclear RNase III enzyme Drosha processes pri-miRNAs into precursor miRNAs (pre-miRNAs), which are subsequently exported to the cytoplasm via exportin 5. In the cytoplasm, pre-miRNAs are further cleaved by another RNase III enzyme (Dicer), generating ~22-nucleotide miRNA duplexes. One strand of the duplex, known as the guide strand, is selectively incorporated into the RNA-induced silencing complex (RISC), which includes Argonaute proteins and the mature miRNA. The opposing strand is typically degraded. Once assembled into RISC, mature miRNAs exhibit increased stability and resistance to degradation (Bartel 2004, 2009) (Fig. 9.1).

Two key features make miRNAs promising biomarkers for PMI estimation. First, due to their involvement in apoptosis, certain miRNAs may increase during the PMI, contributing to cellular degradation (Cianci et al. 2024). Second, miRNAs exhibit exceptional stability compared to other RNA species, particularly in advanced decomposition stages (Rocchi et al. 2020). Their small size and protection within protein and lipid matrices—such as extracellular vesicles, lipoprotein complexes, and RNA-binding proteins—shield them from enzymatic degradation and environmental stressors like extreme temperature and pH. Nonetheless, miRNA stability can be affected by variables including tissue type, environmental conditions, and the extent of putrefaction (Layne et al. 2019; Mayes et al. 2019).

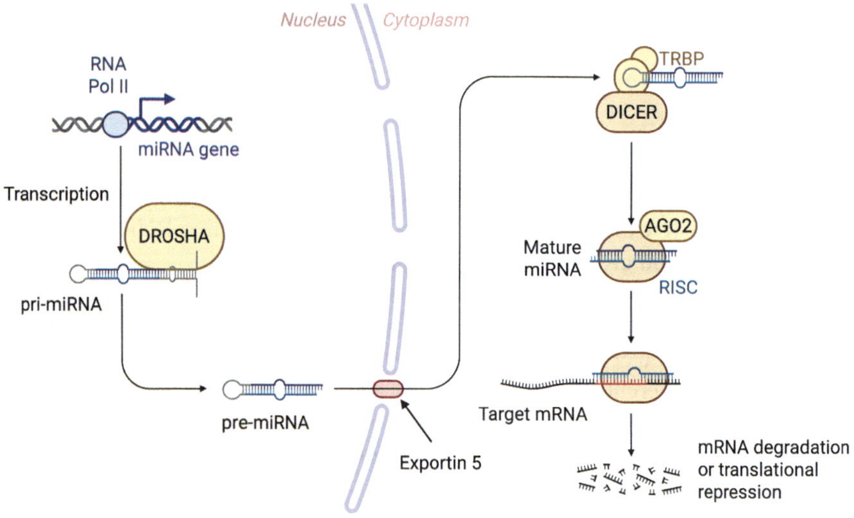

Fig. 9.1 Biogenesis of miRNAs

The methods of election to analyze miRNA are microarray platforms and Next-Generation Sequencing (NGS) techniques, being able to assess several differentially expressed miRNAs (Guardado-Estrada et al. 2023), later validated by Reverse Transcription-Quantitative Polymerase Chain Reaction (RT-QPCR) (Medina-Paz and Zapico 2025). However, many PMI studies evaluated few selected miRNAs, thus, directly applying RT-QPCR, demonstrating an increased sensitivity and reproducibility (Lee et al. 2024a). On the other hand, NGS makes it possible to analyze the structure and expression of known transcripts and even to discover previously unknown ones (Song et al. 2024). This is a major advantage of the technique, as it allows the identification of novel non-coding RNAs.

9.3 Applications of miRNAs to PMI Estimation

The first study on the application of miRNA to PMI estimation was carried out by Li et al. in 2010 (Li 2010). The authors assessed expression levels of two types of RNAs, miR-1-2 and 18S rRNA, in rats' cardiac tissues after different PMIs and in a temperature-controlled environment (25 °C). They found 18S rRNA levels increased up to 96 h after death and decreased after that. In contrast, miR-1-2 was more stable, increasing to 120 h, but declining after this period. This study points out the usefulness of these markers between 96 and 120 h after death.

Wang et al. studied the expression levels of other miRNAs (miR-122, miR-133a, miR-150, miR-195, miR-206) up to 48 h after death in mice liver tissues (Wang et al. 2013). No correlation was found between the expression levels of any of these miRNAs and the early PMI. However, it was demonstrated that the levels of miR-206 and miR-133 significantly decreased 24 h after death, indicating a potential use of these miRNAs for PMI estimation between 24 and 48 h.

Odriozola et al. (2013) assessed the stability of miRNAs in vitreous humor from human corpses, also evaluating the potential impact of the period of the day in which the death occurred on the miRNAs. They found an upregulation of miR-142-5p and miR-541 if deaths occurred at nighttime. However, all miRNAs were stable for at least 24 h, not finding a correlation between miRNA levels and the TSD.

Lv et al. performed a simultaneous analysis of mRNA, miRNAs, 18S rRNA, and U6 snRNA in rats' spleens, PMIs from 0 to 144 h at 25 ± 2 °C, and at 4 ± 2 °C with PMIs between 0 and 312 h (Lv et al. 2014). RT-QPCR was the method of choice to assess the levels of these markers. Applying the geNorm algorithm, the authors found that the miRNAs evaluated (miR-125b and miR-143) were more stable than the other markers, as they were not affected by PMI and temperature; thus, they should be used as control markers. However, the other markers could potentially apply to PMI determination: 18S rRNA, GAPDH2, and ACTB2. The findings of the latter were surprising, as GAPDH2 and ACTB2 are called "housekeeping genes". They are frequently used as endogenous mRNA reference genes in RT-QPCR data analysis in general molecular biology experiments, as they should be stable in all cells. However, according to these results, they exhibited a close relationship with PMI, making them unsuitable as reference genes.

In the same line of research, Li et al. (2014) evaluated the expression levels of 18S-rRNA and miRNAs (miR-1, miR-143, miR-208) in tissue samples from rats (heart, liver, spleen, lung, kidney, brain, back skin, and front leg muscle) during PMIs between 0 and 168 h. RT-QPCR was also applied. At 25 °C in rats' hearts, they found a gradual increase of 18S-rRNA in early PMI and peaked at around 96 h. After that, the levels of 18S-rRNA decreased gradually. miR-1 remained stable during the PMI, and the authors used it to normalize the expression levels of 18S-rRNA. Based on this, the authors proposed a regression model 18S-rRNA/miR-1 to correlate with the PMI, reducing the probability of error of this estimate.

Following the potential influence of circadian rhythm on PMI, Sharam et al. (2015) assessed expression levels of the apoptosis antagonizing transcription factor (AATF, che-1) and its associated miRNA, miR-2909, in various organs from mice (brain, heart, lungs, kidneys, pancreas, spleen, and liver). To evaluate the impact of the circadian rhythm, mice were sacrificed at different times (4 AM, noon, 8 PM), and samples were recovered from 0 to 72 h after death. RT-QPCR was the technique of choice. First, the authors found that AATF and miR-2909 were predominantly expressed in the brain and heart. The results on these tissues indicated that the expression of miR-2909 was stable up to 48 h after death at 8 PM, and 12 h after death at noon. AATF expression was stable up to 24 h after death at 8 PM, and 12 h after death at noon. This study found a significant correlation between the variation in AATF and miR-2909 and the circadian rhythm.

Ma et al. continued with the study of multiple markers to improve PMI estimation and the development of a mathematical model towards this goal (Ma et al. 2015). They used rats' brain samples kept in a controlled environment at three different temperatures (15 ± 1 °C; 25 ± 1 °C; 35 ± 1 °C), and postmortem times from 0 to 144 h. As validation, additional rats were kept at 10 ± 1 °C; 20 ± 1 °C; 30 ± 1 °C, and brain samples were recovered from 10 to 100 h postmortem. RT-QPCR was the technique of choice, assessing the levels of different RNA markers: mRNA markers (β-actin, GAPDH, Ribosomal Protein (RP) S29 mRNA (RPS29)); 5SrRNA; 18SrRNA; U6; miR-9; miR-125b; Let-7a. Their results indicated that, as happened with previous studies (Lv et al. 2014), miR-125 and miR-9 were chosen as control markers based on their high stability with the PMI. β-actin significantly correlates with PMI in all the temperature groups, finding the highest correlation among markers, confirming previous published results (Lv et al. 2014). This work identified a potential, reliable biomarker for the PMI estimation.

Tu et al. also assessed the stability of several RNA biomarkers with the PMI (Tu et al. 2018). Among them are GAPDH, β-actin, RPS18, 5SrRNA, 18SrRNA, U6, miR-122, and miR-133a. The authors also studied circular RNAs (circRNAs). circRNAs are non-coding RNAs characterized by the presence of a covalent bond linking the 3′ and 5′ ends generated by backsplicing, which makes them more stable than linear RNA. As non-coding RNAs, they are involved in gene expression regulation. This study evaluated the expression of five circRNAs: circ-AFF1, circ-Ogdh, LC-Ogdh, Circ-LRP6, and LC-LRP6. This work was performed in tissue samples from mice (heart, liver, and skeletal muscle) with PMIs between 0 and 8 days. RT-QPCR was the method of choice to evaluate this expression. As expected from previous studies, miRNAs and circRNAs were more stable than other RNA biomarkers, ranking them as reference genes accordingly: miR-122, miR-133a, and 18S in heart; LC-Ogdh, circ-AFF1, and miR-122 in liver; miR-133a and circ-AFF1 in skeletal muscle. The authors suggested that this ranking is related to the differential expression of these miRNAs and circRNAs in different tissues. As a result, they proposed that the selection of reference genes should be based on tissue types.

The same group (Tu et al. 2019) in a subsequent study developed specific prediction models for PMI estimation in the same mouse tissues (heart, liver, and skeletal muscle) with TSD between 0 and 8 days. They identified four biomarkers, GAPDH, RPS18, U6, and β-actin to determine PMI and as reference genes miR-122, miR-133a, 18S, LC-Ogdh, and circ-AFF1. Although their accuracy depended on the tissue, finding U6 and RPS18 more suitable target biomarkers in heart and liver tissues, while U6 and β-actin were the best biomarkers in skeletal muscle tissues. These findings confirmed previous results regarding the tissue-specificity and the use of housekeeping genes as potential biomarkers for PMI estimation.

Na (2020) assessed expression levels of miRNAs in human bone samples (patella) from bodies with PMIs ranging from 1 day to 2 years. RT-QPCR was the method of choice and let-7e and miR-16 miRNAs were evaluated, using Ce_miR-39-1 as spike-in control, based on previous studies. A negative correlation was found between miRNA expression and increasing PMI. The expression of let-7a and miR-16 was significantly different between the group with PMIs less than 1 month and the other groups. No other statistical differences were found. This is the first study applying miRNAs to estimate PMI in human remains. Further research is needed to validate these findings.

Martínez-Rivera et al. (2021) designed a rat model to assess expression levels of miRNAs during early PMI (0–24 h). The tissue of election was muscle, and miR-381-3p, miR-23b-3p, and miR-144-3p levels were evaluated by RT-QPCR. They found an upregulation of miR-381-3p at 24 h with respect to the control group; in contrast, miR-23b-3p was downregulated 24 h after death. The authors analyzed the expression of one of the miR-381-3p targets, the EPC1 gene, finding it downregulated at 3 and 12 h after death, keeping stable at 6 and 24 h after death. The authors also tried to develop a mixed effects model to predict the PMI, based on miR-381-3p expression and morphological changes in the rat's brain. This model was a good predictor of the 0-, 3-, 12-, and 24-h PMI, with variability observed at 6 h-PMI. Although these results were promising, additional validation studies are needed to confirm the potential usefulness of miR-381-3p as a biomarker for PMI estimation.

A follow-up study from the previous group (Guardado-Estrada et al. 2023) applied the Affymetrix GeneChip™ miRNA 4.0 microarrays to assess miRNAs expression changes with the PMI in skeletal muscle samples from rats at 0 and 24 h PMI. They found in total 156 dysregulated miRNAs (84 downregulated and 72 upregulated), identifying the most significantly downregulated rno-miR-139-5p and upregulated rno-miR-92b-5p, rno-miR-297, rno-miR-466c-5p, and rno-miR-466-d.

The gene expression of the most significant miRNAs could differentiate the 24-h PMI group from the control group. The authors went beyond these findings and assessed the expression levels of the targets of these dysregulated miRNAs, TGFBR2, SIRT1, and BMF, applying RT-QPCR. They found an increase in the expression of TGFBR2 and BMF 24 h after death, although in the latter was not statistically significant. In contrast, SIRT1 expression was significantly reduced 24 h after death. These genes are involved in autophagy, cell cycle regulation, and low oxygen response, among others, which fit with postmortem processes. This is the first study analyzing the whole miRNome in rat skeletal muscle, identifying potential biomarkers suitable for PMI estimation. Nevertheless, it is important to consider that in this work it only analyzed one time point of PMI, thus the expression patterns of different miRNAs could vary according to other PMIs.

The study of Kim et al. (2021) investigated the degradation profiles of four miRNAs in postmortem blood samples from 28 cadavers with known PMIs, ranging from 16 to 86 h. Two cardiac-specific miRNAs (miR-208b, miR-1) and two non-cardiac-specific miRNAs (Let-7e, miR-16) were analyzed. Blood was collected from the external iliac vein, inferior vena cava, and coronary sinus during autopsy. Samples were categorized by cause of death (cardiac vs. non-cardiac) and whether cardiopulmonary resuscitation (CPR) had been performed. Cardiac-specific miRNAs showed site-dependent variation, while non-cardiac-specific miRNAs remained stable across sites. No significant differences were found between groups (cardiac vs. non-cardiac deaths, CPR vs. non-CPR), nor was there a significant correlation between miRNA levels and PMI. These findings suggest that peripheral venous levels of the selected miRNAs do not reliably correlate with PMI.

The study of Singh et al. assessed miRNAs expression levels on human hearts from traffic accident victims at different postmortem intervals (from 12 to 196 h) (Singh et al. 2023). They used RT-QPCR to evaluate the expression of miRNA-195, miRNA-206, and miRNA-398, and miRNA-1 was chosen as a reference miRNA. They found significant reductions in miRNAs expression at 196 h, particularly miRNA-206 showed significant changes at multiple time intervals. This study points to the usefulness of these markers for PMI estimation up to 196 h.

Lee et al. focused on developing a new method to detect expression levels of miR-133a-5p with the PMI, a molecular beacon probe (Lee et al. 2024b). They used skeletal muscles from mice with PMIs between 0 and 10 days and exposed them to 4 °C and 21 °C. The molecular beacon, thermodynamically stable, was able to detect the miRNA at both temperatures, remaining consistent over time.

The same group assessed the expression levels of three miRNAs (miR-122, miR-133a, and miR-206) in four tissues (brain, liver, heart, and skeletal muscle) from mice exposed to 4 °C and 21 °C during PMIs ranging from 0 to 10 days (Lee et al. 2024a). As a suitable endogenous control, they found RNU6B. At 4 °C, miR-122 levels significantly declined at 8 and 10 days across all tissues, with the liver also showing notable changes at 21 °C. miR-133a exhibited progressive degradation in the heart, muscles, and brain, with marked reductions in the heart and muscles at both temperatures on days 8 and 10. miR-206 levels decreased over time in muscle and liver at 4 °C, but increased in the brain at 21 °C, with no significant changes observed in other tissues. These findings indicated the potential usefulness of these biomarkers for PMI estimation in heart and skeletal muscle.

The miRNAs that have been consistently associated with the PMI include miR-122, miR-133a, miR-206, miR-1, miR-195, miR-125b, miR-143, let-7e, and miR-16. However, it is important to note that miRNAs can be organ-specific, and therefore some of these miRNAs may be more closely related to the tissue being examined. For instance, miR-122 is highly specific to liver tissue, where it has been reported to represent more than 50% of the miRNome in hepatocytes (Locatelli and Pasa 2025). In contrast, miR-1, miR-133a, and miR-206 belong to the group of miRNAs known as myomiRs, which are muscle-specific (Mytidou et al. 2021). These miRNAs are involved in myoblast differentiation and regeneration, as well as in myoblast proliferation and muscle growth (Mytidou et al. 2021). This is important to keep in mind because their presence at a given PMI may be more related to their role in impaired metabolism and structural integrity of these tissues. In addition, miR-195 and miR-16 have been reported to participate in apoptosis and inflammatory responses, pathways that are commonly activated during the PMI (Wang et al. 2015).

It is important to consider that the analysis of miRNAs at the PMI is likely to be more complex than simply evaluating the rate of degradation of these molecules. As in mRNA studies, the expression of miRNAs appears to be dynamic, showing different expression patterns according to the PMI. This is supported by the observation that several miRNAs are overexpressed at early PMI and may be involved in biological pathways related to hypoxia or energy metabolism, among others (Guardado-Estrada et al. 2023). On the other hand, when designing experiments that use miRNAs for PMI estimation, it should be considered that some miRNAs are tissue-specific, and their detection may primarily reflect residual metabolic activity from particular cell populations at the PMI. A promising strategy to identify more precise biomarkers for PMI estimation using miRNAs would be to apply different analytical approaches, such as machine-learning methods, to define miRNome signatures that could also incorporate other molecules (Cheng et al. 2024).

9.4 Conclusions

The microRNAs (miRNAs) are small molecules involved in the regulation of gene expression that, due to their physical properties, can withstand degradation during the postmortem interval (PMI). The presence of several miRNAs at the PMI has been reported in different organs and species. The stability of some of these miRNAs makes them suitable as reference genes for expression studies at the PMI. miR-122, miR-133a, miR-206, miR-1, miR-195, miR-125b, miR-143, let-7e, and miR-16 are among the most extensively studied miRNAs. A summary of these findings is depicted in Table 9.1. Although there is clear evidence of miRNA persistence at the PMI, more studies are needed to better elucidate the molecular mechanisms in which they participate during the postmortem interval.

Table 9.1 miRNAs reported in different species and organs at the postmortem interval.

Study	Analyzed miRNAs	Tissue	Main findings
Li (2010)	miR-1-2	Rats cardiac tissues	Expression increased until 120 h of PMI
Wang (2013)	miR-122, miR-133a, miR-150, miR-195, miR-206	Mice liver tissues	miR-206 and miR-133 decreased at 24 h of PMI
Odriozola et al. (2013)	miR-142-5p and miR-541	Human vitreous humor	Upregulation of miR-142-5p and miR-541 in deaths occurred at nighttime
Lv et al. (2014)	miR-125b; miR-143	Rats spleens	Controls for gene expression
Li et al. (2014)	miR-1, miR-143, miR-208	Heart, liver, spleen, lung, kidney, brain, back skin, front leg muscle tissues from rats	miR-1 stable used to normalize the expression levels of 18S-rRNA in PMI
Sharma et al. (2015)	miR-2909	Brain, heart, lungs, kidneys, pancreas, spleen, and liver tissues from mice	miR-2909 was stable up to 48 h after death at 8 PM, and 12 h after death at noon
Ma et al. (2015)	miR-9; miR-125b; Let-7a	Rat brain samples	miR-125 and miR-9 as controls for gene expression
Tu et al. (2018)	miR-122, and miR-133a	Heart, liver, and skeletal muscle mice samples	miR-122 and miR-133a used as control genes in skeletal muscle, heart and liver
Tu et al. (2019)	miR-122, miR-133a	Heart, liver, and skeletal muscle mice	miR-122 and miR-133a were normalizing reference gene
Na (2020)	let-7e and miR-16	Human bone samples	Negative correlation between miRNA expression and increasing PMI
Martinez-Rivera et al. (2021)	miR-381–3p, miR-23b-3p, and miR-144-3p	Rats muscle	Upregulation of miR-381–3p and downregulation of miR-23b-3p
Guardado-Estrada et al. (2023)	miRNome	Rats muscle	Upregulation of miR-381-3p at 24 h while miR-23b-3p was downregulated
Kim et al. (2021)	miR-208b, miR-1, Let-7e, miR-16	Blood samples from cadaver dead of cardiac and no cardiac disease	No differences
Singh et al. (2023)	miRNA-195, miRNA-206, and miRNA-398	Human heart of dead's by traffic accidents	Reduction of expression of these miRNAs
Lee et al. (2024b)	miR-133a-5p	Mice skeletal muscle	Stable expression through 10 days of PMI

(continued)

Table 9.1 (continued)

Study	Analyzed miRNAs	Tissue	Main findings
Lee et al. (2024a)	miR-122, miR-133a, and miR-206	Mice brain, liver, heart, and skeletal muscle	Downregulation of miR-122 across all tissues, while miR-133a was down regulated in the heart, muscles, and brain

Compliance with Ethical Standards The authors declare that they have no conflict of interest. This chapter is a review of previously published accounts, as such, no animal or human studies were performed.

References

Adserias-Garriga J, Zapico SC (2022) Postmortem interval estimation: new approaches by the analysis of human tissues and microbial communities' changes. Forensic Sci 2(1):163–174. https://www.mdpi.com/2673-6756/2/1/13

Amendt J, Krettek R, Zehner R (2004) Forensic entomology. Naturwissenschaften 91(2):51–65. https://doi.org/10.1007/s00114-003-0493-5

Amendt J, Richards CS, Campobasso CP, Zehner R, Hall MJ (2011) Forensic entomology: applications and limitations. Forensic Sci Med Pathol 7(4):379–392. https://doi.org/10.1007/s12024-010-9209-2

Babapulle CJ, Jayasundera NP (1993) Cellular changes and time since death. Med Sci Law 33(3):213–222. https://doi.org/10.1177/002580249303300306

Bartel DP (2004) MicroRNAs: genomics, biogenesis, mechanism, and function. Cell 116(2):281–297. https://doi.org/10.1016/s0092-8674(04)00045-5

Bartel DP (2009) MicroRNAs: target recognition and regulatory functions. Cell 136(2):215–233. https://doi.org/10.1016/j.cell.2009.01.002

Cheng Y, Xu SM, Santucci K, Lindner G, Janitz M (2024) Machine learning and related approaches in transcriptomics. Biochem Biophys Res Commun 724:150225. https://doi.org/10.1016/j.bbrc.2024.150225

Cianci V, Mondello C, Sapienza D, Guerrera MC, Cianci A, Craco A, Luppino F, Gioffre V, Gualniera P, Asmundo A, Germana A (2024) microRNAs as new biomolecular markers to estimate time since death: a systematic review. Int J Mol Sci 25(17):9207. https://doi.org/10.3390/ijms25179207

Glynn CL (2020) Potential applications of microRNA profiling to forensic investigations. RNA 26(1):1–9. https://doi.org/10.1261/rna.072173.119

Guardado-Estrada M, Cardenas-Monroy CA, Martinez-Rivera V, Cortez F, Pedraza-Lara C, Millan-Catalan O, Perez-Plasencia C (2023) A miRNome analysis at the early postmortem interval. PeerJ 11:e15409. https://doi.org/10.7717/peerj.15409

Guo Y, Liu J, Elfenbein SJ, Ma Y, Zhong M, Qiu C, Ding Y, Lu J (2015) Characterization of the mammalian miRNA turnover landscape. Nucleic Acids Res 43(4):2326–2341. https://doi.org/10.1093/nar/gkv057

Henssge C, Madea B (2004) Estimation of the time since death in the early post-mortem period. Forensic Sci Int 144(2–3):167–175. https://doi.org/10.1016/j.forsciint.2004.04.051

Javan GT, Finley SJ, Abidin Z, Mulle JG (2016a) The thanatomicrobiome: a missing piece of the microbial puzzle of death. Front Microbiol 7:225. https://doi.org/10.3389/fmicb.2016.00225

Javan GT, Finley SJ, Can I, Wilkinson JE, Hanson JD, Tarone AM (2016b) Human thanatomicrobiome succession and time since death. Sci Rep 6:29598. https://doi.org/10.1038/srep29598

Kim SY, Jang SJ, Jung YH, Na JY (2021) Difference in microRNA levels in the post-mortem blood from different sampling sites: a proof of concept. J Forensic Leg Med 78:102124. https://doi.org/10.1016/j.jflm.2021.102124

Layne TR, Green RA, Lewis CA, Nogales F, Dawson Cruz TC, Zehner ZE, Seashols-Williams SJ (2019) microRNA detection in blood, urine, semen, and saliva stains after compromising treatments. J Forensic Sci 64(6):1831–1837. https://doi.org/10.1111/1556-4029.14113

Lee RC, Feinbaum RL, Ambros V (1993) The *C. elegans* heterochronic gene lin-4 encodes small RNAs with antisense complementarity to lin-14. Cell 75(5):843–854. https://doi.org/10.1016/0092-8674(93)90529-y

Lee EJ, Jeong M, Lee H, Je MA, Park K, Lee DG, Xuan X, Kim S, Park S, Kim J (2024a) MiR-122, miR-133a, and miR-206 as potential biomarkers for post-mortem interval estimation. Genes Genomics 46(10):1175–1182. https://doi.org/10.1007/s13258-024-01559-x

Lee EH, Jeong M, Park K, Lee DG, Lee EJ, Lee H, Kim AY, Ahn JW, Woo HJ, Kim S, Lim J, Kim J (2024b) Detection of miR-133a-5p using a molecular beacon probe for investigating postmortem intervals. Noncoding RNA 10(6):58. https://doi.org/10.3390/ncrna10060058

Li W, Ma K, Zhang P, Wang H, Shen Y, Zhou Y, Zhao Z, Ma D, Chen L (2010) Estimation of postmortem interval using microRNA and 18S rRNA degradation in rat cardiac muscle. Fa Yi Xue Za Zhi 26:413–417

Li WC, Ma KJ, Lv YH, Zhang P, Pan H, Zhang H, Wang HJ, Ma D, Chen L (2014) Postmortem interval determination using 18S-rRNA and microRNA. Sci Justice 54(4):307–310. https://doi.org/10.1016/j.scijus.2014.03.001

Lieb JD, Beck S, Bulyk ML, Farnham P, Hattori N, Henikoff S, Liu XS, Okumura K, Shiota K, Ushijima T, Greally JM (2006) Applying whole-genome studies of epigenetic regulation to study human disease. Cytogenet Genome Res 114(1):1–15. https://doi.org/10.1159/000091922

Locatelli C, Pasa TBC (2025) Role of micrornas in the pathophysiology and diagnosis of metabolic dysfunction-associated steatotic liver disease: a bibliometric review. Hepatol Forum 6(4):180–189. https://doi.org/10.14744/hf.2025.17142

Lv YH, Ma KJ, Zhang H, He M, Zhang P, Shen YW, Jiang N, Ma D, Chen L (2014) A time course study demonstrating mRNA, microRNA, 18S rRNA, and U6 snRNA changes to estimate PMI in deceased rat's spleen. J Forensic Sci 59(5):1286–1294. https://doi.org/10.1111/1556-4029.12447

Ma J, Pan H, Zeng Y, Lv Y, Zhang H, Xue A, Jiang J, Ma K, Chen L (2015) Exploration of the R code-based mathematical model for PMI estimation using profiling of RNA degradation in rat brain tissue at different temperatures. Forensic Sci Med Pathol 11(4):530–537. https://doi.org/10.1007/s12024-015-9703-7

Martinez PN, Menendez ST, Villaronga MLA, Ubelaker DH, Garcia-Pedrero JM, Zapico SC (2019) The big sleep: elucidating the sequence of events in the first hours of death to determine the postmortem interval. Sci Justice 59(4):418–424. https://doi.org/10.1016/j.scijus.2019.03.001

Martinez-Rivera V, Cardenas-Monroy CA, Millan-Catalan O, Gonzalez-Corona J, Huerta-Pacheco NS, Martinez-Gutierrez A, Villavicencio-Queijeiro A, Pedraza-Lara C, Hidalgo-Miranda A, Bravo-Gomez ME, Perez-Plasencia C, Guardado-Estrada M (2021) Dysregulation of miR-381-3p and miR-23b-3p in skeletal muscle could be a possible estimator of early postmortem interval in rats. PeerJ 9:e11102. https://doi.org/10.7717/peerj.11102

Mayes C, Houston R, Seashols-Williams S, LaRue B, Hughes-Stamm S (2019) The stability and persistence of blood and semen mRNA and miRNA targets for body fluid identification in environmentally challenged and laundered samples. Leg Med (Tokyo) 38:45–50. https://doi.org/10.1016/j.legalmed.2019.03.007

Medina-Paz F, Zapico SC (2025) Identification of cell death biomarkers in early and late postmortem intervals. In: 30th Congress of the International Society for Forensic Genetics, Santiago de Compostela, Spain

Mytidou C, Koutsoulidou A, Zachariou M, Prokopi M, Kapnisis K, Spyrou GM, Anayiotos A, Phylactou LA (2021) Age-related exosomal and endogenous expression patterns of miR-1,

miR-133a, miR-133b, and miR-206 in skeletal muscles. Front Physiol 12:708278. https://doi.org/10.3389/fphys.2021.708278
Na JY (2020) Estimation of the post-mortem interval using microRNA in the bones. J Forensic Leg Med 75:102049. https://doi.org/10.1016/j.jflm.2020.102049
Odriozola A, Riancho JA, de la Vega R, Agudo G, Garcia-Blanco A, de Cos E, Fernandez F, Sanudo C, Zarrabeitia MT (2013) miRNA analysis in vitreous humor to determine the time of death: a proof-of-concept pilot study. Int J Legal Med 127(3):573–578. https://doi.org/10.1007/s00414-012-0811-6
Rocchi A, Chiti E, Maiese A, Turillazzi E, Spinetti I (2020) MicroRNAs: an update of applications in forensic science. Diagnostics (Basel) 11(1):32. https://doi.org/10.3390/diagnostics11010032
Saber TM, Omran BHF, El Deib MM, El-Sharkawy NI, Metwally MMM, Abd-Elhakim YM (2021) Early postmortem biochemical, histological, and immunohistochemical alterations in skeletal muscles of rats exposed to boldenone undecylenate: forensic implication. J Forensic Leg Med 83:102248. https://doi.org/10.1016/j.jflm.2021.102248
Sharma S, Singh D, Kaul D (2015) AATF RNome has the potential to define post mortem interval. Forensic Sci Int 247:e21–e24. https://doi.org/10.1016/j.forsciint.2014.12.008
Singh P, Ali W, Sandhu S, Mishra S, Singh US, Verma AK, Singh M, Kaleem Ahmad M, Kumari S (2023) Post-mortem interval estimation using miRNAs of road traffic accident cases: a forensic molecular approach. Sci Justice 63(4):485–492. https://doi.org/10.1016/j.scijus.2023.04.011
Song B, Qian J, Fu J (2024) Research progress and potential application of microRNA and other non-coding RNAs in forensic medicine. Int J Legal Med 138(2):329–350. https://doi.org/10.1007/s00414-023-03091-1
Tu C, Du T, Shao C, Liu Z, Li L, Shen Y (2018) Evaluating the potential of housekeeping genes, rRNAs, snRNAs, microRNAs and circRNAs as reference genes for the estimation of PMI. Forensic Sci Med Pathol 14(2):194–201. https://doi.org/10.1007/s12024-018-9973-y
Tu C, Du T, Ye X, Shao C, Xie J, Shen Y (2019) Using miRNAs and circRNAs to estimate PMI in advanced stage. Leg Med (Tokyo) 38:51–57. https://doi.org/10.1016/j.legalmed.2019.04.002
Vass AA (2011) The elusive universal post-mortem interval formula. Forensic Sci Int 204(1–3):34–40. https://doi.org/10.1016/j.forsciint.2010.04.052
Vass AA, Barshick SA, Sega G, Caton J, Skeen JT, Love JC, Synstelien JA (2002) Decomposition chemistry of human remains: a new methodology for determining the postmortem interval. J Forensic Sci 47(3):542–553. https://www.ncbi.nlm.nih.gov/pubmed/12051334
Wang H, Mao J, Li Y, Luo H, Wu J, Liao M, Liang W, Zhang L (2013) MiRNA expression analyze in post-mortem interval (PMI) within 48h. Forensic Sci Int Suppl Ser 4:e190–e191
Wang F, Jiang C, Sun Q, Yan F, Wang L, Fu Z, Liu T, Hu F (2015) miR-195 is a key regulator of Raf1 in thyroid cancer. Onco Targets Ther 8:3021–3028. https://doi.org/10.2147/OTT.S90710
Wescott DJ (2018) Recent advances in forensic anthropology: decomposition research. Forensic Sci Res 3(4):327–342. https://doi.org/10.1080/20961790.2018.1488571
Zapico SC, Menendez ST, Nunez P (2014) Cell death proteins as markers of early postmortem interval. Cell Mol Life Sci 71(15):2957–2962. https://doi.org/10.1007/s00018-013-1531-x

Chapter 10
Future Perspectives on Forensic Epigenetics for DNA Investigative Intelligence

Manfred Kayser

Abstract Forensic interest in epigenetics goes beyond chronological age estimation, tissue type determination, monozygotic twin identification, and postmortem interval estimation discussed in the previous chapters. Here, I provide a perspective on a different forensic area where epigenetics holds promises, namely for DNA investigative intelligence in the context of forensic DNA phenotyping beyond age estimation. Epigenetics may allow predicting a trace donor's appearance, such as regarding age-dependent traits, thereby extending genetic appearance prediction. Epigenetics may permit inferring the geographic region where a trace donor has lived, complementing genetic prediction of biogeographic ancestry. Epigenetics has already started to allow inferring a trace donor's visible habits involving environmental interactions, such as substance intake, adding externally visible information that cannot be genetically predicted. Provided sufficient scientific and technological progress, and when applied in a societally responsible way, epigenetic prediction of externally visible characteristics is expected to widen the spectrum of forensic DNA phenotyping, which together with advancements in genetic prediction, will improve DNA-guided police investigations to help find perpetrators in criminal cases that otherwise cannot be solved.

Keywords Forensic epigenetics · DNA methylation · DNA investigative intelligence · Forensic DNA phenotyping · Appearance · Behavioral traits · Residence

---

M. Kayser (✉)
Department of Pathology and Clinical Bioinformatics, Erasmus MC University Medical Center Rotterdam, Rotterdam, The Netherlands
e-mail: m.kayser@erasmusmc.nl

S. C. Zapico (ed.), *Forensic Science and Epigenetics*, Epigenetics and Human Health 18, https://doi.org/10.1007/978-3-032-23085-0_10

Abbreviations

AUC	Area Under the Curve from a Receiver Operating Characteristic (ROC) curve
CpG	Cytosine nucleotide followed by a Guanine along its $5' \rightarrow 3'$ direction
DNA	Deoxyribonucleic Acid
EWAS	Epigenome-Wide Association Study
FDP	Forensic DNA Phenotyping
GWAS	Genome-Wide Association Study
PRS	Polygenetic Risk Score
SNP	Single-Nucleotide Polymorphism

10.1 Introduction

Currently, forensic epigenetics concerns the use of DNA methylation markers and analysis methods for four main applications, as discussed in the previous chapters. First, the estimation of chronological age of a donor of a crime scene trace to help find unknown perpetrators in cases without suspects (Naue 2023). Second, the inference of the cell or tissue type of a crime scene trace to obtain information on the activity of the DNA-identified trace donor that placed the trace at the crime scene (Vidaki and Kayser 2018). Third, the individual identification of a trace donor who is a monozygotic twin (Kayser 2025). Fourth, the estimation of the time between a person's death and the recovery of the dead body referred to as postmortem interval (Cianci et al. 2024). In this chapter, I describe another area where epigenetics is expected, and in part has already started, to provide forensically relevant information, namely for DNA investigative intelligence in the context of forensic DNA phenotyping (FDP) beyond chronological age prediction.

FDP delivers information about externally visible characteristics of a trace donor (Kayser et al. 2023), while invisible information, such as genetic risks for disease, is not used for this purpose (Zieger 2022). FDP is applied in serious crime cases that lack suspects at the start of the investigation. Or, when all suspects have been excluded from being the trace donor by showing mismatches between their STR-profile obtained from their reference DNA and the STR-profile generated from the trace. Or, when seeing mismatches between the STR profile from the trace and all STR-profiles of convicted criminal offenders stored in the forensic DNA database. FDP outcomes represent externally visible information an eyewitness can see, while in many cases, eyewitnesses are not available, but DNA is and can serve as witness. In cases where human eyewitnesses are available, FDP can help confirm or correct eyewitness statements, which can be error-prone (Albright 2017) due to various reasons such as intrinsic biases eyewitnesses may have, or other reasons why they could not see or remember how a person at or near the crime scene looks like, which does not apply to DNA.

Since forensic DNA profiling is completely comparative, requiring the availability of the STR-profile from the to-be-identified person in addition to the STR-profile of the trace, cases without suspects cannot be solved with standard forensic DNA analysis (Kayser 2025). In the absence of other evidential information about the putative perpetrator, cases without suspects typically become cold cases, sometimes for several decades (Kayser 2017). FDP outcomes can focus police investigation to help find potential suspects, while subsequent individual identification is done via forensic STR-profile matching (Kayser 2025). To further enhance DNA investigative intelligence, FDP can be combined with investigative genetic genealogy and other forms of familial search, such as Y-chromosome analysis to identify male relatives of the unknown male trace donor, thereby increasing the information that guides police investigation to find potential suspects (Kayser 2025).

Currently, genetic information is used for practical applications of FDP, with the exception of chronological age estimated with the help of DNA methylation (Kayser et al. 2023). In the future, epigenetics may allow more externally visible information about an unknown trace donor to become inferable from trace DNA (Vidaki and Kayser 2017). It is envisioned that epigenetic prediction of externally visible characteristics will widen the spectrum of forensic DNA phenotyping, which together with advancements in genetic prediction, will improve DNA-guided police investigations to help find perpetrators in criminal cases that otherwise cannot be solved.

10.2 Epigenetic Inference of Appearance Traits

10.2.1 Age-Depended Facial Trait

While chronological age can be estimated from DNA methylation (Naue 2023), it is well known that some people look older, others younger, for their chronological age. From an investigative perspective, it would be desirable to combine the prediction of how old a trace donor is, with an estimation of how old he/she looks in the face, and use both types of age information for investigative intelligence. Looking younger or older for one's age can run in families, indicating that genetics is involved in how old one looks. Heritability of perceived facial age, i.e., how old one looks in the face, was quantified in twin-studies at 50–60% (Gunn et al. 2009). An early genome-wide association study (GWAS) on perceived facial age determined by multiple observers in thousands of Europeans found associations of DNA variants located in one gene (Liu et al. 2016). Carriers with the risk alleles of a compound marker in the *MC1R* gene looked 2 years older than non-carriers (Liu et al. 2016). While *MC1R* is well known as the only one gene responsible for red hair (Zorina-Lichtenwalter et al. 2019), and one of many genes contributing to light skin (Visconti et al. 2018), the perceived age effect was shown to be independent of pigmentation (Liu et al. 2016), demonstrating the gene's pleiotropic function. A recent enlarged GWAS in hundreds of thousands of Europeans on self-reported perceived age, which is more error-prone than observer-based perceived age, found over 80 independently associated genetic loci, which together explained 14% of perceived age variance (Roberts et al. 2020). The large sample size and the relatively low percentage of genetically explained variance demonstrate the large genetic complexity of perceived facial age.

Since the face changes with age in various aspects based on environmental impact, it is plausible to assume that epigenetics is involved in visible facial age. A proof-of-principle study integrated GWAS data on perceived age with epigenome-wide methylation quantification loci from blood and performed an epigenome-wide Mendelian randomization study on perceived age (Vladimir et al. 2023), reporting hundreds of CpGs potentially causal of perceived facial aging that had little overlap with known epigenetic-based clocks. Recently, the second-generation epigenetic clock VisAgeX was developed to predict visual facial age progression based on methylation microarray data from skin-derived DNA (Bienkowska et al. 2023). For visible facial age, VisAgeX achieved a mean absolute prediction error from cross-validation of 8.20 ± 0.83 years, and in an independent model validation dataset, a significant correlation of $R = 0.84$ between the predicted visual facial age and the assessed age was observed and the mean absolute prediction error was 6.54 years (Bienkowska et al. 2023). However, generating DNA methylation microarrays data from DNA of skin cells obtained from touched crime scene objects is challenging. New technologies for targeting large numbers of CpGs with the sensitivity and accuracy needed for forensic DNA analysis will be required. Moreover, developing epigenetic methods for predicting facial age from other forensically relevant tissue types than skin cells, such as blood or semen, would be desirable too to widen the casework applications to other types of trace materials.

Perceived facial age combines different age-depending facial traits such as wrinkles, pigmented spots, lip heights, and sagging. These traits can also be used separately as indicator of facial age. GWASs on wrinkles, pigmented spots, and sagging eyelids have revealed a few genetic loci, but the phenotypic variance explained by the identified genetic loci remains small (Hamer et al. 2018; Chen et al. 2021; Okuno et al. 2022). Due to the age dependency, epigenetic effects are expected for these traits and may have larger effects than genetic factors. However, epigenome-wide association studies (EWAS) on these traits are pending as of yet. The VisAgeX model for epigenetic prediction of visual age progression applied to the degree of wrinkling revealed a mean absolute prediction error of 12.66 ± 0.94 wrinkle grade, and, in an independent model validation dataset, yielded a significant high correlation between predicted and observed wrinkle grade of $R = 0.86$ and a mean absolute prediction error of 8.81 on the wrinkle grade scale (Bienkowska et al. 2023).

Future progress in epigenetic investigation of visible facial age will show whether accurate and reliable epigenetic prediction of how old a trace donor looks like is achievable. Eventually, it is desired to combined genetic and epigenetic prediction of facial age with epigenetic prediction of chronological age for DNA investigative intelligence in the context of FDP.

10.2.2 Age-Dependent Hair Traits

There are several appearance traits other than facial age that can change with age. One example is head hair, which in some people darkens during puberty from blond in childhood to dark blond or brown, rarely darker, in adulthood. Currently available DNA test systems for hair color prediction, such as HIrisPlex-S (Chaitanya et al. 2018) or Snipper (Sochtig et al. 2015), do not cover this age-dependent hair color darkening. Some of the individuals who experienced age-dependent hair color darkening away from blond, do predict blond as they were as child, and others predict brown as they are as adults. While genetic control of hair color darkening may be expected, but no GWAS is available, the strong age dependency of this trait makes it expected that epigenetics is involved in hair color darkening. However, no EWAS on age-dependent hair color darkening has been published yet. Future E/GWAS may reveal DNA markers suitable for predicting age-dependent hair color darkening. Once they become available, they should be integrated into the available DNA test for hair color prediction to increase the prediction accuracy for those people who did undergo hair color darkening.

Another hair color trait with strong age dependency is hair greying or whitening due to partial or complete loss of pigmentation, which typically happens at advanced age, but under certain circumstances of severe stress can also occur at younger age. As with age-dependent hair color darkening, also hair color loss is not covered with the currently available hair color DNA test systems. Hair greying is under strong genetic control and 90% heritability of grey hair was estimated in twin studies (Gunn et al. 2009). However, a GWAS of various hair traits in thousands of Latin Americans found DNA variants significantly associated with grey hair in only one gene, the well-known pigmentation gene *IRF4*, which is involved in variation in eye, hair, and skin coloration (Adhikari et al. 2016). Gene expression studies on hair greying identified several genes such as the well-known pigmentation genes *TYR*, *TRP1,* and a few others (Choi et al. 2008; Paus et al. 2024). Many more genes are expected to be identified in future enlarged GWASs or expression studies on grey or white hair. A first genetic prediction model for grey hair based on 13 SNP together with age and sex achieved a grey vs non-grey prediction with a cross-validated AUC of 0.87 and AUCs 0.79–0.86 for a 3-category model of hair greying (Pospiech et al. 2020); however, most of the prediction information was provided by age alone and SNPs only explained <10% of hair greying variation. As with hair darkening, the strong age-dependency suggests epigenetics being involved in hair greying too. While no EWAS on hair greying is available yet, it is expected that future research into the epigenetics of hair greying will reveal associated DNA methylation sites, which may allow improving the prediction of grey hair from trace DNA.

A third hair trait with strong age-dependency is hair loss. Several GWASs have been published on hair loss, particularly in men referred to as male pattern baldness, which revealed hundreds of genomic loci (Liu et al. 2025). Several genetic prediction models for MPB were published of which the one based on the largest underlying datasets using a total of over 187,000 samples and 117 predictive SNPs yielded

AUCs ranging from 0.725 to 0.728 for severe, 0.631–0.635 for moderate, 0.598–0.602 for slight, and 0.708–0.711 for no hair loss and external model validation in an early-onset enriched MPB dataset gave AUC of 0.830 for no vs. any hair loss (Chen et al. 2023), leaving room for further improvements by adding more predictive SNPs. The age-dependency of normal onset and early-onset hair loss expects epigenetics to be involved as well, while no EWAS on hair loss has been published yet.

Once epigenetic prediction of age-dependent appearance traits becomes available with progressing science and technology, it will need to be combined with epigenetically predicted chronological age.

10.2.3 Other Appearance Traits

All human appearance traits, particularly those that are not age-dependent, have high heritability's, providing low expectation for large genetically independent epigenetic effects. However, this expectation increases, the lower the heritability is. Body height is an appearance trait with heritability of around 80%. Several GWASs have been carried out on body height, particularly by the GIANT Consortium, of which the most recent and largest in 5.4 million samples from different ancestries identified over 12,000 independent significantly associated SNPs that genetically explained up to 40% height variation in Europeans but 10–20% in other ancestries (Yengo et al. 2022). A genetic prediction model for extremely tall stature in over 10,000 Europeans of which 800 were extremely tall revealed an AUC of 0.79 based on almost 700 SNPs (Yengo et al. 2022).

Regarding epigenetic effects on body height, a recent DNA methylation microarray analysis in over 350 Europeans identified over 11,000 differentially methylated sites that were significantly associated with body height (Wang et al. 2024). Of those, over 850 CpGs achieved in nearly 200 Chinese a cross-validated mean average error of 4.6 cm and a cross-validated correlation of $R^2 = 0.59$ in the case of unknown sex, while in sex-stratified analyses, these values were 4.1 cm and 0.41, respectively, for males and 3.5 cm and 0.38, respectively, for females (Wang et al. 2024). In a study of over 600 Europeans based on DNA methylation microarray analysis, from which almost 1000 CpGs were preselected for height prediction, correlations of R from 0.57 to 0.75 were obtained in the test set, corresponding to height predictions within 5.7 and 6.7 cm (Llobet et al. 2023). However, a larger study in approximately 1400 Europeans found almost no effect of DNA methylation scores on body height with an epigenetically explained height variance of below 1% (Shah et al. 2015).

Another externally visible trait that is shaped by genetic as well as environmental factors and thus is expected to be influenced by epigenetics, is body type or body density as determined by the relationship of body weight to body height, which can be expressed by the Body Mass Index (BMI). Heritability of BMI had been quantified in twin studies to range from 0.47 to 0.90 (Elks et al. 2012). Several GWASs on

BMI have been carried out most notably by the GIANT Consortium, of which the currently largest in 700,000 samples highlighted over 900 SNPs significantly associated with BMI (Yengo et al. 2018). A recent study performed polygenic risk score (PRS) analysis on BMI in over 5 million samples of different continental ancestries based on up to 1.3 million SNPs, which achieved explained variance proportions ranging between 7.2% in Africans and 17.5% in Europeans with a median of 14.0% (Smit et al. 2025).

Regarding epigenetics of BMI, several EWASs on BMI were published too. A recent EWAS meta-analysis of six trans-ethnic cohorts in over 17,000 samples found over 1200 significantly BMI-associated CpGs and an elastic net regression analysis to predict log-normalized BMI identified almost 400 CpGs that explained 32% of BMI variance (Do et al. 2023). Epigenetic predictors for BMI were shown to be largely independent from genetic ones and to have an additive effect on BMI with the genetic ones (Shah et al. 2015) and methylation scores were shown to outperform polygenic risk scores for BMI prediction (Odintsova et al. 2021). Together, these studies show that epigenetic effects on BMI are larger than genetic ones, as maybe expected given the strong environmental impact on BMI.

Based on almost 1000 pre-selected CpGs, a prediction model for BMI revealed correlations of *R* from 0.08 to 0.43 in the model testing dataset, corresponding to a prediction within 3.8–6.8 BMI of the actual BMI values (Llobet et al. 2023). In the same study, waste-to-hip ratio was predicted with correlations of 0.35–0.55 in the test set and a classification of individuals into High and Low gave an AUC of 0.79 with only two DNA methylation markers (Llobet et al. 2023). In a forensically motivated study focusing on minimizing the number of predictive markers, epigenetic prediction of BMI in 700 Koreans with four carefully selected CpGs explained 11% of BMI variance, which was increased to 17% when combining epigenetic with genetic prediction by adding a genetic score from eight SNPs as well as age (Cho et al. 2021). These (epi)genetic markers together with sex and age yielded AUCs between 0.65 and 0.68 based on different BMI cutoffs in a test dataset of over 200 samples (Cho et al. 2021).

Future intensified research on the epigenetics of human appearance traits, with EWASs and epigenetic prediction studies of increased samples size, will show how accurate DNA methylation-based appearance prediction will get and to what degree it can help, on its own and when combined with genetic prediction, to improve forensic DNA phenotyping of appearance.

10.3 Epigenetic Inference of Visible Habits

10.3.1 Tobacco Smoking

The human methylome is largely shaped by environmental impact. Thus, for habits that involve environmental interactions such as via substance intake, there is a large expectation that they influence DNA methylation levels. As far as current knowledge is concerned, of all human habits that involve environmental interactions, tobacco smoking has the largest effect on the human methylome. Several EWASs on smoking have been published and revealed thousands of CpGs with significant smoking association. To give examples of large studies, an earlier EWAS meta-analysis in nearly 16,000 samples from 16 cohorts using DNA methylation microarrays covering 450,000 CpGs identified over 2600 significantly smoking-associated CpGs from over 1400 genes (Joehanes et al. 2016). A recent EWAS using EPIC microarrays targeting 850,000 CpGs, covering 5% of the methylome, in over 15,000 samples from five cohorts reported nearly 13,000 significantly smoking-associated CpGs linked to over 4600 genes (Hoang et al. 2024). In another large recent study in nearly 18,000 samples, DNA methylation markers explained 50.0% of the variance in the pack years of smoking (Chybowska et al. 2025).

Epigenetic prediction analysis revealed AUCs of 0.98 for current vs. never smokers, 0.90 for current vs. former smokers, and 0.85 for former vs. never smokers. Earlier studies also obtained almost perfect discrimination between current and never smokers but modest for former vs. never smokers (Bollepalli et al. 2019). In forensic trace DNA analysis, however, analyzing DNA methylation microarrays is challenging due to the low amount of degraded DNA many trace samples contain. Moreover, most current DNA methylation analysis methods rely on bisulfide conversion, which requires more DNA quantity and reduces DNA quality. To cope with the reduced multiplex capacity of DNA methylation analysis methods suitable for forensic trace analysis, models for DNA methylation-based smoking prediction envisioned for forensic applications were developed by minimizing the number of predictive CpGs. For instance, a study that employed 14 EWASs for marker discovery, and used data from over 4500 samples of 8 cohorts for model building and validation, identified 13 CpGs that yielded AUC of over 0.90 for current smokers, while for never smokers and especially for former smokers, the AUCs were lower with 0.83 and 0.77, respectively (Maas et al. 2019). Due to limited material and financial resources, most models developed for forensic applications suffer from low sample size, such as a few hundred individuals, which makes it uncertain to know how reliable the reported prediction accuracies are. For instance, a smoking prediction model based on 4 CpGs achieved AUCs of 0.8 for smokers versus never-smokers (Pospiech et al. 2024) and other models based on 1–3 CpGs gave AUCs of 0.87 (Ambroa-Conde et al. 2024). While these smoking prediction models were developed based on blood DNA, a recently published model for DNA derived from buccal swabs achieved an AUCs of 0.8 for current versus never smokers (Pospiech et al. 2025). Smoking prediction models with higher AUCs were reported too (Llobet et al. 2023; Alghanim et al. 2018) but were based on even lower sample size, questioning the reliability of the reported accuracies without additional model validations becoming available. The first DNA test systems for smoking prediction were developed using technologies suitable for forensic trace analysis such as pyrosequencing (Alghanim et al. 2018) and more recently targeted MPS (Vidaki et al. 2023).

10.3.2 Alcohol Drinking

Many studies demonstrated the impact of alcohol consumption on the human methylome, and the methylation levels of many genes were found to be influenced by alcohol consumption including alcoholism (Mahna et al. 2018). Regarding epigenetic prediction of alcohol consumption, extremely high AUC values of 0.90–0.99 were reported initially based on models with 5–144 CpGs (Liu et al. 2018). However, when these models were tested in almost 3000 samples independent of those used for model training, the same marker sets achieved much lower AUCs of 0.61–0.83 (Maas et al. 2021), indicating that the previously reported accuracies were likely overestimates due to sample dependencies in the prediction modelling. Newly developed models based on these 5–144 CpG in over 4500 samples, trained and validated in independent datasets, achieved similarly low AUCs of 0.6–0.7 (Maas et al. 2021). Alcohol prediction models based on only 1–3 CpGs yielded AUCs of 0.77–0.80 (Ambroa-Conde et al. 2024), but were based on only a few hundred samples.

More research on the epigenetic impact of substance intake such as tobacco smoking and alcohol drinking, but also other substances (Mahna et al. 2018; Grassi et al. 2025) will show how accurate the DNA methylation-based prediction of habits based involving substance intake will become, and thus how useful it can be in crime scene investigation for FDP purposes.

10.4 Epigenetic Inference of Geographic Residence

Next to appearance prediction and age estimation, a third element of currently practiced FDP is the genetic prediction of biogeographic ancestry. This is typically done via SNPs from the Y-chromosome, mitochondria, and the autosomes, that have restrictive allele frequency distributions and thus allow inferring the geographic region in the world from which the paternal, maternal, and bi-parental biological ancestors of person, such as a trace donor, come from. For a long time, forensic ancestry tests based on DNA technologies suitable for trace analysis suffered from low number of simultaneously analyzable DNA markers in the order of tens of SNPs, which led to ancestry detection on low geographic resolution. The application of targeted MPS allows to increase the number of SNPs analyzed simultaneously to hundreds, maybe thousands of markers. However, compared to high-throughput technologies such as SNP microarrays based on hundreds of thousands of SNPs or whole genome sequencing, the ancestry resolution available with forensic ancestry tests based on targeted MPS is restricted to 5–7 large geographic regions (Kayser et al. 2023). While the uni-parental inheritance of the Y-chromosome and mtDNA makes Y- and mtDNA SNPs very powerful in detecting paternal and maternal biogeographic ancestry, respectively, the lack of recombination does not allow to differentiate between recent and ancient ancestry in persons of mixed ancestry. With autosomal SNPs, bi-parental ancestry components can be quantified and thus recent genetic admixture, which often is externally visible, can be differentiated from ancient admixture, which depending on the admixture scenario may not be visible.

Genetic testing can provide information on the geographic region(s) where the biological ancestors of a trace donor originate from, which is valuable from an investigative perspective. However, being able to infer the region where the trace donor itself has lived, i.e., geographic residence or geolocation, would provide additional geographic information with investigative value to complement genetic ancestry testing. Since environmental factors impact on epigenetics, it can be expected that epigenetics may allow to infer geographic regions of residence from DNA methylation patterns. Indeed, population specificity of human DNA methylation variation has been reported (Fraser et al. 2012; Carja et al. 2017). However, a strong correspondence between population-specific variation detected with DNA methylation and with SNPs was noted too (Carja et al. 2017; Rahmani et al. 2017), implying that geographic information inferred with DNA methylation in fact comes from genetics. That epigenetics holds more geographic information than genetic ancestry is indicated by DNA methylation differences between populations that have a common genetic background but live in different habitats (Fagny et al. 2015), implying the influence of environmental factors on the methylome. A comparative population study using DNA methylation and SNP microarrays found hundreds of CpGs significantly associated with self-declared ethnicity and with genetic ancestry, while shared genomic ancestry accounted for 75% of DNA methylation variance associated with ethnicity (Galanter et al. 2017). Moreover, a significant enrichment of ethnicity-associated CpGs was seen among loci previously associated with environmental exposure (Galanter et al. 2017).

More DNA methylation studies on different populations need to be done that in parallel also study genetic variation to work out population effects provided by epigenetics that cannot be revealed from genetics. The future may see more combined use of genetic testing for biogeographic ancestry and epigenetic testing for geolocation to add geographic information regarding the geographic region where a person lived to the region where its ancestor came from.

10.5 Conclusions

Epigenetics holds great promise to improve DNA investigative intelligence by widening and enhancing forensic DNA phenotyping. Scientific progress is required to catalogue, at best understand, epigenetic links with externally visible characteristics that can describe an unknown trace donor such as age-dependent appearance traits, habits based on substance intake, and geographic residence. Moreover, additional factors such as diet and physical activity leave their foodprints on the human methylome too (Mirisola 2023; Lorenzo et al. 2022; Ostaiza-Cardenas et al. 2025). Intensified future research in lifestyle epigenetics may allow the inference of diet and lifestyle factors with DNA methylation markers from trace material. With scientific progress in identifying more and more DNA methylation markers with investigative value, technical developments are needed to simultaneously analyze large numbers of CpGs with the accuracy, reliability, sensitivity, and robustness required for forensic trace analysis. If applied in a societally responsible way, it is expected that combining genetic with epigenetic prediction of externally visible traits from crime scene DNA will improve DNA investigative intelligence to better help solve serious criminal cases that lack suspects and otherwise cannot be solved.

Compliance with Ethical Standards Writing of this paragraph did not consume any funding other than the author's employment by Erasmus University Medical Center Rotterdam that is acknowledged. The author declares that he has no conflict of interest regarding the presented information. This chapter is a review of previously published accounts and an outlook into the future, as such, no animal or human studied were performed.

References

Adhikari K, Fontanil T, Cal S, Mendoza-Revilla J, Fuentes-Guajardo M, Chacon-Duque JC et al (2016) A genome-wide association scan in admixed Latin Americans identifies loci influencing facial and scalp hair features. Nat Commun 7:10815

Albright TD (2017) Why eyewitnesses fail. Proc Natl Acad Sci USA 114(30):7758–7764

Alghanim H, Wu W, McCord B (2018) DNA methylation assay based on pyrosequencing for determination of smoking status. Electrophoresis 39(21):2806–2814

Ambroa-Conde A, Casares de Cal MA, Gomez-Tato A, Robinson O, Mosquera-Miguel A, de la Puente M et al (2024) Inference of tobacco and alcohol consumption habits from DNA methylation analysis of blood. Forensic Sci Int Genet 70:103022

Bienkowska A, Raddatz G, Sohle J, Kristof B, Volzke H, Gallinat S et al (2023) Development of an epigenetic clock to predict visual age progression of human skin. Front Aging 4:1258183

Bollepalli S, Korhonen T, Kaprio J, Anders S, Ollikainen M (2019) EpiSmokEr: a robust classifier to determine smoking status from DNA methylation data. Epigenomics 11(13):1469–1486

Carja O, MacIsaac JL, Mah SM, Henn BM, Kobor MS, Feldman MW, Fraser HB (2017) Worldwide patterns of human epigenetic variation. Nat Ecol Evol 1(10):1577–1583

Chaitanya L, Breslin K, Zuniga S, Wirken L, Pospiech E, Kukla-Bartoszek M et al (2018) The HIrisPlex-S system for eye, hair and skin colour prediction from DNA: introduction and forensic developmental validation. Forensic Sci Int Genet 35:123–135

Chen Y, Andre M, Adhikari K, Blin M, Bonfante B, Mendoza-Revilla J et al (2021) A genome-wide association study identifies novel gene associations with facial skin wrinkling and mole count in Latin Americans. Br J Dermatol 185(5):988–998

Chen Y, Hysi P, Maj C, Heilmann-Heimbach S, Spector TD, Liu F, Kayser M (2023) Genetic prediction of male pattern baldness based on large independent datasets. Eur J Hum Genet 31(3):321–328

Cho S, Lee EH, Kim H, Lee JM, So MH, Ahn JJ, Lee HY (2021) Validation of BMI genetic risk score and DNA methylation in a Korean population. Int J Legal Med 135(4):1201–1212

Choi YJ, Yoon TJ, Lee YH (2008) Changing expression of the genes related to human hair graying. Eur J Dermatol 18(4):397–399

Chybowska AD, Bernabeu E, Yousefi P, Suderman M, Hillary RF, Clark R et al (2025) A blood- and brain-based EWAS of smoking. Nat Commun 16(1):3210

Cianci V, Mondello C, Sapienza D, Guerrera MC, Cianci A, Craco A et al (2024) microRNAs as new biomolecular markers to estimate time since death: a systematic review. Int J Mol Sci 25(17):9207

Do WL, Sun D, Meeks K, Dugue PA, Demerath E, Guan W et al (2023) Epigenome-wide meta-analysis of BMI in nine cohorts: examining the utility of epigenetically predicted BMI. Am J Hum Genet 110(2):273–283

Elks CE, den Hoed M, Zhao JH, Sharp SJ, Wareham NJ, Loos RJ, Ong KK (2012) Variability in the heritability of body mass index: a systematic review and meta-regression. Front Endocrinol (Lausanne) 3:29

Fagny M, Patin E, MacIsaac JL, Rotival M, Flutre T, Jones MJ et al (2015) The epigenomic landscape of African rainforest hunter-gatherers and farmers. Nat Commun 6:10047

Fraser HB, Lam LL, Neumann SM, Kobor MS (2012) Population-specificity of human DNA methylation. Genome Biol 13(2):R8

Galanter JM, Gignoux CR, Oh SS, Torgerson D, Pino-Yanes M, Thakur N et al (2017) Differential methylation between ethnic sub-groups reflects the effect of genetic ancestry and environmental exposures. elife 6. https://doi.org/10.7554/eLife.20532.001

Grassi S, Costantino A, Dimitrova A, Croce EB, Iasi F, Puggioni A et al (2025) Drug-induced epigenetic alterations: a set of forensic toxicological fingerprints?—a systematic review. Genes (Basel) 16(10):1129

Gunn DA, Rexbye H, Griffiths CE, Murray PG, Fereday A, Catt SD et al (2009) Why some women look young for their age. PLoS One 4(12):e8021

Hamer MA, Pardo LM, Jacobs LC, Deelen J, Uitterlinden AG, Slagboom E et al (2018) Facial wrinkles in Europeans: a genome-wide association study. J Invest Dermatol 138(8):1877–1880

Hoang TT, Lee Y, McCartney DL, Kersten ETG, Page CM, Hulls PM et al (2024) Comprehensive evaluation of smoking exposures and their interactions on DNA methylation. EBioMedicine 100:104956

Joehanes R, Just AC, Marioni RE, Pilling LC, Reynolds LM, Mandaviya PR et al (2016) Epigenetic signatures of cigarette smoking. Circ Cardiovasc Genet 9(5):436–447

Kayser M (2017) Forensic use of Y-chromosome DNA: a general overview. Hum Genet 136(5):621–635

Kayser M (2025) Forensic genetics in the omics era. Nat Rev Genet 27(2):170–186

Kayser M, Branicki W, Parson W, Phillips C (2023) Recent advances in forensic DNA phenotyping of appearance, ancestry and age. Forensic Sci Int Genet 65:102870

Liu F, Hamer MA, Deelen J, Lall JS, Jacobs L, van Heemst D et al (2016) The MC1R gene and youthful looks. Curr Biol 26(9):1213–1220

Liu C, Marioni RE, Hedman AK, Pfeiffer L, Tsai PC, Reynolds LM et al (2018) A DNA methylation biomarker of alcohol consumption. Mol Psychiatry 23(2):422–433

Liu Y, Tosti A, Wang ECE, Heilmann-Heimbach S, Aguh C, Jimenez F et al (2025) Androgenetic alopecia. Nat Rev Dis Primers 11(1):73

Llobet MO, Johansson A, Gyllensten U, Allen M, Enroth S (2023) Forensic prediction of sex, age, height, body mass index, hip-to-waist ratio, smoking status and lipid lowering drugs using epigenetic markers and plasma proteins. Forensic Sci Int Genet 65:102871

Lorenzo PM, Izquierdo AG, Rodriguez-Carnero G, Fernandez-Pombo A, Iglesias A, Carreira MC et al (2022) Epigenetic effects of healthy foods and lifestyle habits from the Southern European Atlantic diet pattern: a narrative review. Adv Nutr 13(5):1725–1747

Maas SCE, Vidaki A, Wilson R, Teumer A, Liu F, van Meurs JBJ et al (2019) Validated inference of smoking habits from blood with a finite DNA methylation marker set. Eur J Epidemiol 34(11):1055–1074

Maas SCE, Vidaki A, Teumer A, Costeira R, Wilson R, van Dongen J et al (2021) Validating biomarkers and models for epigenetic inference of alcohol consumption from blood. Clin Epigenetics 13(1):198

Mahna D, Puri S, Sharma S (2018) DNA methylation signatures: biomarkers of drug and alcohol abuse. Mutat Res Rev Mutat Res 777:19–28

Mirisola MG (2023) The nutriepigenome. Genes (Basel) 14(11):1997

Naue J (2023) Getting the chronological age out of DNA: using insights of age-dependent DNA methylation for forensic DNA applications. Genes Genomics 45(10):1239–1261

Odintsova VV, Rebattu V, Hagenbeek FA, Pool R, Beck JJ, Ehli EA et al (2021) Predicting complex traits and exposures from polygenic scores and blood and buccal DNA methylation profiles. Front Psychol 12:688464

Okuno R, Inoue Y, Hasebe Y, Igarashi T, Kawagishi-Hotta M, Yamada T, Hasegawa S (2022) Genome-wide association studies in Japanese women identified genetic loci associated with wrinkles and sagging. Exp Dermatol 31(9):1411–1420

Ostaiza-Cardenas J, Tobar AC, Costa SC, Calero DS, Lopez-Carrera A, Bermudez FG, Orellana-Manzano A (2025) Epigenetic modulation by life-style: advances in diet, exercise, and mindfulness for disease prevention and health optimization. Front Nutr 12:1632999

Paus R, Sevilla A, Grichnik JM (2024) Human hair graying revisited: principles, misconceptions, and key research frontiers. J Invest Dermatol 144(3):474–491

Pospiech E, Kukla-Bartoszek M, Karlowska-Pik J, Zielinski P, Wozniak A, Boron M et al (2020) Exploring the possibility of predicting human head hair greying from DNA using whole-exome and targeted NGS data. BMC Genomics 21(1):538

Pospiech E, Rudnicka J, Noroozi R, Pisarek-Pacek A, Wysocka B, Masny A et al (2024) DNA methylation at AHRR as a master predictor of smoke exposure and a biomarker for sleep and exercise. Clin Epigenetics 16(1):147

Pospiech E, Pisarek-Pacek A, Herda K, Wysocka B, Sitek A, Spolnicka M et al (2025) Epigenetic predictor of smoking status in buccal cells. Toxicol Appl Pharmacol 502:117415

Rahmani E, Shenhav L, Schweiger R, Yousefi P, Huen K, Eskenazi B et al (2017) Genome-wide methylation data mirror ancestry information. Epigenetics Chromatin 10:1

Roberts V, Main B, Timpson NJ, Haworth S (2020) Genome-wide association study identifies genetic associations with perceived age. J Invest Dermatol 140(12):2380–2385

Shah S, Bonder MJ, Marioni RE, Zhu Z, McRae AF, Zhernakova A et al (2015) Improving phenotypic prediction by combining genetic and epigenetic associations. Am J Hum Genet 97(1):75–85

Smit RAJ, Wade KH, Hui Q, Arias JD, Yin X, Christiansen MR et al (2025) Polygenic prediction of body mass index and obesity through the life course and across ancestries. Nat Med 31(9):3151–3168

Sochtig J, Phillips C, Maronas O, Gomez-Tato A, Cruz R, Alvarez-Dios J et al (2015) Exploration of SNP variants affecting hair colour prediction in Europeans. Int J Legal Med 129(5):963–975

Vidaki A, Kayser M (2017) From forensic epigenetics to forensic epigenomics: broadening DNA investigative intelligence. Genome Biol 18(1):238

Vidaki A, Kayser M (2018) Recent progress, methods and perspectives in forensic epigenetics. Forensic Sci Int Genet 37:180–195

Vidaki A, Planterose Jimenez B, Poggiali B, Kalamara V, van der Gaag KJ, Maas SCE et al (2023) Targeted DNA methylation analysis and prediction of smoking habits in blood based on massively parallel sequencing. Forensic Sci Int Genet 65:102878

Visconti A, Duffy DL, Liu F, Zhu G, Wu W, Chen Y et al (2018) Genome-wide association study in 176,678 Europeans reveals genetic loci for tanning response to sun exposure. Nat Commun 9(1):1684

Vladimir K, Perisic MM, Storga M, Mostashari A, Khanin R (2023) Epigenetics insights from perceived facial aging. Clin Epigenetics 15(1):176

Wang Z, Fu G, Ma G, Wang C, Wang Q, Lu C et al (2024) The association between DNA methylation and human height and a prospective model of DNA methylation-based height prediction. Hum Genet 143(3):401–421

Yengo L, Sidorenko J, Kemper KE, Zheng Z, Wood AR, Weedon MN et al (2018) Meta-analysis of genome-wide association studies for height and body mass index in approximately 700000 individuals of European ancestry. Hum Mol Genet 27(20):3641–3649

Yengo L, Vedantam S, Marouli E, Sidorenko J, Bartell E, Sakaue S et al (2022) A saturated map of common genetic variants associated with human height. Nature 610(7933):704–712

Zieger M (2022) Forensic DNA phenotyping in Europe: how far may it go? J Law Biosci 9(2):lsac024

Zorina-Lichtenwalter K, Lichtenwalter RN, Zaykin DV, Parisien M, Gravel S, Bortsov A, Diatchenko L (2019) A study in scarlet: MC1R as the main predictor of red hair and exemplar of the flip-flop effect. Hum Mol Genet 28(12):2093–2106

GPSR Compliance

The European Union's (EU) General Product Safety Regulation (GPSR) is a set of rules that requires consumer products to be safe and our obligations to ensure this.

If you have any concerns about our products, you can contact us on ProductSafety@springernature.com

In case Publisher is established outside the EU, the EU authorized representative is:

Springer Nature Customer Service Center GmbH
Europaplatz 3
69115 Heidelberg, Germany

Batch number: 10755216

Printed by Printforce, the Netherlands